SEMANTIC TABLEAU TESTS

Test for Argument Validity (and Implication)

$$p_1$$

$$p_n \quad\Big| \quad q$$

Closed: Valid (Argument); **Implied** (Implication)
Open: Invalid (Argument); **Not Implied** (Implication)

Test for Equivalence

$$p \quad\Big|\quad q \qquad\qquad q \quad\Big|\quad p$$

Closed: p implies q   **Closed: q implies p**
Both Close: p and q equivalent

Test for Logical Truth

$$\Big| \quad p$$

Closed: Tautologous
Open: Contingent or Contradictory

Test for Contradictoriness or Satisfiability

$$p \quad\Big|$$

Closed: Contradictory
Open: Satisfiable

SEMANTIC TABLEAU RULES

~L (Negation Left)

$$\checkmark \quad {\sim}p \quad\Big|\quad p$$

~R (Negation Right)

$$p \quad\Big|\quad {\sim}p \;\checkmark$$

·L (Conjunction Left)

$$\checkmark \quad p \cdot q \quad\Big|$$
$$p$$
$$q$$

·R (Conjunction Right)

$$p \cdot q \;\checkmark$$
$$p \qquad q$$

vL (Disjunction Left)

$$\checkmark \quad p \vee q \quad\Big|$$
$$p \qquad q$$

vR (Disjunction Right)

$$\Big|\quad p \vee q \;\checkmark$$
$$p$$
$$q$$

⊃L (Conditional Left)

$$\checkmark \quad p \supset q \quad\Big|$$
$$p \qquad q$$

⊃R (Conditional Right)

$$p \quad\Big|\quad p \supset q$$
$$q$$

≡L (Biconditional Left)

$$\checkmark \quad p \equiv q \quad\Big|$$
$$p \qquad\qquad p$$
$$q \qquad\qquad q$$

≡R (Biconditional Right)

$$\Big|\quad p \equiv q \;\checkmark$$
$$p \qquad\qquad p$$
$$q \qquad\qquad q$$

SIMPLE LOGIC

SIMPLE LOGIC

Daniel Bonevac

The University of Texas at Austin

Under the general editorship of

Robert C. Solomon

The University of Texas at Austin

New York   Oxford

OXFORD UNIVERSITY PRESS

Oxford University Press

Oxford   New York
Auckland   Bangkok   Buenos Aires   Cape Town
Chennai   Dar es Salaam   Delhi   Hong Kong   Istanbul   Karachi
Kolkata   Kuala Lumpur   Madrid   Melbourne   Mexico City   Mumbai   Nairobi
São Paulo   Shanghai   Singapore   Taipei   Tokyo   Toronto

ISBN 978-0-19-515502-0

Printing number: 9 8

Printed in the United States of America
on acid-free paper

for
Molly
and
Melanie

PREFACE

Simple Logic is a comprehensive introduction to logic. Logic has played a central role in higher education since its origin in the writings of Aristotle. It continues to occupy an important position in contemporary university curricula for much the same reason it occupied such a position in ancient academies, medieval centers of learning, and Enlightenment universities—its object of study, reasoning, is fundamental to all intellectual activity and to most other human endeavors. The aim in this book is to introduce readers to the traditional areas of logic: (1) "informal logic," concerned with language, communication, and fallacies; (2) "formal logic," including the theory of the syllogism and modern symbolic techniques; and (3) "inductive logic," the study of empirical reasoning.

ORGANIZATION

In its structure, this book follows the now-common pattern for introductory logic texts established by John Stuart Mill, Morris Cohen, Ernest Nagel, and Irving Copi. Chapter 1 introduces the basic concepts of logic. Chapter 2 discusses emotive language and definitions. Chapter 3 is a comprehensive look at informal fallacies. Chapters 4 and 5 present traditional, Aristotelian logic, while Chapters 6, 7, and 8 cover propositional logic, developing truth tables, semantic tableaux, and natural deduction to evaluate arguments. Chapter 9 introduces predicate logic, and Chapters 10, 11, and 12 cover inductive logic, discussing generalizations, analogies, causal inferences, and explanations.

NEW FEATURES

There are two chief enemies in the logic classroom: boredom and confusion. *Simple Logic* has been designed to defeat both.

First, boredom. Logic can be fun. Yet most students—and even most philosophers—do not find the subject so intrinsically appealing. Typically, logic is as interesting as the examples, arguments, and problems to which it is applied. For that reason, this book's goal has been finding examples and exercises that are fun. They illumine the points in the text but also intrigue the reader in their own right. Students will want to read this book just for the examples. Some are philosophical; whenever possible, famous definitions and arguments from philosophical classics (western and non-western) have been included. Some are from important historical documents and speeches, but most of the examples are from popular sources, and many are humorous. (The exercises in Chapter 1 are a good illustration.)

Second, confusion. This is indeed simple logic; material for an introductory logic course has been streamlined as much as possible, eliminating needless complexities. Most logic texts spend too much time on details that students will not remember a year later. Some details are necessary, of course, but many can be eliminated by designing logical systems carefully.

Students are provided with step-by-step help where they need it, while the discussion is limited where most students can pick up the needed skills easily. Chapter 8 reflects this most clearly and thus differs most significantly from other introductory logic books, even those using the same proof method. The text takes students through each rule step-by-step, giving an English argument that motivates the rule, explaining the rule itself, explaining why it is sound, explaining how to use it, and then using it in a proof.

In the interests of simplicity and elegance, semantic tableaux are presented for propositional and monadic predicate logic. Based on E. W. Beth's tableaux, the method bears much similarity to Richard Jeffrey's truth trees (which are also covered in the *Instructor's Manual* for instructors who prefer them). The method is very easy to teach and to learn; it directly mirrors the semantics of the logical operators. Tableaux provide a ready test for validity and other logical properties in both sentential and predicate logic. (Of course, in full predicate logic, the method is not a decision procedure.) Semantic tableaux are even simpler than truth trees, for tableaux have the subformula property—on a tableau, formulas are always decomposed into their subformulas.

The deduction system is that of Irving Copi's *Introduction to Logic*. It uses only simple rules, that is, rules stating that formulas of certain kinds can be deduced from formulas of other kinds. There are, consequently, no subordinate proofs. In the last two sections of Chapter 8, we extend the proof method to categorical as well as hypothetical proofs and introduce a limited indirect proof method that makes many proofs even simpler.

Some complications have been relegated to the footnotes, which are intended primarily for instructors, not students. Footnotes are also used to explain why occasional differences occur from other authors of introductory logic texts and to present historical background. Especially in the more traditional areas of logic, the history of the subject sheds light on the topic, as well as being interesting in its own right. It is hoped the treatment of syllogistic logic is more elegant than many others, but also more historically sensitive. At the end of Chapter 5, having explained the standard methods using Venn diagrams and rules, the text presents a variant of Aristotle's method of reduction. This not only gives students practice on immediate inference but also introduces the idea of a deduction system in a very restricted and simple setting.

This book is thoroughly modular. Instructors can pick and choose topics freely, without making later sections unintelligible, except in cases of obvious dependence. (For example, the presentation of natural deduction in predicate logic presupposes the presentation of natural deduction in propositional logic.) Instructors omitting topics and instructors accustomed to another book should find no unpleasant surprises.

SUPPLEMENTS

Accompanying this book are the *Instructor's Manual* with solutions. The *Instructor's Manual* presents solutions to problems not already solved in the back of the book, with the exception of a few essay-type exercises that are omitted for obvious reasons. It also presents a brief summary of each section, with all key definitions and rules. It is intended it to be an ideal in-class resource for instructors, who can use it to structure their presentations, as well as to provide illustrations of rules and concepts. Interested instructors should contact their Harcourt Brace representative for more information.

A Simple Logic Web site provides access to many on-line resources for both instructors and students. Start your tour of this site at the Harcourt Brace home page—www.hbcollege.com.

ACKNOWLEDGEMENTS

I owe a special debt of gratitude to Andrew Schwartz, who contributed to this book in many ways. His comments on a draft of the manuscript led to many improvements, especially in Chapters 4, 6, and 9; his insistence on clarity and consistency led me to revise very substantially the book's discussions of validity and of the object/metalanguage distinction. He also revised my draft of Chapter 10, sharpening the treatment of several key issues.

During the development stage of the book, the following people provided valuable input: Betty J. Baker, Brookdale Community College; Richard Burnor, University of Toledo; Randall Dipert, U.S. Military Academy; Bill Garland, University of the South; Harold Greenstein, SUNY–Brockport; Janet Grouchy, Louisiana State University; Christopher Hoffman, George Washington University; Hugh Hunt, Kennesaw State College; Greg Landini, University of Iowa; Bonnie Paller, California State University–Northridge; Michael Principe, Middle Tennessee State University; Daniel Rothbart, George Mason University; Edward Sherline, University of Wyoming; David Stern, University of Iowa; Robert Weingard, Rutgers University; and Daniel Wueste, Clemson University.

ABOUT THE AUTHOR

Daniel Bonevac is Professor and Chair of the Department of Philosophy at the University of Texas at Austin. He was educated at Haverford College and the University of Pittsburgh, where he received his Ph.D. in 1980 under the direction of Wilfrid Sellars. His first book, *Reduction in the Abstract Sciences*, received Johnsonian Prize in Philosophy from *The Journal of Philosophy*. He is also the author of *Deduction* and *The Art and Science of Logic*, editor of *Today's Moral Issues*, and coeditor of *Beyond the Western Tradition* and *Understanding Non-Western Philosophy*. His articles on metaphysics, the philosophy of mathematics, and philosophical logic have appeared in a wide array of philosophical journals. Recipient of several grants from the National Science Foundation, he serves on the editorial board of the *American Philosophical Quarterly*.

CONTENTS

PART II   ARISTOTELIAN LOGIC

PART III   SYMBOLIC LOGIC

SIMPLE LOGIC

LOGIC AND LANGUAGE

CHAPTER 1

REASONING

Logic is the study of reasoning. Aristotle (384–322 B.C.) founded logic as a system of principles upon which all other knowledge rests. Logic pertains to all subjects; people can reason about anything. Sometimes the reasoning is good. Sometimes it is not. People use logic to tell the difference.

Reasoning well is more than an academic exercise. Gathering information, making decisions, and carrying out plans all require reasoning. Good reasoning leads to accurate information, good decisions, and appropriate plans, whereas bad reasoning leads to inaccuracies, bad decisions, and misguided plans. Part I of this book discusses features of good reasoning and sources of bad reasoning. Part II presents traditional Aristotelian logic. Part III presents the core of modern deductive logic. And Part IV presents the fundamentals of induction, the kind of reasoning used frequently in the sciences as well as everyday life.

Logic is not the study of how people *do* reason but how they *should* reason. It does not describe the psychology of reasoning; it prescribes ways of supporting conclusions—that is, for showing reasoning to be good. Arithmetic describes the rules for addition rather than the psychological process of addition. Just so, logic describes not the psychological process of reasoning but the rules for correct reasoning. Logic describes an ideal that actual reasoning sometimes fails to reach.

◼ 1.1 PREMISES AND CONCLUSIONS

Arguments are bits of reasoning in language. Frequently, we think of arguments as conflicts. Sometimes, however, we speak of a politician arguing for the passage of a bill, a lawyer arguing a case, or a reader of Shakespeare arguing that *King Lear* is better than *Hamlet*. In this sense, arguments are attempts to establish a conclusion.

An argument begins with some assertions trying to justify a thesis. The initial assertions of the argument are its **premises;** the thesis the argument tries to justify is its **conclusion.**

> An *argument* is a finite string of statements, called *premises*, together with another statement, the *conclusion*, which the premises are taken to support.

The statements that make up the argument come in a particular order, whether the argument is spoken, written, or encoded in a computer language. For our purposes in this text, however, the order in which an argument presents its premises rarely makes a difference. So, we generally will not worry about order of presentation. But it is important that the string of premises be finite. If the premises never end, the conclusion is never established.

Arguments consist of **statements,** sentences that can be true or false.[1] Many ordinary sentences, including almost every one in this book, fall into this category. Grammatically, they are declarative, in the indicative mood. They say something about the way the world is, correctly or incorrectly.

Commands, for example, are different. *Shut the door* can be appropriate or inappropriate, irritating or conciliatory, and friendly or hostile, but it cannot be true or false.[2] Questions, too, are neither true nor false: Consider *What time is it?* or *What is the capital of North Dakota?* Interjections—*Hello, Goodbye, Ouch!, Alright!*—and exhortations *(Let's go!, Let us pray)* are likewise neither true nor false.

A statement is true or false in a particular context: as used on a particular occasion, by a particular speaker, to a particular audience, in a given circumstance, as part of a given discourse. Whether *I love you* is true, for example, clearly depends on who you and I are, when the sentence is spoken, and so on. Statements are true or false only in a context.

---

[1]This is Aristotle's definition: "Every sentence is significant . . . but not every sentence is a statement-making sentence, but only those in which there is truth and falsity" (*De Interpretatione* 4, 17a1–3).

[2]We use italics to quote English expressions. (Most logic books use quotation marks for this purpose. If many things are being quoted, however, quotation marks make the text look cluttered. We find italics more readable.) Thus,
   Shut the door
refers to the English sentence
   Shut the door.
If John says, "Shut the door," he *uses* the sentence; if he says, "*Shut the door* is a command rather than a statement," he *mentions* it. We use italics to indicate that we are mentioning rather than using an expression.
   We use italics for two other purposes: (a) for words being defined, and (b) for emphasis. Context, we hope, makes clear our intentions in each case.

Some logicians therefore speak of **propositions,** which they take statements used in particular contexts to express.[3] Anyone so inclined can substitute *proposition* for *statement* throughout the rest of this book. But because statements are less abstract than propositions, and because context rarely enters directly into what follows, we will continue to speak of statements and treat them as having truth values.

Statements may contain other statements. Consider the traditional aphorism

(1) When the going gets tough, the tough get going.

Example (1) itself is a statement. Within it, furthermore, are two other statements: *The going gets tough* and *The tough get going*. Example (1) results from combining these two shorter statements. The number of premises in an argument, therefore, is not always one less than the number of periods, exclamation points, and so on. Indeed, an entire argument, with premises and conclusion, can appear within a single sentence.

The following example of a simple argument is Abraham Lincoln's explanation of why he did not expect to marry:

(2) I have come to the conclusion never again to think of marrying, and for this reason: I can never be satisfied with anyone who would be blockhead enough to marry me.

We can represent Lincoln's argument as

(3) I can never be satisfied with anyone who would be blockhead enough to marry me.

∴ I shall never again think of marrying.

When we write an argument "officially," in **standard form,** we list the premises in the order in which they are given, and then list the conclusion. In addition, we preface the conclusion with the symbol ∴, which means "therefore." So, in official representations, conclusions always come last. This is not true in natural language, as Lincoln's argument shows. Conclusions may appear at the beginning, in the middle, or at the end of arguments, if they are stated at all.

---

[3]Traditionally, propositions are (a) expressed by statements with the same meaning in the same context; (b) the basic things that are true or false; and (c) the things people believe, know, think, doubt, etc.

Some philosophers use *statement* as equivalent to *proposition*. We invite those so inclined to take *statement* that way in this book. We stay with Aristotle's conception of a statement as a sentence that can be true or false, however, for sentences have a well-understood structure, but propositions do not. We can speak readily of one statement containing another, for example, but whether one proposition can contain another depends on one's theory of propositions.

Another simple argument is from French essayist Joseph Joubert (1754–1824):

(4) Nothing that is proved is obvious; for what is obvious shows itself and cannot be proved.

This argument, too, starts with its conclusion, and then introduces premises to justify it. In standard form, the argument becomes:

(5) What is obvious shows itself and cannot be proved.

∴ Nothing that is proved is obvious.

The premises of an argument are meant to support the conclusion. The relation of intended support distinguishes arguments from other uses of language. We can recognize arguments, then, by recognizing when some statements are offered in support of others. We can do this most easily, in turn, if we can distinguish premises from conclusions. But how can we pick out the conclusion of an argument? In (4), the word *for* indicated that a premise was being introduced. In English, various words and phrases can signal the premises or the conclusion of an argument.

CONCLUSION INDICATORS

therefore

thus

hence

so

consequently

it follows that

in conclusion

as a result

then

must

accordingly

this implies that

this entails that

we may infer that

PREMISE INDICATORS

because

as

for

since

given that

for the reason that

These words and phrases have other uses; they are not always premise or conclusion indicators.

(6) John has been depressed since Mary left him.

is not an argument; *since* here expresses not a logical relation but a relation in time. Still, these expressions can, and often do, serve as indicators because they can attest to relations of support among the statements of an argument. The single English sentence

(7) Since state legislatures will reapportion voting districts after the next census, control of those legislatures will be hotly contested.

presents a simple argument. The word *since* indicates that we should take what follows it,

(8) State legislatures will reapportion voting districts after the next census,

as a premise, supporting the conclusion

(9) Control of those legislatures will be hotly contested.

Premise indicators often signal not only that one or more statements are premises but that a certain statement is a conclusion. *Since,* for example, exhibits a relation of support between the statements it links together. Its occurrence in (7) points out not only that the statement immediately following it is a premise but also that (9) is a conclusion. Earlier, we saw that *for* usually indicates that the statements following it are premises and that the statement preceding it is the conclusion.

Indicators provide important clues to the structure of arguments. Often, however, no explicit indicators appear. Sometimes the conclusion is not even stated. In such cases, we must ask what the author is trying to establish. Some arguments are quite clear about what their conclusions are, but others are hard to analyze. Consider, for example, this argument from the French essayist Voltaire (1694–1778):

(10) If it were permitted to reason consistently in religious matters, it is clear that we all ought to become Jews, because Jesus Christ was born a Jew, lived a Jew, and died a Jew, and because he said that he was accomplishing and fulfilling the Jewish religion.

Voltaire seems to be arguing for the conclusion that we all ought to become Jews. The word *because* indicates that the rest of the argument consists of a list of premises. But Voltaire, a satirist, is really aiming not at this conclusion but at another. Everything he says is supposed to follow from the hypothetical *if it were permitted to reason consistently in religious matters*. This is a clue that the argument that we ought to become Jews is part of a larger argument. Voltaire does not state its conclusion. But he apparently is trying to show that it is not permitted to reason consistently in religious matters. The conclusion of the smaller argument—*we all ought to become Jews*—is an observation few among his intended audience of Christians would accept, even though, according to Voltaire, their own doctrine commits them to it.

P R O B L E M S

Which of the following passages contain arguments? Identify any conclusions you find.

1. Crime is common. Logic is rare. Therefore it is upon the logic rather than upon the crime that you should dwell. (Sir Arthur Conan Doyle)

2. Children make the most desirable opponents in Scrabble as they are both easy to beat and fun to cheat. (Fran Lebowitz)

3. Since we take an average of 45,000 car trips over the course of a lifetime, say statisticians, the chance of being in a serious accident is nearly one in two. (Jane Stein)

4. If there hadn't been women we'd still be squatting in a cave eating raw meat, because we made civilization in order to impress our girlfriends. (Orson Welles)

5. We owe a lot to Thomas Edison—if it wasn't for him, we'd be watching television by candlelight. (Milton Berle)

6. Since happiness consists in peace of mind, and since durable peace of mind depends on the confidence we have in the future, and since that confidence is based on the science we should have of the nature of God and the soul, it follows that science is necessary for true happiness. (Gottfried Leibniz)

7. Cats are smarter than dogs. You can't get eight cats to pull a sled through snow. (Jeff Valdez)

8. One has to belong to the intelligentsia to believe things like that; no ordinary man could be such a fool. (George Orwell)

9. Do not love your neighbor as yourself. If you are on good terms with yourself, it is an impertinence; if on bad, an injury. (George Bernard Shaw)

10. I should take little comfort in a world without books, but reality is not to be found in them because it is not there whole. (Marguerite Yourcenar)

11. It is possible to own too much. A man with one watch knows what time it is; a man with two watches is never quite sure. (Lee Segall)

12. Every luxury must be paid for, and everything is a luxury. . . . (Cesare Pavese)

13. Life does not agree with philosophy: there is no happiness that is not idleness, and only what is useless is pleasurable. (Anton Chekhov)

14. Ireland set out to crack down on alcohol-related traffic accidents. A spokesman for the Automobile Association in Dublin said it's time to stop blaming accidents on motorists: "In many cases the pedestrian is to blame. Often, he is lying prone in the roadway." *(Esquire)*

15. It is absurd to bring back a runaway slave. If a slave can survive without a master, is it not awful to admit that the master cannot live without the slave? (Diogenes of Sinope)

16. Man being a reasonable, and so a thinking creature, there is nothing more worthy of his being than the right direction and employment of his thoughts; since upon this depends both his usefulness to the public, and his own present and future benefits in all respects. (William Penn)

17. It's fun politically to attack the rich on behalf of the poor—the Robin Hood syndrome. But if you're serious about helping the poor, you'd better give economic security to those in the middle, so they're paying the necessary taxes. You're just not going to be able to finance all social programs from tax collections from the rich. (Gary Hart)

18. Science demands an uncompromising approach to reality— measurement and experiment surrounded by theory, yielding

careful definition of degrees of certainty. Science also demands independent investigator initiative, diversity, dialogue, and, above all, *freedom* of debate and discussion. Our culture does endeavor to cherish such values in the humanities and in politics as well, despite . . . frustration with scientific hedging. And that is, of course, my point. The manifest power of science to explain and predict phenomena has made science appear to dominate *all* values, and even the human condition. Nevertheless, the opposite more clearly describes the condition. Culture dominates science. (E. E. David, Jr.)

19. An Iron Curtain is being drawn down over their front. We do not know what lies behind it. It is vital, therefore, that we reach an understanding with Russia now before we have mortally reduced our armies and before we have withdrawn into our zones of occupation. (Winston Churchill)

20. . . . astrology was progressive. Astrology differed in asserting the continuous, regular force of a power at a distance. The influences of heavenly bodies on the events on earth it described as periodic, repetitious, *invisible* forces like those that would rule the scientific mind. (Daniel J. Boorstin)

21. Stunned by his defeat, Dewey later said he felt like the man who woke up to find himself inside a coffin with a lily in his hand and thought: "If I'm alive, what am I doing here? And if I'm dead, why do I have to go to the bathroom?" (Paul F. Boller, Jr.)

22. Everything is new. And we are living among events so singular that old people have no more knowledge of them, are no more habituated to them, and have no more experience of them than young people. We are all novices, because everything is new. (Joseph Joubert)

23. If we take in hand any volume; of divinity or school metaphysics, for instance; let us ask, *Does it contain any abstract reasoning concerning quantity or number?* No. *Does it contain any experimental reasoning concerning matter of fact and existence?* No. Commit it then to the flames; for it can contain nothing but sophistry and illusion. (David Hume)

24. The pure sciences express results of comparison exclusively; comparison is not a conceivable effect of the order in which outer impressions are experienced—it is one of the house-born portions of our mental structure; therefore, the pure sciences form a body of propositions with whose genesis experience has nothing to do. (William James)

1.2 RECOGNIZING ARGUMENTS

An argument tries to establish something. The premises of an argument are meant to support the conclusion. The relation of intended support distinguishes arguments from other uses of language. We can recognize arguments, then, by recognizing when some statements are offered in support of others.

The simplest way to recognize a relation of intended support—often called an **inferential relation** or **claim**—is to look for premise or conclusion indicators: words such as *therefore, since, thus,* and so forth. When indicators are used to mark a premise or conclusion, the inferential claim is **explicit.** Many arguments, however, lack indicators. In such cases, the inferential claim is **implicit.**

The first two arguments we considered in section 1.1 make explicit inferential claims. (Indicators are in italics.)

(2) I have come to the conclusion never again to think of marrying, and *for this reason*: I can never be satisfied with anyone who would be blockhead enough to marry me.

(4) Nothing that is proved is obvious; *for* what is obvious shows itself and cannot be proved.

But arguments frequently make implicit inferential claims. Consider this argument from French essayist Michel de Montaigne (1533–1592):

(11) When all is summed up, a man never speaks of himself without loss; his accusations of himself are always believed; his praises never.

The first statement is the conclusion; the others are intended to support it. They give reasons why speaking of yourself is a losing proposition. If you criticize yourself, people will believe you; if you praise yourself, they will not. In either case, you lose.

Inferential claims are not easy to recognize. Reasonable interpreters may disagree. One must judge the intentions of an author who has provided no explicit guides, and sometimes this is not easy. Even when premise or conclusion indicators are present, it can be difficult to tell whether some statements are meant to support others, for, as we have seen, indicator words and phrases usually have other noninferential uses.

Recognizing arguments is much easier, therefore, if one can recognize other, noninferential relations between bits of language. Most relations between statements in a discourse are not inferential. The theory of discourse relations is still young. But here are some basic kinds of noninferential relations between statements.

NARRATIVE

A **narrative** or **report** is a group of statements that tells a story. It relays information about a course of events. Statements in a narrative generally proceed in temporal order. A simple example:

(12) John fell down. He got up, dusted himself off, and went on his way.

This portrays four events in the order in which they occurred. A well-known example of a narrative is Robert Southey's (1774–1843) "The Story of the Three Bears" ("A tale which may content the minds of learned men and grave philosophers," according to its motto, quoted from playwright George Gascoyne [1539–1577]). Here is just one paragraph, relating a sequence of nine events:

(13) So first she tasted the porridge of the Great, Huge Bear, and that was too hot for her; and she said a bad word about that. And then she tasted the porridge of the Middle Bear, and that was too cold for her; and she said a bad word about that, too. And then she went to the porridge of the Little, Small, Wee Bear, and tasted that; and that was neither too hot, nor too cold, but just right; and she liked it so well, that she ate it all up; but the naughty old Woman said a bad word about the little porridge-pot, because it did not hold enough for her.

DESCRIPTION

A **description** is a group of statements about a situation. It relays information about some circumstance. Here is a simple example:

(14) John was wearing a plaid shirt, untucked on one side, and blue pants that had seen better days. His backpack fell at an angle across his back. His shoulders were slumped slightly forward, and his face bore a look of such intensity that no one would dare to interrupt his train of thought.

Note that this describes a single situation, not a sequence of events. A famous description is Samuel Taylor Coleridge's (1772–1834) portrait of Xanadu in *Kubla Khan*:

(15) Where Alph, the sacred river, ran
Through caverns measureless to man
Down to a sunless sea.
So twice five miles of fertile ground
With walls and towers were girdled round:

And there were gardens bright with sinuous rills,
Where blossomed many an incense-bearing tree;
And here were forests ancient as the hills,
Enfolding sunny spots of greenery.

The statements in a description are related chiefly by being about the same person, place, thing, or situation.

ASSOCIATION

Statements may also be linked only in that they have the same or similar topics. In such cases, they are loosely associated as falling under some general category. Consider, for example, the Beatitudes from Jesus' Sermon on the Mount (*Matthew* 5:3–11):

(16) Blessed are the poor in spirit, for theirs is the kingdom of heaven.

Blessed are those who mourn, for they shall be comforted.

Blessed are the meek, for they shall inherit the earth.

Blessed are those who hunger and thirst for righteousness, for they shall be satisfied.

Blessed are the merciful, for they shall obtain mercy.

Blessed are the pure in heart, for they shall see God.

Blessed are the peacemakers, for they shall be called sons of God.

Blessed are those who are persecuted for righteousness' sake, for theirs is the kingdom of heaven.

Blessed are you when men revile you and persecute you and utter all kinds of evil against you falsely on my account.

Or this, the first of Confucius' *Analects:*

(17) Is it not a joy to learn and put into practice what has been learned? Is it not delightful to have friends coming from afar? And is he not a superior man who does not mind that others do not recognize him?

In both cases, it is not easy to explain the **association.** Spelling out the topic of the Beatitudes or of *Analects* 1.1 takes one to the heart of Christian and Confucian ethics, respectively. But we recognize the statements in these examples as linked by association not only because of their parallel grammatical structure but because each is connected to the others primarily by being related topically. Changing the order of statements linked by association usually makes little difference to the meaning of the discourse. That is not true of most other discourse relations.

ELABORATION

A group of statements elaborates a statement if it develops, expands upon, or gives further details about the topic of that statement. A simple example:

> (18) John fell down. He tumbled, head over heels, and, like a rag doll tossed by a careless child, crumpled as he struck the unforgiving earth.

A famous example of an **elaboration** is William Wordsworth's (1770–1850) sonnet composed on Westminster Bridge, September 3, 1802:

> (19) Earth has not anything to show more fair:
> Dull would he be of soul who could pass by
> A sight so touching in its majesty:
> This city now doth, like a garment, wear
> The beauty of the morning; silent, bare,
> Ships, towers, domes, theatres, and temples lie
> Open unto the fields, and to the sky;
> All bright and glittering in the smokeless air.
> Never did sun more beautifully steep
> In his first splendour, valley, rock, or hill;
> Ne'er saw I, never felt, a calm so deep!
> The river glideth at his own sweet will:
> Dear God! the very houses seem asleep;
> And all that mighty heart is lying still!

The last eleven lines of the sonnet are a description of London at dawn, intended to elaborate the dual claims made in the first three lines.

ILLUSTRATION

A group of statements illustrates a statement if it gives an example of what that statement discusses. All the numbered passages in this section are **illustrations** of kinds of noninferential discourse relations. Here is an example from Thomas J. Peters and Robert Waterman's *In Search of Excellence:*

> (20) . . . what's called "foot-in-the-door" research demonstrates the importance of incrementally acting our way into major commitment. For instance, in one experiment, in Palo Alto, California, most subjects who initially agreed to put a *tiny* sign in their front window supporting a cause (traffic safety) subsequently agreed to display a billboard in their front yard, which required letting outsiders dig sizable holes in the lawn. On the other hand, those not asked to take the first

small step turned down the larger one in ninety-five cases out of a hundred.

Phrases such as *for instance, for example, a case in point,* and so on indicate that what is to come illustrates the statement or statements just made. But illustrations often occur without telltale phrases:

(21) Even more than the scientists—Dalton, Davy, and Faraday—the technocrats came from nowhere and had nothing given to them except what they earned with their hands. George Stephenson began as a cowherd; Telford, a shepherd's son, as a stonemason. Alexander Naysmith started as an apprentice coach painter. (Paul Johnson)

Johnson's mention of Stephenson, Telford, and Naysmith illustrates his point about the democratization of science in the nineteenth century. One could take him as using the three to support his generalization; one could see (21), that is, as an argument. This is often true of illustrations or anecdotes—they may be construed as examples or as supporting evidence.

EXPLANATION

A group of statements explains a statement if it is intended to increase understanding of it. Many, but not all, **explanations** are **causal:** They explain a fact or event by pointing to an event causing it. A simple example:

(22) John fell down. Mary pushed him.

The second statement explains the first. Mary's push came before and caused John's fall. The temporal order is thus the opposite of narrative. In a narrative, statements proceed more or less in temporal order. Explanations, in contrast, often introduce circumstances and events that precede the event or fact being explained.

Not all explanations deal with events. Sometimes generalizations explain other generalizations. Here, for example, Jerome E. Groopman, professor at Harvard Medical School, explains why mad cow disease is so difficult to fight:

(23) How might a mad cow's prion [a kind of protein] cause our brain to degenerate into a sponge? It turns out that we have normal prion proteins in our brains, and the TSE [mad cow disease] prion may induce our normal prions to change into a diseased form and precipitate. Our immune system has evolved to protect us against microbial invaders by first recognizing them as foreign and then fighting to expel them from the body. But the TSE agent is likely a prion protein whose composition is identical to our own, so our immune

defenses fail to recognize it as alien. This makes a vaccine remote, since the TSE prion looks like a friend until it settles into the brain and reveals itself as a deadly foe.

This passage contains two explanations. The first is contained in the first two sentences. How does mad cow disease turn a human brain into a sponge? Diseased prions change human prions and cause them to precipitate. The second explanation is contained in the remainder of the passage. Why can't our immune system repel the disease? And why is a vaccine difficult to devise? The answer to both questions is the same: Our immune system cannot distinguish diseased from healthy prions.

Explanations are similar to arguments in many ways. They even use many phrases that can function as premise and conclusion indicators. But there are important differences. Arguments use premises to establish a conclusion. Explanations seek to answer the questions How does this happen? and Why is this the case? Arguments seek to answer the question Why believe this?

As we shall see in Chapter 3, the premises of an argument are evidence for the conclusion. The premises are accepted facts; the conclusion is not yet accepted. The point of arguing is to get the conclusion accepted. In an explanation, the thing to be explained is the accepted fact; the statements that do the explaining are hypotheses that may not yet be accepted. Arguments and explanations are thus opposite in what Aristotle called the order of knowledge.

In summary, an argument is an inferential relation. The conclusion is inferred from other statements, which are taken to support or provide evidence for it.

P R O B L E M S

Which of the following contain arguments? Identify the conclusions of the arguments you find. Identify the noninferential relations involved in passages without arguments.

1. You can't have everything. Where would you put it? (Stephen Wright)

2. Scientists discovered a Bechstein's bat, Britain's rarest. As they began to celebrate, it was eaten by a cat. *(The Economist)*

3. Woman to umbrella vendor, in the pouring rain: "I don't want that one. It's wet." *(New York)*

4. Since the politician never believes what he says, he is surprised when others believe him. (Charles De Gaulle)

5. Everyone has talent at 25. The difficulty is to have it at 50. (Edgar Degas)

6. I'm not sure a bad person can write a good book. If art doesn't make us better, then what on earth is it for? (Alice Walker)

7. I never want to be the best at anything. Anybody who wants to be the best at anything in the world must spend 80 percent of their waking time on it. And you lose the pursuit of other things. (Warren Avis)

8. ABC sportscaster Chip Cipolla has a friend whose car radio was heisted three times. Not being a clunk, the friend invested in a clip-on, removable radio. In going to dinner this week the friend unclipped the brand new gizmo and carted it into the restaurant with him. When he came out he had the radio nice and safe, but his car had been stolen. (Cindy Adams)

9. The important thing to remember about the statist society is not that it *doesn't* work, but that it *can't*. No central planner—even one as smart as Bill Clinton—can match the sheer genius of the marketplace. (Theodore J. Forstmann)

10. I ate the chips, not stuffing them rapidly down my craw with both hands as I usually do, but slowly munching them one by one, allowing myself to savor the unique texture and listening to my individual teeth crunching them sensuously to a pulp. Being, at that time, a driven, overstressed, Type-A personality, I had never before allowed myself to stop and actually taste my food. It was a staggering revelation. "God," I thought, "this stuff is terrible." (Lewis Grossberger)

11. In 1960 there were 365 paid lobbyists of the Senate. Today, there are over 40,000, or approximately 400 lobbyists per Senator. (Institute for Policy Innovation)

12. Deficits are bad—but not because they necessarily raise interest rates. They are bad because they encourage political irresponsibility. They enable our representatives in Washington to buy votes at our expense without having to vote explicitly for taxes to finance the largess. The result is a bigger government and a poorer nation. (Milton Friedman)

13. Nothing is cheap which is superfluous, for what one does not need is dear at a penny. (Plutarch)

14. To the question, How conservative are you?, the American people have given the answer that they talk a very much more conservative game than they are prepared to see played. They

have a voracious appetite for government. They just have a negligible willingness to pay for it. (George Will)

15. Once upon a time there were Three Bears, who lived together in a house of their own, in a wood. One of them was a Little, Small, Wee Bear; and one was a Middle-sized Bear, and the other was a Great, Huge Bear. (Robert Southey)

16. Marty Venker [then a Secret Service agent] was on duty in the White House one day when he heard a panic button go off—alerting the Secret Service to a president in distress.

    Hand on his gun, the agent rushed to the scene of the emergency: an Oval Office bathroom. After "practically knocking down the door," Marty says, he found Carter inside, zipping up his pants. Apparently the commander-in-chief mistook the bathroom's panic button for a toilet flush mechanism. (Richard Johnson)

17. The first time I played [in an old-timers game], I got two quick outs in the inning and couldn't get the third one. There were 11 straight people up. Nine runs and I still didn't get them out. They finally had to come and get me.

    Somebody asked our catcher, Joe Garagiola, how hard I was throwing. He said, "I don't know. I haven't caught one yet." (Jim Bunning)

18. Oxford University was offered a $34 million grant to build a "world-class school of business management"—and turned it down. The gift was from a respectable Saudi Arabian billionaire. Business was deemed too distasteful a subject to impose on students seeking wisdom and enlightenment. (James Russell)

19. The late Carleton Coon [said] that cooking was "the decisive factor in leading man from a primarily animal existence into one that was more fully human." While this is a tempting proposition for the food historian, it is based on the premise that humans could not get down to developing a culture until they were released from the need to spend hours every day chewing tough meat—whereas any deskbound sandwich eater today knows that working and chewing tough meat are by no means mutually exclusive occupations. (Reay Tannahill)

20. Actually it was not until I was forty that I was able to go into a room and say to myself, "What do I think of these people?" Before that, I had always thought, "What do these people think of me?" When I became forty, I said to myself, "You are either a whole person now, or you never will be. Believe in yourself." (Brooke Astor)

21. Quintessential New York kids conversation—this one between two 7-year old boys—overheard by Rhoda Cahan at a private school on the Upper East Side, where she was picking up her son:

    First boy: I found a condom on the patio.
    Second boy: What's a patio? *(New York Times)*

22. Patron to waiter: "How is the fish?"
    Waiter to patron: "What do you mean how is the fish? It's dead."
    (Anthony Haden-Guest)

23. Statistics indicate that most fatal [cycle] accidents occur within 25 miles of home. (The logical response to that, of course, is to move away.) (Grace Butcher)

24. I am not the type who wants to go back to the land; I am the type who wants to go back to the hotel. This state of affairs is at least partially due to the fact that nature and I have so little in common. We don't go to the same restaurants, laugh at the same jokes or, most significant, see the same people. (Fran Lebowitz)

25. "Uh, Jean?"
    "Yes?"
    "Are you awake?"
    "Yes."
    "What the hell is that?"
    "What?"
    "That funny glare outside?"
    (Pause.) "The moon, Fran." (E. Jean Carroll)

26. The Congressional Quarterly Researcher lists the top problems in public schools as identified by teachers in 1940: talking out of turn, chewing gum, making noise, cutting in line, dress code infraction, and littering.

    And in 1990: drug abuse, alcohol abuse, pregnancy, suicide, rape, robbery, assault. *(The Wall Street Journal)*

27. How do we know that if one sees someone drowning, mauled by beasts, or attacked by robbers, one is obligated to save him? From the verse, "Do not stand by while your neighbor's blood is shed" [*Leviticus,* 19:16]. (Babylonian Talmud, *Sanhedrin* 73a)

28. Forty-four-year-old Daniel Yannielli scaled a 10-foot concrete wall beside an elephant cage at the Honolulu Zoo in Hawaii, stripped to his underwear, and played his harmonica for Empress, the Indian elephant inside. Yannielli was arrested and charged with cruelty to animals. (Associated Press, as reprinted in *National Lampoon*)

29.   . . . a guest politely turned to a woman seated to the right of him. "Did I get your name correctly?" the guest said. "Is your name Post?"
"Yes," was the response.
"Is it Emily Post?" he said.
"Yes," she replied.
He said, "Are you the world renowned authority on manners?"
She said, "Yes," then added: "Why do you ask?"
"Because," he said, "you have just eaten my salad." (Clark Clifford)

30.   General Wild Bill Donovan, [WWII] head of the OSS [Office of Strategic Services, forerunner of the CIA], turned to me. "I want you to find out how the British coordinate all their sources of information at the top level. Needless to say, if you are caught, we shall completely disown you."
It was a five-minute ride to the British War Office. The Director of Military Intelligence asked me, "Well, Fitz, what can I do for you?"
"The Yanks want me to spy on you, Sir."
"Splendid!" he said. "What do they wish to know?"
So I told him. He glanced at his wristwatch: "Come back in one hour and I'll have the charts ready."
An hour later I looked at the administrative chart upon his desk.
"I hope it's enough," said the D.M.I.
"It looks so to me, Sir."
At five to five General Donovan examined it.
"Excellent," he said. "Excellent. But how on earth . . . ?"
"Do I have to tell you, Sir?"
"No, FitzGibbon. Of course not." (Constantine FitzGibbon)

31.   Six hundred million years ago our ancestors were worms, ten thousand years ago they were savages. Both these periods are negligible compared with our possible future. Provided, therefore, that man has a future lasting for more than a few million years we can at once say our descendants may, for anything we can see to the contrary, excel us a great deal more than we excel worms or jellyfish. (J. B. S. Haldane)

32.   At a dinner honoring French President François Mitterand, he and his wife and Nancy and I walked into the State Dining Room. As was customary, everyone was to stand until Nancy led François to her table and I led Mrs. Mitterand to mine. Nancy and François headed for their table, but Mrs. Mitterand stood frozen. I whispered, "We're supposed to go over to the other

side." But she wouldn't move. She said something to me very quietly in French, which I didn't understand. Then she repeated it, and I shook my head. I still didn't know what she was saying. Suddenly an interpreter ran up to us and said, "She's telling you that you're standing on her gown." (Ronald Reagan)

33. [My wife] Bette and I were driving outside of Beijing. We came to this Buddhist temple, and the head monk came out and greeted us, and his eyes lit up when he heard that I was the American ambassador and that my wife was a famous author. He said, "Would you do this temple a great honor and favor for our future visitors, to guide and instruct them? Would you write something in English?" We said we'd be delighted. In fact, we were quite flattered because this is an honor reserved for emperors and great poets.

    He came back after a brief absence, carrying two wooden plaques and said, "Now to guide and instruct our future visitors, would you write in English on this plaque the word 'Ladies,' and on this plaque the word 'Gentlemen'?" (Winston Lord)

34. And on the pedestal these words appear:
"My name is Ozymandias, king of kings;
Look on my works, ye Mighty, and despair!"
Nothing beside remains. Round the decay
Of that colossal wreck, boundless and bare
The lone and level sands stretch far away. (Percy Bysshe Shelley)

35. Nothing can be made to be of interest to the reader that was not first of vital concern to the writer. Each writer's prejudices, tastes, background and experience tend to limit the kinds of characters, actions and settings he can honestly care about, since by the nature of our mortality we care about what we know and might lose (or have lost), dislike that which threatens what we care about, and feel indifferent toward that which has no visible bearing on our safety or the safety of the people and things we love. Thus no two writers get aesthetic interest from exactly the same materials. Mark Twain, saddled with a cast of characters selected by Henry James, would be quick to maneuver them all into wells. (John Gardner)

36. Throughout the ferocious barrage of the previous week the Germans had hidden in their bunkers, tormented by the incessant concussions [of] artillery fire. But they survived. As the barrage lifted, machine-guns were hurriedly hauled out of safekeeping and mounted, and through the smoke the Germans peered out on an astounding sight: successive waves of men plodding steadily forward, as if on parade. A German soldier writing after

the war recalled, with evident incredulity: "[The British] came on at a steady, easy pace as if expecting to find nothing alive in our front trenches." When the Germans opened fire the slaughter was immense. The machine-guns cut down the lines of advancing soldiers like hay before a scythe. (Martin Marix Evans)

37. At the end of the twentieth century, English is more widely spoken and written than any other language has ever been. It has become the first truly global language.

   The statistics of English are astonishing. Of all the world's languages, it is arguably the richest in vocabulary. The compendious *Oxford English Dictionary* lists about 500,000 words. According to traditional estimates, neighboring German has a vocabulary of about 185,000 words and French fewer than 100,000. Three-quarters of the world's mail, and its telexes and cables, are in English. So are more than half the world's technical and scientific periodicals: It is the language of technology from Silicon Valley to Shanghai. English is the medium for 80 percent of the information stored in the world's computers. Nearly half of all business deals in Europe are conducted in English. (Robert McCrum, William Cran, and Robert MacNeil)

38. An astronaut wants to cool off from the pressures and publicity of NASA, so he drives up to a little bar in College Station. The bartender keeps staring at him and then finally says, "Are you one of those astronauts?"

   The astronaut answers yes, but he wants to get away from the publicity and really doesn't want to talk about it. The bartender says, "Well, we got a couple of engineers who've been building their own spaceship out back. Why don't you take a look?"

   The astronaut doesn't want to, but he agrees to see it. It's a mess of beer bottles, cans, and tamale husks. The bartender says, "You fellas at NASA think you're something, going to the moon. But we designed this old spaceship to go to the sun." The astronaut shakes his head and says, "I hate to tell you this, but this thing will be incinerated before you get within a few hundred miles of the sun."

   The bartender says, "We got that all figured out. We're gonna go at night." (Herman Gollub)

39. Bill Clinton was a law professor at the University of Arkansas. Rarely was he in his office. Frequently his classes were cancelled. He might be seen sitting around the law school lounge, chatting and joshing. He was almost never seen using the law library. He filed grades late. In 1974 he did not post

grades for his spring criminal procedure class until the end of the summer. Only six months passed after Clinton graduated from Yale Law School before [he] launched his campaign for a seat in the House of Representatives. (R. Emmett Tyrrell, Jr.)

40. It is worse to steal from the many than to steal from an individual; for one who steals from an individual can appease him by returning the theft: one who steals from the many, however, cannot [since he does not even know all the people from whom he stole]. (Tosefta, *Baba Kamma* 10:14)

41. Life is culturally biased. We live twice as long as people in some of the poorer parts of the world, not because we are more deserving, individually smarter or otherwise more meritorious, but simply because we had the dumb luck to be born into a culture which produces cures and preventions for deadly diseases that have ravaged the human race for centuries. (Thomas Sowell)

42. High incomes in business and the professions are draining "talent" from teaching and government, says Derek Bok, ex-president of Harvard. [But] even if we could raise teachers' salaries 10 percent or 15 percent, the schools' basic problems would remain. The quality of new teachers might improve somewhat. But schools would still have to cope with the social problems of family breakdown. And academic standards would remain lax because, although we say we want good schools, we don't want them to flunk our kids. Higher teachers' salaries can't cure hypocrisy. (Robert J. Samuelson)

43. A pretty woman with a gun makes a man nervous. At the White House Correspondents dinner, a female FBI agent opened her purse for the Secret Service detail, revealing her revolver, badge, and credentials. In her black lace dress, her blonde hair swept up, the agent said, "I'm on call for the Oklahoma bombing." Handing back her gun, the Secret Service man asked, "Do you really know how to use this thing?" She smiled and said she would be happy to meet him Monday on the firing range. (*The Washingtonian*)

44. A New York woman who visited Switzerland recently discovered that import chic is not limited to Americans. She went to a department store in Basel thinking it would be the ideal place to buy a Swiss army knife at a favorable price. She got two surprises. The first was that the store did not sell any of those famous red-handled devices, thick with blades for every imaginable purpose. The second surprise was what the store was selling instead—a black-handled device, thick with blades and

emblazoned in bold white letters, "U.S. ARMY KNIFE." *(New York Times)*

45. Ah, love, let us be true
To one another! for the world, which seems
To lie before us like a land of dreams,
So various, so beautiful, so new,
Hath really neither joy, nor love, nor light,
Nor certitude, nor peace, nor help for pain;
And we are here as on a darkling plain
Swept with confused alarms of struggle and flight,
Where ignorant armies clash by night. (Matthew Arnold)

46. Following the publication of his book *Winesburg, Ohio*, Sherwood Anderson found out that Theodore Dreiser lived on the same block. He decided to pay his fellow author a call. When Dreiser came to the door, Anderson nervously introduced himself. "I am Sherwood Anderson," he said. "I thought I would come to see you." "Oh, hello," said Dreiser, and shut the door in his face.

Anderson later received a note of apology; Dreiser had simply been embarrassed. Dreiser threw a stag party in Anderson's honor. Later in the evening, the doorbell rang. When Dreiser answered it he found a fair-haired young man standing on his stoop holding a bottle. "I am Scott Fitzgerald," the man said. "I have brought you this champagne." "Thanks," said Dreiser, and shut the door in his face. *(New York)*

47. St. Peter is at his post, greeting heaven's new arrivals and assigning them living quarters. First in line is the Pope. St. Peter directs him to heaven's equivalent of one of those $6-a-night roadside motel rooms. No phone, no TV—nothing. Next in line is a lawyer. St. Peter assigns him to a two-bedroom suite. Sauna—the works. The third man can't restrain his curiosity. "St. Peter, forgive me . . . I mean, the *Pope*, for God's sake! And some lawyer?" He gestures weakly at their respective accommodations.

"My son," St. Peter replies calmly, "we have 75 popes up here. We've never had a lawyer." (Andrew Tobias)

48. Sometimes I go home and I can't sleep.

I'm physically tired, but I can't sleep.

And it's getting worse.

I lie in bed and think about things.

Like Brian, who got blown up by a bomb at JFK. He was only 26.

The Croatians thought that blowing up a cop would get them their freedom.

You know what his wife said when they told her? "What's a Croatian?"

Nineteen years on the job, I've seen too much. (Jill Freedman)

49. A slumber did my spirit seal;
I had no human fears;
She seemed a thing that could not feel
The touch of earthly years.

No motion has she now, no force;
She neither hears nor sees;
Rolled round in earth's diurnal course,
With rocks, and stones, and trees. (William Wordsworth)

50. Two letters have passed between these parties, letters which are admitted to be in the handwriting of the defendant, and which speak volumes indeed. . . . The next has no date whatsoever, which in itself is suspicious. "Dear Mrs. B., I shall not be home until to-morrow. Slow coach." And then follows this very remarkable expression. "Don't trouble yourself about the warming-pan." The warming-pan! Why, gentlemen, who *does* trouble himself about a warming-pan? When was the peace of mind of man or woman broken or disturbed by a warming-pan, which is in itself a harmless, a useful, and I will add, gentlemen, a comforting article of domestic furniture? Why is Mrs. Bardell so earnestly entreated not to agitate herself about this warming-pan, unless (as is no doubt the case) it is a mere cover for hidden fire—a mere substitute for some endearing word or promise, agreeably to a preconcerted system of correspondence, artfully contrived by Pickwick with a view to his contemplated desertion. . . ? (Charles Dickens)

1.3 EXTENDED ARGUMENTS

Arguments in natural language can be complicated. A lawyer arguing for the innocence of a client, for instance, typically offers many specific arguments in presenting the case. The lawyer may argue that a piece of evidence is inadmissible, that results from a lab test are ambiguous, that the client could not have reached the scene of the crime by the time it was committed, and so on. All these smaller arguments form part of the larger argument for the client's innocence.

We can divide arguments, then, into two groups: **extended** or **complex** arguments, which contain other arguments, and **simple** arguments, which do not. Because extended arguments are good only if the simple arguments within them are good, we will usually analyze simple arguments. In fact, we will so often focus on simple arguments that we will generally drop the adjective *simple* and speak simply of *arguments*. In this section, however, we focus on extended arguments.

Extended arguments contain other arguments. Analyzing them, therefore, requires not only identifying premises and conclusions but also determining which premises are being taken to support which conclusions. Premise and conclusion indicators still provide the most important clues to the structure of an extended argument. Because authors and speakers do not always make relations of support clear, however, analyzing an extended argument often requires interpretation.

Extended arguments may consist of several simple arguments in sequence. They may contain other extended arguments. And they may consist of a list of premises, followed by several conclusions stated at once. Typically, the conclusion of one part of an extended argument serves as a premise for another part.

Consider an extended argument that the ancient skeptic Sextus Empiricus (ca. A.D. 150–225) offered to challenge our usual notion of change:

(24) If Socrates died, he died either when he was living or when he was dead. But he did not die while living; for assuredly he was living, and as living he had not died. Nor when he was dead; for then he would be twice dead. Therefore Socrates did not die.

The argument as a whole tries to establish the startling conclusion that Socrates did not die. It moreover includes two subarguments. Signaling them is the word *for*, which, as in the earlier argument from Joubert, tends to follow a conclusion and introduce a premise supporting it. In this argument, the first *for* introduces premises supporting the intermediate conclusion that Socrates did not die while living:

(25) Assuredly, Socrates was living.

As living, he had not yet died.

∴ Socrates did not die while living.

The second *for* introduces premises to support the conclusion that Socrates did not die while he was dead:

(26) If Socrates died while he was dead, he would be twice dead.

∴ Socrates did not die while he was dead.

The conclusions of these subarguments act as premises in the larger argument that Socrates never died:

(27) If Socrates died, he died either when he was living or when he was dead.

He did not die while living.

He did not die while he was dead.

∴ Socrates did not die.

Understanding any argument, simple or extended, requires recognizing conclusions. Analyzing an extended argument requires two additional steps. First, it is important to determine whether premises are offered in support of a given conclusion *jointly* or *independently*.

> Premises are taken to support a conclusion *independently* when each premise alone purports to be sufficient to establish the conclusion.
>
> Premises are taken to support a conclusion *jointly* when they purport to be sufficient to establish the conclusion only when taken together.

This is an important distinction. If the premises support the conclusion jointly, the falsity of any one premise brings down the argument. If they support it independently, however, the falsity of a premise does not condemn the argument as a whole. Second, it is important to realize that more than one conclusion may be drawn from a string of premises.

The premises in Sextus's arguments support their respective conclusions jointly. Consider, for example, the larger argument (27) within which the others are embedded. The conclusion *Socrates did not die* is not established by any premise taken all by itself. Only when taken together do the premises lend any support to the conclusion.

For an example of independent support for a conclusion, consider this outline of a debate case for the proposition that methods of environmental control should be changed (a case one of us offered at the Connecticut state tournament years ago):

(28) 1. Current methods of pollution control create serious monitoring problems. (We cannot tell when laws are being broken.)

   2. Current methods of pollution control create serious enforcement problems. (Even when we can tell, we cannot take effective action.)

   3. Current methods of pollution control create serious administrative problems. (Our efforts are costly and counterproductive.)

These contentions are meant to support the conclusion independently. Any one is meant to provide sufficient reason for accepting the conclusion. Debate cases are often structured to provide independent support, so, in theory, at least, winning on one premise is enough to win the debate.

. An excellent illustration of some of these possibilities occurs at the beginning of Sir Arthur Conan Doyle's first Sherlock Holmes story. Dr. Watson had recently been married and had resumed his medical practice. He had not seen Holmes for some time and stopped by to see him. Holmes looked Watson over "in his singular introspective fashion" and said:

> "And in practice again, I observe. You did not tell me that you intended to go into harness."
>
> "Then how do you know?"
>
> "I see it, I deduce it. How do I know that you have been getting yourself very wet lately, and that you have a most careless and clumsy servant girl?"
>
> "My dear Holmes," said I, "this is too much. You would certainly have been burned, had you lived a few centuries ago. It is true that I had a country walk on Thursday and came home in a dreadful mess; but, as I have changed my clothes, I can't imagine how you deduce it. As to Mary Jane, she is incorrigible and my wife has given her notice; but there again I fail to see how you work it out."
>
> He chuckled to himself and rubbed his long nervous hands together.
>
> "It is simplicity itself," said he; "my eyes tell me that on the inside of your left shoe, just where the firelight strikes it, the leather is scored by six almost parallel cuts. Obviously they have been caused by someone who has very carelessly scraped round the edges of the sole in order to remove crusted mud from it. Hence, you see, my double deduction that you had been out in vile weather, and that you had a particularly malignant boot-slitting specimen of the London slavey. As to your practice, if a gentleman walks into my rooms smelling of iodoform, with a black mark of nitrate of silver upon his right forefinger, and a bulge on the side of his top-hat to show where he had secreted his stethoscope, I must be dull indeed, if I do not pronounce him to be an active member of the medical profession."[4]

Holmes here draws three conclusions: (1) that Watson has resumed his medical practice; (2) that Watson has been out in vile weather recently; and (3) that Watson has a careless servant girl. To understand the structure of the extended argument he offers, we must not only recognize these conclusions but also identify the premises supporting each.

The last two conclusions are supported by the same premises (in what Holmes calls a "double deduction"):

---

[4]Sir Arthur Conan Doyle, "A Scandal in Bohemia," *The Adventures of Sherlock Holmes* (Secaucus: Castle Books, 1980; originally published in *The Strand,* 1891), pp. 11–12.

(29) The leather on the inside of Watson's left shoe is scored by six parallel cuts.

The cuts were caused by careless scraping of crusted mud.

∴ Watson had been out in vile weather.

(30) The leather on the inside of Watson's left shoe is scored by six parallel cuts.

The cuts were caused by careless scraping of crusted mud.

∴ Watson had a careless servant girl.

In both cases, Holmes relies on unstated premises or intermediate conclusions: in (29), that Watson's shoes had recently been crusted with mud and that shoes become crusted with mud when their wearer goes out in vile weather; in (30), that a servant girl had scraped Watson's shoes. Arguments that rely on unstated premises are called **enthymemes** (or **elliptical,** or **enthymematic**). In each case, the premises offer joint support for the conclusion; Holmes needs both the cuts and his causal hypothesis to obtain the conclusions.

Holmes's inference that Watson had resumed his medical practice is more straightforward:

(31) You smell of iodoform.

You have a black mark of nitrate of silver upon your right forefinger.

There is a bulge on the side of your top-hat from your stethoscope.

∴ You are an active member of the medical profession.

Are the premises here meant to offer joint or independent support of the conclusion? Probably independent support; any one would be a sign that Watson had returned to active medical practice. Taken together, of course, they offer even stronger support.

P R O B L E M S

Each of the following passages contains a simple or extended argument. Analyze its structure.

1. When you negotiate with people who take hostages you are obliged, in the negotiation, to give something. It may be just a little, it may be a lot, but you have to give something. Once you have given something, the kidnapper gains from his action. So what is his normal and spontaneous reaction? He does it again, thinking that it is a way of obtaining what he cannot obtain by other means.

So you get caught in a process. Naturally you can get maybe two, three or four hostages freed. But you immediately give the kidnapper an inducement to seize another three, four, five or six. So it is an extraordinarily dangerous and irresponsible process. That is why I do not negotiate. (Jacques Chirac)

2. But we are convinced that the American national purpose must at some point be fixed. If it is redefined—or even subject to redefinition—with every change of administration in Washington, the United States risks becoming a factor of inconstancy in the world. The national tendency to oscillate between exaggerated belligerence and unrealistic expectation will be magnified. Other nations—friends or adversaries—unable to gear their policies to American steadiness will go their own way, dooming the United States to growing irrelevance. (Henry Kissinger and Cyrus Vance)

3. The "housing crisis" agitating Mayor Koch and Governor Cuomo is actually a product of local attempts to suspend economic law. Everybody who isn't a tenant admits how destructive the city's World War II rent controls have been, but tenants have the most votes. An extralegal free market in abandoned factory-loft conversions compensated somewhat, but City Hall's first instinct was to prohibit most conversions. Anyone fool enough to want to build new apartments has to pay double the standard price for concrete because of a Mafia stranglehold on the commodity. Vast areas of New York City are a wasteland of abandoned housing and rubble. *(The Wall Street Journal)*

4. It is absurd to call Pope John Paul II a traditionalist. Seldom has there been a more future-oriented pope, such a visionary activist. It would also be wrong to call him a "progressive"; the "progressives" hate him. The Pope criticizes both traditionalists and progressives.

    True, the progressives call everybody to their right "conservative." If by a "neoconservative" we mean a nontraditionalist who criticizes the illusions of progressives, then the Pope is a neoconservative. (Michael Novak)

5. . . . Mars would be the next logical niche for human expansion in the universe. Why Mars? Clearly, Mars will have priority in any manned solar system exploration program because it offers the least severe environment for humans. Due to its atmosphere, its accessible surface, its probable availability of water and its relatively moderate temperatures . . . it is the most hospitable of all the planets other than earth.

Moreover, Mars resources include materials that could be adapted to support human life, including air, fuels, fertilizers, building materials and an environment that could grow food. . . . (James M. Beggs)

6. Computer makers must recognize that the old marketing rule is still golden: Listen to your customers. What corporate computer customers say they want is hardware and software that will allow them to tie their entire organization together in a true information network. Before the industry can give them that, individual manufacturers must agree on the uniform standards under which computers will "talk" to one another. This will be a complex effort. But as long as makers delay, customer frustration rises. *(Business Week)*

7. At this critical point in history, American immigration policies are in a shambles.

   Our borders are totally out of control. Our border patrol apprehends 3,000 illegal immigrants per day, 1.2 million per year. And two get in for every one caught; those caught just try again. There has been a 3,000 percent increase in apprehensions since 1965 with only a 50 percent increase in manpower.

   Not just our borders, but our whole immigration apparatus is out of control: 400,000 more people fly into the U.S. than fly out every year. INS believes that 30 percent of the persons granted permanent residence each year on the basis of family ties are making fraudulent claims. (Richard D. Lamm)

8. The specific case against business schools is that they have neglected certain skills and outlooks that are essential to America's commercial renaissance while inculcating values that can do harm. The traditional strength of business education has been to provide students with a broad view of many varied business functions—marketing, finance, production, and so forth. But like sociology and political science, business training has gotten all wrapped up in mathematical models and such ideas as can be boiled down to numbers. This shift has led schools to play down two fundamental but hard-to-quantify business imperatives: creating the conditions that will permit the design and production of high-quality goods, and waging the constant struggle to inspire, cajole, discipline, lead and in general persuade employees to work in a common cause. (James Fallows)

9. In 1983 government agents in South Florida seized some six tons of cocaine and 850 tons of marijuana (which tends to come by the boatload). In 1985 the figures were twenty-five and 750 tons respectively. In other words, seizures of cocaine, a poten-

tially lethal drug, have quadrupled while seizures of marijuana, a substance that looks benign in comparison, have fallen off. It is generally believed that the amounts of drugs seized reflects the amounts coming in. Thus, almost certainly, more cocaine is being imported now than ever before. . . .

Why the sudden abundance of cocaine? The government's strategy in the war on drugs may be partly to blame. The heightened risk of interdiction has prompted smugglers to favor drugs that are compact and expensive, like cocaine, over drugs that are bulky and relatively cheap, like marijuana. *(Atlantic Monthly)*

10. Being good liberals themselves, they had no ground in principle by which to justify indefinite Israeli rule over a rebellious Palestinian population. Nor could they answer the contention that continued Israeli occupation of the territories would ultimately erode the Jewishness of the state or transform it from a democracy into another South Africa. The only argument they could rely on was security: the argument that Israeli withdrawal in favor of a Palestinian state run by the PLO posed so great a danger to the "body" of Israel that, for the time being and for the foreseeable future, it had to take precedence over the danger to Israel's "soul" admittedly posed by continued occupation. (Norman Podhoretz)

11. He does not want to overturn the Soviet system; he wants to strengthen it. To paraphrase Churchill from another context, Gorbachev did not become general secretary to preside over the demise of the Communist Party. We have an interest in the success of his reforms only to the extent that they change the system to make it less threatening to our security and interests. We should applaud glasnost and perestroika but not pay for them, for if his reforms do not irrevocably alter Soviet foreign policy we will be subsidizing the threat of our own destruction. (Richard Nixon)

12. Those of a more conservative bent, in contrast, believe that the recent changes are little more than a public relations campaign aimed at getting naive Americans to make unilateral concessions that will allow Castro to weather Cuba's current crisis. They argue that U.S. policy toward Cuba had worked and continues to work. Castro is no longer seen abroad as a charismatic revolutionary hero, but rather as a ruthless dictator and an abuser of human rights who has ruined Cuba. . . .

In either case, the United States cannot lose. If Cuba becomes more open economically and politically, that is good for the United States. And if the Soviets have to keep bankrolling Cuba, that is better than what they believe the more liberal policy would lead to—the American subsidy of a Cuba that remains under Castro's control and militarily allied with the Soviet Union. (Susan Kaufman Purcell)

13. A struggle for existence inevitably follows from the high rate at which all organic beings tend to increase. Every being, which during its natural lifetime produces several eggs or seeds, must suffer destruction during some period of its life, and during some season or occasional year; otherwise, on the principle of geometric increase, its numbers would quickly become so inordinately great that no country could support the product. Hence, as more individuals are produced than can possibly survive, there must in every case be a struggle for existence, either one individual with another of the same species, or with the individuals of distinct species, or with the physical conditions of life. It is the doctrine of Malthus applied with manifold force to the whole animal and vegetable kingdoms; for in this case there can be no artificial increase in food, and no prudential restraint from marriage. Although some species may now be increasing, more or less rapidly, in numbers, all cannot do so, for the whole world would not hold them. (Charles Darwin)

14. . . . Holmes [was] still carrying with him the stone which he had picked up in the wood.
"This may interest you, Lestrade," he remarked, holding it out. "The murder was done with it."
"I see no marks."
"There are none."
"How do you know, then?"
"The grass was growing under it. It had lain there only a few days. There was no sign of a place whence it had been taken. It corresponds with the injuries. There was no sign of any other weapon." (Sir Arthur Conan Doyle)

15. "From the first, two facts were very obvious to me, the one that the lady had been quite willing to undergo the wedding ceremony, the other that she had repented of it within a few minutes of returning home. Obviously something had occurred during the morning, then, to cause her to change her mind. What could that something be? She could not have spoken to anyone when she was out, for she had been in the company of the bridegroom. Had she seen someone, then? If she had, it must be someone from America, because she had spent so short a time in this country that she could hardly have allowed anyone to acquire so deep an influence over her that the mere sight of him

would induce her to change her plans so completely. You see we
have already arrived, by a process of exclusion, at the idea that
she might have seen an American. Then who could this American be, and why should he possess so much influence over her?
It might be a lover; it might be a husband. Her young womanhood had, I knew, been spent in rough scenes, and under strange
conditions. So far I had got even before ever I heard Lord St.
Simon's narrative. When he told us of a man in a pew, of the
change in the bride's manner, of so transparent a device for obtaining a note as the dropping of a bouquet, of her resort to her
confidential maid, and of her very significant allusion to claim-
jumping, which in miners' parlance means taking possession of
that which another person has a prior claim to, the whole situation became absolutely clear. She had gone off with a man, and
the man was either a lover or was a previous husband, the
chances being in favor of the latter." (Sir Arthur Conan Doyle)

The following are well-known philosophical arguments. Identify them
as simple or extended, and analyze their structure, identifying premises
and conclusions.

16. If all ideas are determined; and to be determined is to be false;
    then it follows that truth is impossible. (Kwasi Wiredu)

17. If the sceptic's position is that all utterances are devoid of mean-
    ing, then this itself cannot be an utterance. For if it is, it falsifies
    itself. (B. K. Matilal)

18. Whatsoever we imagine is finite. Therefore there is no idea or
    conception of anything we call infinite. (Thomas Hobbes)

19. Let us weigh the gain and the loss in wagering that God is. Let
    us estimate these two chances. If you gain, you gain all; if you
    lose, you lose nothing. Wager, then, without hesitation that He
    is. (Blaise Pascal)

20. To this war of every man against every man, this also is conse-
    quent; that nothing can be unjust. The notions of right and
    wrong, justice and injustice, have there no place. Where there
    is no common power, there is no law; where no law, no injustice.
    (Thomas Hobbes)

21. Is the arrow-maker less benevolent than the maker of armour of
    defense? And yet the arrow-maker's only fear is lest men should
    not be hurt, and the armour-maker's only fear is lest men should
    be hurt. So it is with the priest and the coffin maker. The choice
    of profession, therefore, is a thing in which great caution is
    required. (Mencius)

22. Marx and Engels are therefore on the horns of a dilemma. If all philosophical thinking is ideological, then their own philosophical thinking is ideological and, by their hypothesis, false. If, on the other hand, their philosophical thinking is not ideological then the philosophical thinking of the bourgeoisie is not necessarily ideological. (Kwasi Wiredu)

23. . . . I am not at all afraid of the arguments of the Academicians, who say, What if you are deceived? For if I am deceived, I am. For he who is not, cannot be deceived; and if I am deceived, by this same token I am. . . . certainly I am not deceived in this knowledge that I am. And, consequently, neither am I deceived in knowing that I know. For, as I know that I am, so I know this also, that I know. (Augustine)

24. By Matter, therefore, we are to understand an inert, senseless substance, in which extension, figure, and motion do actually subsist. But it is evident from what we have already shown, that extension, figure, and motion are only ideas existing in the mind, and that an idea can be like nothing but another idea, and that consequently neither they nor their archetypes can exist in an unperceiving substance. Hence, it is plain that the very notion of what is called Matter or corporeal substance, involves a contradiction in it. (George Berkeley)

25. It is a question of fact, whether the perceptions of the senses be produced by external objects, resembling them: How shall this question be determined? By experience surely, as all other questions of a like nature. But here experience is, and must be entirely silent. The mind has never anything present to it but the perceptions, and cannot possibly reach any experience of their connection with objects. The supposition of such a connection is, therefore, without any foundation in reasoning. (David Hume)

26. All men have a mind which cannot bear to see the sufferings of others. . . . if men suddenly see a child about to fall into a well, they will without exception experience a feeling of alarm and distress. They will feel so, not as a ground on which they may gain the favour of the child's parents, nor as a ground on which they may seek the praise of their neighbours and friends, nor from a dislike to the reputation of having been unmoved by such a thing. From this case we may perceive that the feeling of commiseration is essential to man. . . . (Mencius)

27. If *a* is identical with *b*, whatever is true of the one is true of the other, and either may be substituted for the other in any proposition without altering the truth or falsehood of that proposition.

Now George IV wished to know whether Scott was the author of *Waverley;* and in fact Scott *was* the author of *Waverley,* Hence we may substitute *Scott* for the author of *Waverley,* and thereby prove that George IV wished to know whether Scott was Scott. Yet an interest in the law of identity can hardly be attributed to the first gentleman of Europe. (Bertrand Russell)

28. The contrary of every matter of fact is still possible; because it can never imply a contradiction, and is conceived by the mind with the same facility and distinctness, as if ever so conformable to reality. That the sun will not rise tomorrow is no less intelligible a proposition, and implies no more contradiction than the affirmation, that it will rise. We should in vain, therefore, attempt to demonstrate its falsehood. Were it demonstratively false, it would imply a contradiction, and could never be distinctly conceived by the mind. (David Hume)

29. Nor can we doubt that God does well even in the permission of what is evil. For He permits it only in the justice of His judgment. And surely all that is just is good. Although, therefore, evil, in so far as it is evil, is not a good; yet the fact that evil as well as good exists, is a good. For if it were not a good that evil should exist, its existence would not be permitted by the omnipotent God, who without doubt can as easily refuse to permit what He does not wish, as bring about what He does wish. (Augustine)

30. We may remark, then, that every virtue or excellence both brings into good condition the thing of which it is the excellence and makes the work of that thing be done well; e.g. the excellence of the eye makes both the eye and its work good; for it is by the excellence of the eye that we see well. Similarly the excellence of the horse makes a horse both good in itself and good at running and at carrying its rider and at awaiting the attack of the enemy. Therefore, if this is true in every case, the virtue of man also will be the state of character which makes a man good and which makes him do his own work well. (Aristotle)

31. Whatsoever therefore is consequent to a time of war, where every man is enemy to every man, the same consequent to the time wherein men live without other security than what their own strength and their own invention shall furnish them withal. In such condition there is no place for industry, because the fruit thereof is uncertain: and consequently no culture of the earth; no navigation, nor use of the commodities that may be imported by sea; no commodious building; no instruments of moving and removing such things as require much force; no

knowledge of the face of the earth; no account of time; no arts; no letters; no society; and which is worst of all, continual fear, and danger of violent death; and the life of man, solitary, poor, nasty, brutish, and short. (Thomas Hobbes)

32. Soc. As thus: he who sees knows, as we say, that which he sees; for perception and sight and knowledge are admitted to be the same.

    Theaet. Certainly.

    Soc. But he who saw, and has knowledge of that which he saw, remembers, when he closes his eyes, that which he no longer sees.

    Theaet. True.

    Soc. And seeing is knowing, and therefore not-seeing is not-knowing?

    Theaet. Very true.

    Soc. Then the inference is, that a man may have attained the knowledge, of something, which he may remember and yet not know, because he does not see; and this has been affirmed by us to be a monstrous supposition.

    Theaet. Most true.

    Soc. Thus, then, the assertion that knowledge and perception are one, involves a manifest impossibility?

    Theaet. Yes. (Plato)

33. That moral virtue is a mean, then, and in what sense it is so, and that it is a mean between two vices, the one involving excess, the other deficiency, and that it is such because its character is to aim at what is intermediate in passions and in actions, has been sufficiently stated. Hence also it is no easy task to be good. For in everything it is no easy task to find the middle, e.g. to find the middle of a circle is not for every one but for him who knows; so, too, any one can get angry—that is easy—or give or spend money; but to do this to the right person, to the right extent, at the right time, with the right motive, and in the right way, that is not for every one, nor is it easy; wherefore goodness is both rare and laudable and noble. (Aristotle)

34. The "people" who exercise the power are not always the same people with those over whom it is exercised; and the "self-government" spoken of is not the government of each by himself, but of each by all the rest. The will of the people, moreover, practically means the will of the most numerous or the most active part of the people; the majority, or those who succeed in making themselves accepted as the majority; the people,

consequently may desire to oppress a part of their number; and precautions are as much needed against this as against any other abuse of power. The limitation, therefore, of the power of government over individuals loses none of its importance when the holders of power are regularly accountable to the community, that is, to the strongest party therein. (John Stuart Mill)

35. Again, it is a proper office of public authority to guard against accidents. If either a public officer or any one else saw a person attempting to cross a bridge which had been ascertained to be unsafe, and there were no time to warn him of his danger, they might seize him and turn him back, without any real infringement of his liberty; for liberty consists in doing what one desires, and he does not desire to fall into the river. Nevertheless, when there is not a certainty, but only a danger of mischief, no one but the person himself can judge of the sufficiency of the motive which may prompt him to incur the risk: in this case, therefore (unless he is a child, or delirious, or in some state of excitement or absorption incompatible with the full use of the reflecting faculty), he ought, I conceive, to be only warned of the danger; not forcibly prevented from exposing himself to it. (John Stuart Mill)

36. . . . after the constant conjunction of two objects—heat and flame, for instance, weight and solidity—we are determined by custom alone to expect the one from the appearance of the other. This hypothesis seems even the only one which explains the difficulty, why we draw, from a thousand instances, an inference which we are not able to draw from one instance, that is, in no respect, different from them. Reason is incapable of any such variation. The conclusions which it draws from considering one circle are the same which it would form upon surveying all the circles in the universe. But no man, having seen only one body move after being impelled by another, could infer that every other body will move after a like impulse. All inferences from experience, therefore, are effects of custom, not of reasoning. (David Hume)

37. When we look about us towards external objects, and consider the operation of causes, we are never able, in a single instance, to discover any power or necessary connection; any quality, which binds the effect to the cause, and renders the one an infallible consequence of the other. We only find, that the one does actually, in fact, follow the other. The impulse of one billiard ball is attended with motion in the second. This is the whole that appears to the outward senses. The mind feels no sentiment or inward impression from this succession of objects. Consequently,

there is not, in any single, particular instance of cause and effect, anything which can suggest the idea of power or necessary connection. (David Hume)

38. Now, as in the ideas of God there is an infinite number of possible universes, and as only one of them can be actual, there must be a sufficient reason for the choice of God, which leads Him to decide upon one rather than another. And this reason can be found only in the fitness [convenance], or in the degrees of perfection, that these worlds possess, since each possible thing has the right to aspire to existence in proportion to the amount of perfection it contains in germ. Thus the actual existence of the best that wisdom makes known to God is due to this, that His goodness makes Him choose it, and His power makes Him produce it. Now this connection or adaptation of all created things to each and of each to all, means that each simple substance has relations which express all the others, and, consequently, that it is a perpetual living mirror of the universe. (G. W. Leibniz)

Philosophers have advanced many arguments to prove that there is a God. Here are twelve famous arguments for God's existence. The first is from French philosopher René Descartes (1596–1650); the next three are from St. Anselm of Canterbury (1033–1109). They are versions of the **ontological argument.** The next five are from St. Thomas Aquinas (1225–1274); the last three, from the Persian philosopher Avicenna (980–1037), the Spanish and Morrocan philosopher Averroës (1126–1198), and the German philosopher G. W. F. Leibniz (1646–1716). All but one are versions of the **cosmological argument.** (Number 47 is a version of the **teleological argument.**) Analyze the structure of these arguments, identifying premises and conclusions.

39. Although it is not necessary that I happen upon any thought of God, nevertheless as often as I think of a being first and supreme—and bring forth the idea of God as if from the storehouse of my mind—I must of necessity ascribe all perfections to it, even though I do not at that time enumerate them all, nor take note of them one by one. This necessity plainly suffices so that afterwards, when I consider that existence is a perfection, I rightly conclude that a first and supreme being exists.

40. Even the Fool . . . is forced to agree that something, the greater than which cannot be thought, exists in the intellect, since he understands this when he hears it, and whatever is understood is in the intellect. And surely that, the greater than which cannot be thought, cannot exist in the intellect alone. For even if it exists solely in the intellect, it can be thought to exist in reality, which

is greater. If, then, that, the greater than which cannot be thought, exists in the intellect alone, this same being, than which a greater cannot be thought, is that than which a greater can be thought. But surely this is impossible. Therefore, there can be absolutely no doubt that something, the greater than which cannot be thought, exists both in the intellect and in reality.

41. Certainly, this being so truly exists that it cannot even be thought not to exist. For something can be thought to exist that cannot be thought not to exist, and this is greater than whatever can be thought not to exist. Hence, if that, the greater than which cannot be thought, can be thought not to exist, then that, the greater than which cannot be thought, is not the same as that, the greater than which cannot be thought, which is absurd. Therefore, something, the greater than which cannot be thought, exists so truly that it cannot even be thought not to exist.

42. You exist so truly, Lord my God, that You cannot even be thought not to exist. And this is as it should be. For, if a mind could think of something better than You, the creature would rise above its creator and judge its creator, and that is completely absurd. In fact, everything else, except You alone, can be thought not to exist. You alone, then, of all things most truly exist, and therefore of all things possess existence to the highest degree; for anything else does not exist as truly, and possesses existence to a lesser degree.

43. The first and most obvious way is based on change. Certainly, our senses show us that some things in the world are changing. Now anything changing is changed by something else. For nothing changes except what can but does not yet have some actuality; something that causes change has that actuality already. For to cause change is to bring into being what was before only potential, and only something that already is can do this. Thus, fire, which is actually hot, causes wood, which can be hot, to become actually hot, and so causes change in the wood. Now it is impossible for the same thing to be simultaneously actually F and potentially F, though it can be actually F and potentially G: The actually hot cannot at the same time be potentially hot, though it can be potentially cold. It is therefore impossible that something undergoing a change cause itself to undergo that very change. It follows that anything changing must be changed by something else. If this other thing is also changing, it is being changed by another thing, and that by another. Now this does not go on to infinity, or else there would be no first cause of the change and, consequently, no other changes. The intermediate causes will not

produce change unless they are affected by the first change, just as a stick does not move unless moved by a hand. Therefore, it is necessary to arrive at some first cause of change, itself changed by nothing, and this all understand to be God.

44. The second way is based on the nature of causation. In the observable world causes are to be found ordered in series; we never observe, or even could observe, something causing itself, for this would mean it preceded itself, and this is impossible. Such a series of causes, however, must stop somewhere. For in all series of causes, an earlier member causes an intermediate, and the intermediate a last (whether the intermediate be one or many). If you eliminate a cause you also eliminate its effects. Therefore there can be neither a last nor an intermediate cause unless there is a first. But if the series of causes goes on to infinity, and there is no first cause, there would be neither intermediate causes nor a final effect, which is patently false. It is therefore necessary to posit a first cause, which all call 'God'.

45. The third way depends on what is possible and necessary, and goes like this. We observe in things something that can be, and can not be, for we observe them springing up and dying away, and consequently being and not being. Now not everything can be like this, for whatever can not be, once was not. If all things could not be, therefore, at one time there was nothing. But if that were true there would be nothing even now, because something that does not exist can be brought into being only by something that already exists. So, if there had been nothing, it would have been impossible for anything to come into being, and there would be nothing now, which is patently false. Not all things, therefore, are possible but not necessary; something is necessary. Now what is necessary may or may not have its necessity caused by something else. It is impossible to go on to infinity in a series of necessary things having a cause of their necessity, just as with any series of causes. It is therefore necessary to posit something which is itself necessary, having no other cause of its necessity, but causing necessity in everything else.

46. The fourth way is based on the gradation observed in things. For some things are found to be more good, true, and noble, and other things less. But 'more' and 'less' describe varying degrees of approximating the maximum; for example, things are hotter and hotter the more they approach the hottest. Something, therefore, is the best and truest and noblest of things, and consequently exists to the highest degree; for Aristotle says that the truest things exist to the highest degree. Now when many things

have a common property, the one having it most fully causes the others to have it. Fire, the hottest of all things, causes the heat in all other things, to use Aristotle's example. Therefore, something causes all other things to be, to be good, and to have any other perfections, and this we call 'God'.

47. The fifth way is based on the rule-governed character of nature. The ordering of actions toward an end is observed in all bodies obeying natural laws, even when they lack awareness. For their behavior hardly ever varies, and will practically always turn out well; this shows that they truly tend toward a goal, and do not merely hit it by accident. Nothing, however, that lacks awareness tends toward a goal, except under the direction of someone aware and intelligent. The arrow, for example, requires an archer. All things in nature, therefore, are directed toward a goal by someone intelligent, and this we call 'God'.

48. Whatever has being must either have a reason for its being, or have no reason for it. If it has a reason, then it is contingent. . . . If on the other hand it has no reason for its being in any way whatsoever, then it is necessary in its being. This rule having been confirmed, I shall now proceed to prove that there is in being a being which has no reason for its being.

Such a being is either contingent or necessary. If it is necessary, then the point we sought to prove is established. If on the other hand it is contingent, that which is contingent cannot enter upon being except for some reason which sways the scales in favour of its being and against its not-being. If the reason is also contingent, then there is a chain of contingents linked one to the other, and there is no being at all; for this being which is the subject of our hypothesis cannot enter into being so long as it is not preceded by an infinite succession of beings, which is absurd. Therefore contingent beings end in a Necessary Being.

49. Possible existents must of necessity have causes which precede them, and if these causes again are possible it follows that they have causes and that there is an infinite regress; and if there is an infinite regress there is no cause, and the possible will exist without a cause, and this is impossible. Therefore the series must end in a necessary cause, and in this case this necessary cause must be necessary through a cause or without a cause, and if through a cause, this cause must have a cause and so on infinitely, and if we have an infinite regress here, it follows that what was assumed to have a cause has no cause, and this is impossible. Therefore the series must end in a cause necessary without a cause, i.e. necessary by itself, and this necessarily is the necessary existent.

50. . . . *nothing happens without a sufficient reason;* that is to say, that nothing happens without its being possible for him who should sufficiently understand things, to give a reason sufficient to determine why it is so and not otherwise. . . . Now this sufficient reason for the existence of the universe cannot be found *in the series of contingent things,* that is, of bodies and of their representation in souls; for matter being indifferent in itself to motion and to rest, and to this or another motion, we cannot find the reason of motion in it, and still less of a certain motion. And although the present motion which is in matter, comes from the preceding motion, and that from still another preceding, yet in this way we make no progress, try as we may; for the same question always remains. Thus it must be that the sufficient reason, which has no need of another reason, be outside this series of contingent things and be found in a substance which is the cause, or which is a necessary being, carrying the reason of its existence within itself; otherwise we still should not have a sufficient reason in which we could rest. And this final reason of things is called *God.*

 ## 1.4 Validity and Strength

Some arguments are good; others are not. What distinguishes good from bad arguments? What makes a good argument good?

A good argument links its premises to its conclusion in the right way. A special connection exists between the premises and the conclusion. To see what this special connection is, consider an argument that has true premises and a true conclusion but is nevertheless bad:

(32) Aristotle was Greek.

Reno is west of San Diego.

∴ The Berlin Wall fell in 1989.

The facts cited in the premises have little to do with the conclusion. The Berlin Wall might have stood into 1990 if a few things had happened differently; if more had happened differently, it might never have been built at all. None of the events in question has anything to do with ancient philosophy or U.S. geography. That is, the conclusion of this argument could have turned out false, even when the premises were true. The truth of the premises does nothing to guarantee the truth of the conclusion. This is the mark of a **deductively invalid** argument: Its premises could all be true in a circumstance in which its conclusion is false.

In a **deductively valid** argument, the truth of the premises does guarantee the truth of the conclusion. If the premises are all true, then the conclusion has to be true. Consider, for example, the argument:

(33) All Canadians are North Americans.

Jeff is a Canadian.

∴ Jeff is a North American.

In any circumstance in which the premises of this argument are true, the conclusion must be true as well. It is impossible to conceive of a state of affairs in which, while all Canadians are North Americans, Jeff is a Canadian but not a North American. In a deductively valid argument, the truth of the premises guarantees the truth of the conclusion. Or, to say the same thing, if the conclusion of a deductively valid argument is false, at least one premise must also be false.

> An argument is *deductively valid* if and only if its premises cannot all be true while its conclusion is false.

A deductively valid argument can have true premises and a true conclusion, some (or all) false premises and a false conclusion, or false premises and a true conclusion. But no deductively valid argument has true premises and a false conclusion.

SOME DEDUCTIVELY VALID ARGUMENTS

TRUE PREMISES, TRUE CONCLUSION	FALSE PREMISES, FALSE CONCLUSION	FALSE PREMISES, TRUE CONCLUSION
You can read.	You are a duck.	You are a duck.
All readers are mortal.	All ducks eat dirt.	All ducks can read.
∴ You are mortal.	∴ You eat dirt.	∴ You can read.

Each argument is deductively valid. In each case, it would be impossible for the premises all to be true but the conclusion false. How could it be true that you are a duck, and true that all ducks eat dirt, but false that you eat dirt? Whether the premises and conclusion are actually true or false makes little difference to the validity of the argument. We evaluate deductive validity *as if* the premises were true. What matters is that *if* the premises are true, the conclusion cannot be false.

Deductively valid arguments are only one species of good argument. Other arguments are **inductively strong** (or **reliable**). Although the truth of the premises of such an argument does not guarantee the truth of its conclusion, it does make the truth of the conclusion probable. Consider, for example, this argument:

(34) Every crow that has been observed is black.

∴ All crows are black.

It is possible for the premise to be true while the conclusion is false. There may be white crows that nobody has ever seen. So, the argument is deductively invalid. Nevertheless, the premise lends some support to the conclusion. The argument is inductively strong; how strong depends on how many crows have been observed, among other things. Inductively reliable arguments are extremely important in both scientific and everyday reasoning. They constitute the subject matter of Part III of this book. Until then, we will focus on deductive validity.

Generally, a good argument should not only be valid or inductively strong but also have true premises. In most circumstances, people want to learn or say true things; they want to come to true conclusions. Arguments with false premises, however, lead to true conclusions only accidentally. Using premises we know to be false, we may be able to convince someone, but this is sophistry, not an ideal of argumentative success.

A **sound** argument has true premises and is deductively valid. Furthermore, since, in any valid argument, the truth of the premises guarantees the truth of the conclusion, it also has a true conclusion.

> An argument is *sound* if and only if (a) it is valid and (b) all its premises are true.

The analogous concept for inductive arguments is **cogency:**

> An argument is *cogent* if and only if (a) it is inductively strong and (b) all its premises are true.

Most of this book focuses not on soundness or cogency but on validity or strength. First, validity is obviously a crucial component of soundness. Strength is a crucial component of cogency. We cannot evaluate whether an argument is sound, for example, without first determining whether it is valid. Second, evaluating soundness or cogency requires judging the actual truth or falsehood of premises. This is no business of a logician. Determining truth or falsehood requires a knowledge of the subject matter of the argument.

Third, although we usually want to argue from true premises, many useful arguments start from false ones. Some arguments try to show that a certain statement is false by using it as a premise to reach an outrageous or absurd conclusion. Others adopt a premise purely as a hypothesis to see what would follow if it were true. Aristotle first realized how important such arguments are; he characterized them as having **dialectical,** rather than **demonstrative,** premises.

P R O B L E M S

Evaluate the following arguments as valid or invalid. If the argument is invalid, describe a circumstance in which the premises would be true but the conclusion would be false. Do any of the invalid arguments nevertheless make their conclusions probable in your opinion?

1. If the rain continues, there will be a real danger of floods. The rain will continue. Therefore, flooding will be a real danger.

2. If Sally has pneumonia, she needs penicillin and lots of rest. Sally does need penicillin and lots of rest. So, Sally has pneumonia.

3. The Republicans in Congress will agree to the cut only if the president announces his support first. The president won't announce his support first, so, the Congressional Republicans won't agree to the cut.

4. If chicken is overcooked, it is dry. This chicken is not overcooked. Therefore, this chicken is not dry.

5. I have already said that he must have gone to King's Pyland or to Mapleton. He is not at King's Pyland, therefore he is at Mapleton. (Sir Arthur Conan Doyle)

6. Nobody saw what happened. If nobody witnessed it, nobody can testify. If nobody can testify, you can't be convicted. So, you can't be convicted.

7. We will let you out of the lease only if you pay us two months' rent. You can't pay us two months' rent. So, we won't let you out of the lease.

8. If Socrates died, he died either while he was living or while he was dead. But he did not die while living; moreover, he surely did not die while he was already dead. Hence, Socrates did not die. (Sextus Empiricus)

9. A man cannot serve both God and Mammon. But if a man does not serve Mammon, he starves; if he starves, he can't serve God. Therefore, a man cannot serve God.

10. Modern physics asserts that there is no such thing as absolute motion. If this is correct, then there is no such thing as absolute time, and our ordinary notions of time are wrong. So, either our ordinary ideas about time or modern physics are mistaken.

11. Either we ought to philosophize or we ought not. If we ought, then we ought. If we ought not, then also we ought (to justify this view). Hence in any case we ought to philosophize. (Aristotle)

12. All logic problems are easy. Everything that isn't easy gives me a headache. Therefore, some things that give me a headache aren't logic problems.

13. All flying horses are quick and clever; all flying horses live forever. And sad but true, all horses die. It follows that no horses fly.

14. Some illegal acts go unpunished. All blatantly wrong acts are punished. Therefore, some illegal acts are not blatantly wrong.

15. All who do not remember the past are condemned to repeat it. No one condemned to repeat the past looks forward to the future with eagerness. So, everyone who eagerly looks forward to the future remembers the past.

16. Lori is unhappy with some people who didn't write thank-you notes. Lori will send presents next year to everyone with whom she's happy. Therefore, some people who didn't write thank-you notes won't get presents from Lori next year.

17. Anyone who is not an idiot can see that Jake is lying. Some people in this room can't tell that Jake is lying. Hence, some people in this room are idiots.

18. Genevieve befriends anyone who has been treated unfairly. But she doesn't befriend some really obnoxious people. So, some really obnoxious people receive fair treatment.

19. No mammals but bats can fly. Every commonly kept house pet is a mammal, but none are bats. So nothing that can fly is a commonly kept house pet.

20. Nothing stupid is difficult. Everything you can do is stupid; anything that isn't difficult, I can do better than you. So anything you can do, I can do better.

21. For while every man is able to judge a demonstration (it would not deserve this name if all those who consider it attentively were not convinced and persuaded by it), nevertheless not every man is able to discover demonstrations on his own initiative, nor to present them distinctly once they are discovered, if he lacks leisure or method. (G. W. F. Leibniz) [Therefore, not all those who can judge demonstrations can discover them.]

22. 1. You had chalk between your left finger and thumb when you returned from the club last night. 2. You put chalk there when you play billiards to steady the cue. 3. You never play billiards

except with Thurston. 4. You told me four weeks ago that Thurston had an option on some South African property which would expire in a month, and which he desired you to share with him. 5. Your cheque-book is locked in my drawer, and you have not asked for the key. 6. You do not propose to invest your money in this manner. (Sir Arthur Conan Doyle)

23. "I see that you are professionally rather busy just now," said he, glancing very keenly across at me.

"Yes, I've had a busy day," I answered. "It may seem very foolish in your eyes," I added, "but really I don't know how you deduced it."

Holmes chuckled to himself.

"I have the advantage of knowing your habits, my dear Watson," said he. "When your round is a short one you walk, and when it is a long one you use a hansom. As I perceive that your boots, although used, are by no means dirty, I cannot doubt that you are at present busy enough to justify the hansom."

"Excellent!" I cried.

"Elementary," said he. (Sir Arthur Conan Doyle)

 ## 1.5 IMPLICATION AND EQUIVALENCE

A concept closely related to validity is **implication.** The verb *imply* has various uses in English. Here we discuss just one highly specialized use. We can express the idea that an argument is valid by saying that its conclusion *follows from* its premises. Equivalently, we can say that its premises *imply* or *entail* its conclusion. At least part of what we mean, in either case, is that the truth of the premises guarantees the truth of the conclusion. If the premises are true, the conclusion has to be true, too. Implication, then, is very similar to validity. But validity is a property of arguments; implication is a relation between statements and sets of statements.

A set of statements implies a given statement just in case the truth of that statement is guaranteed by the truth of all the members of the set.

> A set of statements *implies* a statement if and only if, whenever every statement in the set is true, that statement must also be true.

If an argument is valid, the set consisting of its premises implies its conclusion.

We can also speak of a single statement implying another statement.

> One statement *implies* another if and only if, whenever the first is true, the second must be true as well.

One statement implies another, that is, if and only if the truth of the former guarantees the truth of the latter. In every circumstance in which the first is true, the second must be true as well.

Consider these two pairs of statements:

(35) a. Alice was born in New York but now lives in Texas.
     b. Alice was born in New York.

(36) a. Bill will spend his vacation skiing or sailing.
     b. Bill will spend his vacation sailing.

Statement (35)a implies (35)b. It is impossible to conceive of a situation in which it is true that Alice was born in New York but now lives in Texas and false that Alice was born in New York. In such a circumstance, Alice would have to have been born and not been born in New York; the statement *Alice was born in New York* would have to be both true and false at the same time. There are no such circumstances; no statement can be both true and false at the same time. So, the truth of (35)a guarantees the truth of (35)b. Does the truth of (36)a similarly guarantee the truth of (36)b? Obviously not. Imagine that Bill spends his vacation skiing, never going near the water. In this situation, (36)a is true, but (36)b is false. So, (36)a does not imply (36)b.

If two statements imply each other, they must be true in exactly the same circumstances. In such a case, we say that they are **equivalent.**

> One statement is *equivalent* to another if and only if it is impossible for them to disagree in truth value.

Equivalent statements must be true in the same circumstances and false in the same circumstances. There could be no situation in which one would be true while the other would be false. Thus, equivalence amounts to implication in both directions. A is equivalent to B if and only if A implies B and B implies A.

To make this more concrete, consider four more statements:

(37) a. No dollar bills are purple.
     b. Nothing purple is a dollar bill.

(38) a. All bears are animals.
     b. All animals are bears.

The statements in (37) are equivalent. Any circumstance in which no dollar bills are purple is one in which nothing purple is a dollar bill and vice versa. Both statements say that nothing is both purple and a dollar bill. In (38), however, the statements are obviously not equivalent. All bears are animals, so (38)a is true. But not all animals are bears, so (38)b is false. The real world is a case in which these statements disagree in truth value.

P R O B L E M S

Consider the statements in each pair: Are they equivalent? If not, does either statement imply the other?

1. (a) I'll do the dishes.   (b) Frank and I will do the dishes.

2. (a) Geraldine went to Jamaica this year.   (b) Geraldine went to Jamaica or Trinidad this year.

3. (a) Paul and Kate are both from Cleveland.   (b) Paul is from Cleveland, and so is Kate.

4. (a) Donna and Miguel won't both win reelection.   (b) Neither Donna nor Miguel will win reelection.

5. (a) Either Dr. Jones or Dr. Smith will see you.   (b) If Dr. Jones sees you, Dr. Smith won't.

6. (a) If a fetus is a human being, then abortion is wrong.   (b) If a fetus isn't a human being, then abortion is permissible.

7. (a) Xenia will graduate only if she improves her GPA.   (b) If Xenia doesn't improve her GPA, she won't graduate.

8. (a) If Orrin retires next year, we will promote Edna.   (b) We won't promote Edna unless Orrin retires next year.

9. (a) If either Barry or Jane drops out of the race, Yvonne will reap a huge windfall.   (b) If Barry drops out of the race, Yvonne will reap a huge windfall.

10. (a) If Sandra and Harold leave by 11:00, I'll be amazed.   (b) If Harold leaves by 11:00, I'll be amazed.

11. (a) Irma knows that Carl has been seeing Nina.   (b) Carl has been seeing Nina, and Irma knows it.

12. (a) Quincy and Lou will agree only if you do.   (b) Quincy will agree only if you do; the same holds for Lou.

13. (a) Fred will drop the suit if you settle for $50,000.   (b) Fred will drop the suit only if you settle for $50,000.

14. (a) Everyone who can answer this question is clever.   (b) Everyone who is clever can answer this question.

15. (a) No French wines are inexpensive.   (b) No inexpensive wines are French.

16. (a) Several cabinet ministers are Socialists.   (b) Several Socialists are cabinet ministers.

17. (a) All linebackers are strong and agile.   (b) All linebackers are strong, and all are agile.

18. (a) No Communists favor both a free market and individual liberties.   (b) No Communists favor a free market, and no Communists favor individual liberties.

19. (a) Tracy ran up the stairs.   (b) Tracy ran.

20. (a) You didn't have to leave the ring.   (b) You could have stayed in the ring.

21. (a) Julian ate.   (b) Julian ate something.

22. (a) Andy and Roxanne gave me a blow-by-blow description.   (b) Andy gave me a blow-by-blow description, and so did Roxanne.

23. (a) It is not true that at most two people can identify Pat.   (b) At least three people can identify Pat.

24. (a) Karen knows Alice if anyone does.   (b) Either Karen knows Alice, or nobody does.

25. (a) None but the brave deserve the fair.   (b) All the brave deserve the fair.

26. (a) Only the good die young.   (b) All who die young are good.

Consider the statement "If Ralph fails the final exam, he will fail the course." What follows from this, together with the information listed?

27. Ralph will fail the final exam.

28. Ralph will fail the course.

29. Ralph won't fail the final exam.

30. Ralph won't fail the course.

Consider the statement "The patient will die unless we operate immediately." What follows from this, together with the information listed?

31. The patient will die.

32. The patient will not die.

33. We will operate immediately.

34. We won't operate immediately.

Consider the statement "You may have soup or eggroll." What follows from this?

35. You may have soup.

36. You may have both soup and eggroll.

Consider the statement "If a fetus is a person, it has a right to life." Which of the following statements follow from this? Which imply it?

37. A fetus is a person.

38. If a fetus has a right to life, then it is a person.

39. A fetus has a right to life only if it is a person.

40. A fetus is a person only if it has a right to life.

41. If a fetus isn't a person, it doesn't have a right to life.

42. If a fetus doesn't have a right to life, it isn't a person.

43. A fetus has a right to life.

44. A fetus isn't a person only if it doesn't have a right to life.

45. A fetus doesn't have a right to life only if it isn't a person.

46. A fetus doesn't have a right to life unless it is a person.

47. A fetus isn't a person unless it has a right to life.

48. A fetus is a person unless it doesn't have a right to life.

49. A fetus has a right to life unless it isn't a person.

Consider this statement from IRS publication 17, *Your Federal Income Tax:*

> If you are single, you must file a return if you had gross income of $3,560 or more for the year.

What follows from this, together with the information listed?

50. You are single with an income of $2,500.

51. You are married with an income of $2,500.

52. You are single with an income of $25,000.

53. You are married with an income of $25,000.

54. You are single but do not have to file a return.

55. You are married but do not have to file a return.

56. You have an income of $4,500 but do not have to file a return.

57. An old joke: Mutt says, "See you later." Jeff answers, "Not if I see you first." Suppose that both statements are true. What follows?

58. Lao-Tzu said, "Real words are not vain, Vain words not real." Are these two statements equivalent? If not, in what circumstances could one be true while the other is false?

59. From an episode of *The Simpsons:*

> Bart: If Lisa stays home, I stay home.
>
> Lisa: If Bart stays home, I'm going to school.

Suppose that both statements are true. What follows?
   In the episode, it turns out that Bart goes to school while Lisa stays home. That implies that one of the preceding statements is false. Which?

A party invitation says: "Please come to a party this Friday at the Ashers' house at 8 P.M. Respond only if you can't come." Do you abide by this request if you

60. Respond and do not come?

61. Respond and come?

62. Do not respond and come?

63. Do not respond and do not come?

64. Donora, Pennsylvania, used to greet visitors with a road sign saying, "Donora. The nicest town on earth, next to yours." Could this be true for every visitor? If so, what does that imply about Donora?

 ## 1.6 FORM AND INVALIDITY

The methods of this book provide ways of showing arguments to be good. Some also provide ways of showing arguments to be bad. But the most powerful ways of showing arguments to be invalid are intuitive. The methods to come largely make arguments more precise in certain contexts.

An argument is valid if the truth of the premises guarantees the truth of the conclusion. To show that an argument is invalid, therefore, one needs to show that the premises could all be true while the conclusion is false.

There are two ways of doing this. The first is simply to describe such a situation. That is, we can show an argument to be invalid by depicting a possible circumstance in which the premises were all true but the conclusion false. We call this the **direct** method. Consider the paradigm of invalidity we considered a few pages ago:

(32)   Aristotle was Greek.

Reno is west of San Diego.

∴ The Berlin Wall fell in 1989.

It is easy to depict a situation in which the premises are true but the conclusion false. Suppose that ancient history and U.S. geography are unchanged but, then, that the Berlin Wall collapsed just a few weeks later than it actually did. The situation that results is one in which Aristotle was Greek, Reno is west of San Diego, but the Berlin Wall fell in 1990, not 1989.

To consider a more interesting argument:

(39)   If the patient's liver suffers irreparable harm, her skin will turn yellow.

The patient's skin is turning yellow.

∴ The patient's liver has suffered irreparable harm.

This argument has some inductive force; jaundice is evidence of liver damage. Nevertheless, it is not deductively valid. It exemplifies a fallacy known as **affirming the consequent.** To see that it is invalid, we might conceive of a situation in which the patient's skin turns yellow for other reasons, for example, because of temporary malfunctioning of the liver or a deficiency of vitamin D.

Another well-known fallacy is **denying the antecedent,** exemplified here:

(40)   If no one testifies against him, Joey will continue to run the mob.

Someone will testify against him.

∴ Joey won't continue to run the mob.

This argument fails. If no one testifies, the first premise says, Joey will continue to run the mob. That tells us nothing at all about what will happen if someone does testify. Maybe Joey will be convicted and will no longer run the mob. But maybe Joey will nevertheless be acquitted. Maybe Joey will be convicted but will continue to run the mob from his prison cell. In either case, the premises would be true while the conclusion was false.

Finally, consider a different form of argument:

(41)   Everyone in the family had heard about the rumor.

Everyone in Springfield had heard about it.

∴ Some people in the family live in Springfield.

The truth of the conclusion is not guaranteed. We can easily depict a situation in which everyone in the family had heard about the rumor and everybody in Springfield had heard about it without there being any overlap between the two groups. (Perhaps a member of the family committed some indiscretion while visiting Springfield, for example. Or perhaps the Springfield newspaper did a story on the rumor.)

The simplest way to show an argument to be invalid, then, is to describe a situation in which its premises are true and its conclusion is false. This method can be difficult to apply, however, because people may disagree about whether the situation described is really possible. Aristotle devised another method, based on the idea of **form.** It is often called the method of **counterexamples.**

> To show that an argument is invalid, produce another argument of the same form with true premises and a false conclusion.

Throughout this book, we will analyze arguments in terms of form. Consider the argument about jaundice we just examined:

(39)   If the patient's liver suffers irreparable harm, her skin will turn yellow.

The patient's skin is turning yellow.

∴ The patient's liver has suffered irreparable harm.

This argument has something in common with each of the following:

(42)   a. If Jane gets an A on the exam, she'll pass the course.

Jane will pass the course.

∴ Jane will get an A on the exam.

b. If the sky fills with very dark clouds, it will rain.

It will rain.

∴ The sky will fill with very dark clouds.

c. If the United States is in South America, it is in the Western Hemisphere.

The United States is in the Western Hemisphere.

∴ The United States is in South America.

Each has the form

(43)   If p, then q

q

∴ p

In general, an argument has a certain form if it can be produced by uniformly substituting terms or statements for the letters in the form. If an argument has a certain form, it is said to be an **instance** of that form.

Aristotle was perhaps the first logician to see that validity is **formal:**

An argument form is valid if and only if every instance of it is valid.

An argument form is invalid if and only if some instance of it is invalid.

If an argument form is invalid, in fact, some instance is bound to have true premises and a false conclusion. One may show that an argument form is invalid, then, by displaying an instance of it with true premises and a false conclusion.

Argument (42)c shows that affirming the consequent, argument form (43), is invalid. Anything that is in South America is in the Western Hemisphere, so the first premise of (42)c is true. And the United States is in the Western Hemisphere. But the United States is not in South America. An instance of (43) is invalid, so (43) itself is invalid. The method of counterexamples allows us to conclude that the other arguments of that form—(39), (42)a, and (42)b—are also invalid.

But we must be careful. What is the link between the validity of an argument and the validity of an argument form?

An argument is valid if and only if it has a valid form.

An argument is invalid if and only if it does not have a valid form—that is, if and only if no form of it is valid.

If an argument instantiates a valid form, it is valid. All we need to do to demonstrate validity, then, is to exhibit a valid form.

But invalidity is more difficult. An argument that instantiates an invalid form might nevertheless be valid. Here, for example, are two arguments that instantiate the form (43) but are nevertheless valid:

(44) a. If pigs grunt, pigs grunt.

Pigs grunt.

∴ Pigs grunt.

b. If the United States is the United States, it is in the Western Hemisphere.

The United States is in the Western Hemisphere.

∴ The United States is the United States.

Both result from substituting statements for the letters in (43). So, both instantiate that invalid form. But (44)a is valid. The truth of the premise *Pigs grunt* guarantees the truth of the conclusion *Pigs grunt*. The argument instantiates not only the invalid form (43) but also the valid form

(45) If p, then p

p

∴ p

Similarly, (44)b is valid because it is necessarily true that the United States is the United States; the conclusion can never be false, no matter what the premises are. It instantiates the valid form

(46) If a = a, then p

p

∴ a = a

So, exhibiting an invalid form does not in itself show an argument to be invalid.

In both (44)a and (44)b, it is easy to see what has gone wrong. The arguments have more structure than the form (43) displays. The method of counterexamples requires not merely that we find an argument with true premises and a false conclusion that shares a form with the original argument, but also that we find one that shares its *specific form:* the most explicit form we can devise, displaying the most structure. The invalidity of (42)c allows us to deduce the invalidity not only of the form (43), affirming the consequent, but also of any argument that has that as its most explicit form.

To apply the method of counterexamples to (40), we first identify its form as explicitly as possible:

(40) If no one testifies against him, Joey will continue to run the mob.

Someone will testify against him.

∴ Joey won't continue to run the mob.

(47) If p, then q

It is not the case that p

∴ It is not the case that q

Now we look for an argument having that form with true premises and a false conclusion:

(48) If the United States is in South America, it is in the Western Hemisphere.

The United States is not in South America.

∴ The United States is not in the Western Hemisphere.

The premises are true, but the conclusion is false. We may infer that (48), (40), and the form (47), denying the antecedent, are invalid.

Finally, consider the invalid argument

(41) Everyone in the family had heard about the rumor.

Everyone in Springfield had heard about it.

∴ Some people in the family live in Springfield.

We identify its form as explicitly as possible:

(49) All F are R

All S are R

∴ Some F are S

We then construct an argument of the same form with true premises and a false conclusion:

(50) All people born in France were born in Europe.

All people born in Spain were born in Europe.

∴ Some people born in France were born in Spain.

The invalidity of (50) shows the invalidity of (49) and (41).

To summarize the method of counterexamples: We may show that an argument is invalid if we (1) identify its specific form and (2) find an argument of the same form with true premises and a false conclusion.

P R O B L E M S

Use the direct and the counterexample methods to show that these arguments are invalid. For the counterexample method, display the form used.

1. The economy will stabilize only if the government institutes a currency board. The government will institute a currency board. So, the economy will stabilize.

2. Eric will get a job unless he doesn't apply everywhere he can. He won't apply everywhere he can. Therefore he won't get a job.

3. If a fetus is a person, it has a right to life. So, if a fetus isn't a person, it doesn't have a right to life.

4. All students who like to study philosophy enjoy abstract thinking. Nobody with a practical turn of mind likes to study philosophy. So, no students with a practical turn of mind enjoy abstract thinking.

5. Some politicians learn to tell lies as a matter of course. No politician who does that can maintain the people's trust over the long term. So, some politicians who are able to maintain the long-term trust of the people do not learn to tell lies as a matter of course.

6. Schools and hospitals are analogous in some ways to prisons. Prisons are means of social control. So, schools and hospitals are means of social control.

7. There could be lightning without thunder. So, there could be thunder without lightning.

8. Nothing is better than liberty. Prison life is better than nothing. Therefore prison life is better than liberty.

9. The meat I bought yesterday was raw. The meat we ate tonight was the meat I bought yesterday. So, tonight we ate raw meat.

10. Saying that women earn less than men because women interrupt careers to have children, work fewer hours, obtain fewer degrees in technical areas than men, and the like of course serves the interests of successful men in the workforce who do not want their own privileges to be challenged. So, those are not the real reasons for women's reduced earnings.

11. If I would make any proposition whatsoever, then by that I would have a logical error. But I do not make a proposition; therefore I am not in error. (Nagarjuna)

12. Every art and every inquiry, and similarly every action and pursuit, is thought to aim at some good; and for this reason the good has rightly been declared to be that at which all things aim. (Aristotle)

13. Now each man judges well the things he knows, and of these he is a good judge. And so the man who has been educated in a subject is a good judge of that subject, and the man who has received an all-round education is a good judge in general. Hence, a young man is not a proper hearer of lectures on political science; for he is inexperienced in the actions that occur in life. . . . (Aristotle)

14. And having remarked that there was nothing at all in the statement, "I think, therefore I am," which assures me of having thereby made a true assertion, excepting that I see very clearly that to think it is necessary to be, I came to the conclusion that I might assume, as a general rule, that the things which we conceive very clearly and distinctly are all true. . . . (Rene Descartes)

15. No reason can be given why the general happiness is desirable, except that each person, so far as he believes it to be attainable, desires his own happiness. This, however, being a fact, we have not only all the proof which the case admits of, but all which it is possible to require, that happiness is a good: that each person's happiness is a good to that person, and the general happiness, therefore, a good to the aggregate of all persons. (John Stuart Mill)

CHAPTER 2

LANGUAGE

Language serves many purposes and functions. Logic mostly concerns statements, which describe the world and can be true or false. But people use language to do many things, including:

1. Ask questions: What was the capital of Assyria?
2. Attack someone: You idiot!
3. Express feelings: That is outrageous! I am so angry I could spit!
4. Give advice: I'd try a kitchen towel dipped in milk.
5. Give commands: Go directly to jail. Do not pass Go! Do not collect $200.
6. Greet someone: Hello. How do you do?
7. Make bets: I bet the Cowboys will beat Miami.
8. Make choices: I'll take a chocolate and vanilla swirl.
9. Make promises: I promise that I'll be there by eight.
10. Make proposals: I move that this be referred to the Finance Committee.
11. Name: Her name is Melanie Elizabeth.
12. Tell stories: Once upon a time there were three Bears.
13. Tell jokes: He said, "I don't have to outrun the bear. I just have to outrun you."

To reason well, we must distinguish these uses. We must be clear about how language is being used. If we confuse uses of language, we may commit mistakes.

(1) "How do you do?"

"Well, things haven't been going well lately. I broke up with my boyfriend. My cat is sick. Someone filed a lawsuit against me last week. I put on five pounds over the holidays and can't seem to get them off."

Here, a simple greeting has been mistaken for a question seeking information.

> (2) Proud parents, to nurse: "Her name is Melanie Elizabeth."
>   Nurse: "I'm sorry, but that's not true. The birth certificate still
>   has no name."

Here, an act of naming has been confused with a statement of fact.

In this chapter, we will consider the most important sources of confusion in language—confusions of meaning. We will also consider definitions as possible cures.

■ 2.1 REASON AND EMOTION

Arguments are designed to establish conclusions. They are thus means of persuasion. Not all means of persuasion, however, are logical. Someone may try to persuade others through force or emotion as well as by reason. Many arguments mix emotional appeals with appeals to evidence. There is nothing illegitimate about this. Emotions are often based on rational judgments and can reinforce them. But emotion can also conflict with reason. Such conflicts have inspired much great art and literature, from the Roman poet Catullus's anguished love elegies to Shakespeare's *Othello* and Mr. Spock and Commander Data on *Star Trek*. Emotions can lead us toward the truth or away from it. To evaluate arguments rationally, therefore, it is important to keep logical and emotional or other considerations distinct.

A masterpiece of persuasion is Pope Urban II's speech to the Council of Clermont, in his native French rather than Latin, in 1095 to begin the First Crusade. Note the mix of emotional and rational appeals:

> (3) Oh, race of Franks, race beyond the mountains, race beloved
>   and chosen by God (as is clear from many of your works), set
>   apart from all other nations by the situation of your country,
>   as well as by your Catholic faith and the honor you render to
>   the holy Church: to you our discourse is addressed, and for
>   you our exhortations are intended. We wish you to know what
>   a serious matter has led us to your country, for it is the im-
>   minent peril threatening you and all the faithful that has
>   brought us hither.

This introduction is roughly equivalent to "Fellow Frenchmen, we have a problem." But it appeals to the listeners' ethnic and national pride and to their religious faith. It singles out the listeners as having a special role to play in what follows. Notice that the problem is "a serious matter," an "imminent peril threatening you and all the faithful." Urban continues:

(4) From the confines of Jerusalem and from the city of Constantinople a grievous report has gone forth and has been brought repeatedly to our ears; namely, that a race from the kingdom of the Persians, an accursed race, a race wholly alienated from God, 'a generation that set not their heart aright, and whose spirit was not steadfast with God,' has violently invaded the lands of those Christians and has depopulated them by pillage and fire. They have led away a part of the captives into their own country, and a part they have killed by cruel tortures. They have either destroyed the churches of God or appropriated them for the rites of their own religion. They destroy the altars, after having defiled them with their uncleanness. . . .

Here Urban spells out the nature of the problem: The Persians have captured Christian territory and churches. Again there is a combined racial and religious appeal. The capture is described in charged language: "violently invaded . . . depopulated . . . pillage and fire . . . killed by cruel tortures . . . destroyed . . . defiled. . . ." To note that it is charged is not to say that it is inappropriate. The facts may be that bad. But it is important to isolate the factual basis for drawing conclusions.

(5) On whom, therefore, rests the labor of avenging these wrongs and of recovering this territory, if not upon you—you, upon whom, above all other nations, God has conferred remarkable glory in arms, great courage, bodily activity, and strength to humble the heads of those who resist you? Let the deeds of your ancestors encourage you and incite your minds to manly achievements—the glory and greatness of King Charlemagne, and of his son Louis, and of your other monarchs, who have destroyed the kingdoms of the Turks and have extended the sway of the holy church over lands previously pagan. Let the holy sepulcher of our Lord and Savior, which is possessed by the unclean nations, especially arouse you, and the holy places which are now treated with ignominy and irreverently polluted with the filth of the unclean. Oh most valiant soldiers and descendants of invincible ancestors, do not degenerate, but recall the valor of your ancestors.

Urban's point here is, in essence, "You could take it back." But the contrast between the Franks, "most valiant soldiers and descendants of invincible ancestors," with their "remarkable glory in arms, great courage, bodily activity, and strength" and the "unclean" Turks, who "irreverently polluted" the Holy Land with "filth," has great emotional power. Expressed in neutral language, Urban's appeal amounts to

(6) Fellow Frenchmen, we have a problem. The Turks have captured areas of the Middle East from Christian inhabitants, who are suffering as a result. You could take the land back.

But this would hardly have motivated thousands to embark on a difficult and dangerous military journey to fight a powerful adversary for someone else's land.

Advertising frequently combines cognitive and emotional appeals. Consider these ad lines, taken from a single issue of the *Wall Street Journal:*

Spectacular Savings, Switzerland's Finest Watches, Elegant Price Cutters since 1973

Own part of an island. Destined to become the crown jewel of historic West Galveston Island, Laffite's Cove is a waterfront community laced with canals and masterfully planned by The Woodlands Corporation, a nationally recognized developer.

Valet Parking: As terrifying to the driver of a pre-owned Lexus as it is to the driver of a new one.

Own the road. The CF-62. Rugged. Portable. Fully loaded.

This fall, Compaq will introduce a radical new idea in home computing. Computers actually designed for the home.

The Japanese Market is opening! Now is the time for M&A in Japan.

"I think this is the *greatest* thing that I have ever seen offered from a broker." Lind On-Line is just what futures traders have been looking for. In fact, this e-mail comment from a Lind On-Line user is typical. We're not surprised by the rave reviews because Lind On-Line is the most complete PC-based futures trading support you'll find anywhere!

Each ad gives some evidence to persuade potential customers to buy a product or use a service. But each also uses emotional appeals as subtle as an elegant writing style or as obvious as an exclamation point. Emotionally loaded phrases in these ads include *spectacular, finest, elegant, destined, crown jewel, historic, laced, masterfully planned, nationally recognized, terrifying, rugged, fully loaded, radical, greatest,* and *rave reviews.*

Again, there is nothing intrinsically wrong with emotional appeals. They are important tools of persuasion. They can reinforce reasoning. But it is important to base conclusions on the facts, not merely on emotions that may be generated independently of them. The truth of the premises of a valid argument guarantees the truth of the conclusion. But the emotional pitch of the premises guarantees nothing.

PROBLEMS

For each of the following terms, suggest some emotionally loaded equivalents.

1. lawyer

2. elderly person

3. politician

4. salesperson

5. change

Identify emotionally loaded terms and phrases in these advertisements and suggest emotionally neutral replacements. Restructure each in the form of an argument.

6. NHL Hockey
   fast
   powerful!
   action
   NHL: Coolest Game on earth (National Hockey League)

7. Why are millions of children less afraid of the dark? We can shed some light on that. As America's largest generator of low cost electricity we have a 99% record of reliable service. And that means a lot of night lights stay on and a lot of kids get a good night's sleep. Not to mention their parents. (Southern Company)

8. We recognized a long time ago that managing enterprise systems and applications requires more than just the latest products. To be successful, you need a total solution—the best technology and the resources, skills, and expertise to help make enterprise systems management a reality. (Tivoli Systems)

9. The Web Power Curve. You're either on it. Or you're out of it. The #1 Web hosting provider for Fortune's top 1000 companies will show you the way to the top.

   For unparalleled expertise, America's most successful companies have chosen GTE Internetworking as their preferred Web hosting provider. GTE has chosen Compaq ProLiant™ servers to deliver outstanding performance, reliability and manageability required to support its customers' NT-based, mission-critical Web sites.

   We are the engineers of opportunity at the heart of Internet commerce projected to grow from $8 billion now to $327 billion by 2002, on an Internet that will multiply by 100 times. Our commitment is to help you get on, stay on and move up the Web power curve.

GTE Internetworking serves this challenging future with proven performance, technological finesse, and a gift for innovation. (GTE)

10. Think your corporate strategy is ironclad? Think again. The skills and knowledge that made you successful in an age of chivalry may only slow you down in the age of hypercompetition. The armor that was an advantage is now a burden. In this dynamic time, true security comes from learning and flexibility.

    Wharton Executive Education will suit you up to win. Each year, more than 9,000 executives come to Wharton to discover sharp new tools to aid their companies and their careers. While we can't take you to the Grail, we can lead you to some of the latest wisdom about leadership, management, marketing and finance so you can break out of your old mindset. With a mix of more than 200 programs, we'll help you hone your skills for the competitive contests that lie ahead.

    Come to Wharton. Chivalry may be dead, but you don't have to be. (The Wharton School, University of Pennsylvania)

Identify emotionally loaded terms and phrases in these passages and suggest emotionally neutral replacements.

11. Subject to the three-fold yoke of ignorance, tyranny, and vice, the American people have been unable to acquire knowledge, power, or [civic] virtue. The lessons we received and the models we studied, as pupils of such pernicious teachers, were most destructive. We have been ruled more by deceit than by force, and we have been degraded more by vice than by superstition. Slavery is the daughter of Darkness: an ignorant people is a blind instrument of its own destruction. Ambition and intrigue abuse the credulity and experience of men lacking all political, economic, and civic knowledge; they adopt pure illusion as reality; they take license for liberty, treachery for patriotism, and vengeance for justice. This situation is similar to that of the robust blind man who, beguiled by his strength, strides forward with all the assurance of one who can see, but, upon hitting every variety of obstacle, finds himself unable to retrace his steps. (Simon Bolivar, 1819)

12. The political system of the allied powers [Russia, France, and Spain] is essentially different in this respect from that of America. This difference proceeds from which exists in their respective governments; and to the defense of our own which has been achieved by the loss of so much blood and treasure, and matured by the wisdom of their most enlightened citizens and under

which we have enjoyed unexampled felicity, this whole nation is devoted. We owe it, therefore, to candor and to the amicable relations existing between the United States and those powers, to declare that we should consider any attempt on their part to extend their system to any portions of this hemisphere as dangerous to our peace and safety. (James Monroe, The Monroe Doctrine, 1823)

13. We have conquered our liberty and we shall maintain it. We offer to bring this inestimable blessing to you, for it has always been rightly ours, and only by a crime have our oppressors robbed us of it. We have driven out your tyrants. Show yourselves free men and we will protect you from their vengeance, their machinations, or their return. . . . You are, from this moment, brothers and friends; all are citizens, equal in rights, and all are alike called to govern, to serve, and to defend your country. (*Proclamation on Spreading the French Revolution*, 1792)

14. I proceeded, filled with indignation and chagrin, to the Council of the Elders. I besought them to carry their noble designs into execution. I directed their attention to the evils of the nation, which were their motives for concealing those designs. They concurred in giving me new proofs of their unanimous good will. (Napoleon Bonaparte, 1799)

15. The Communists disdain to conceal their views and aims. They openly declare that their ends can be attained only by the forcible overthrow of all existing social conditions. Let the ruling classes tremble at a Communistic revolution. The proletarians have nothing to lose but their chains. They have a world to win.

    Workingmen of all countries, unite! (Karl Marx and Friedrich Engels, *The Communist Manifesto*, 1848)

16. The rioting and agitation in the capitals and in many localities of our empire fills our heart with great and deep grief. The welfare of the Russian Emperor is bound up with the welfare of the people and its sorrows are his sorrows. The turbulence which has broken out may confound the people and threaten the integrity and unity of our Empire.

    The great vow of service by the Tsar obligates us to endeavor, with all our strength, wisdom, and power, to put an end as quickly as possible to the disturbance so dangerous to the Empire. . . .

    We call on the true sons of Russia to remember their duties toward their country, to assist in combatting these unheard-of disturbances, and to join us with all their might in reestablishing quiet and peace in the country. (Tsar Nicholas II, *The October Manifesto*, 1905)

17. The history of recent years, and in particular the painful events of the 28th June last [1914; the day Gavrilo Princip assassinated the Archduke Francis Ferdinand], have shown the existence of a subversive movement with the object of detaching a part of the territories of Austria–Hungary from the Monarchy. The movement which had its birth under the eye of the Serbian Government has gone so far as to make itself manifest on both sides of the Serbian frontier in the shape of acts of terrorism and a series of outrages and murders. (Austrian Ultimatum to Serbia, July 23, 1914)

18. . . . since April of last year the Imperial Government had somewhat restrained the commanders of its undersea craft in conformity with its promise then given to us that passenger boats should not be sunk, and that due warning would be given to all other vessels which its submarines might seek to destroy, when no resistance was offered or escape attempted, and care taken that their crews were given at least a fair chance to save their lives in their open boats. The precautions taken were meagre and haphazard enough, as was proved in distressing instance after instance in the progress of the cruel and unmanly business, but a certain degree of restraint was observed. The new policy has swept every restriction aside. Vessels of every kind, whatever their flag, their character, their cargo, their destination, their errand, have been ruthlessly sent to the bottom without warning and without thought of help or mercy of those on board, the vessels of friendly neutrals along with those of belligerents. Even hospital ships and ships carrying relief to the sorely bereaved and stricken people of Belgium, though the latter were provided with safe conduct through the proscribed areas by the German Government itself and were distinguished by unmistakable marks of identity, have been sunk with the same reckless lack of compassion or of principle. (Woodrow Wilson, April 2, 1917)

19. I have, myself, full confidence that if all do their duty, if nothing is neglected, and if the best arrangements are made, as they are being made, we shall prove ourselves once again able to defend our Island home, to ride out the storm of war, and to outlive the menace of tyranny, if necessary for years, if necessary alone. At any rate, that is what we are going to try to do. That is the resolve of His Majesty's Government—every man of them. That is the will of Parliament and of the nation. The British Empire and the French Republic, linked together in their cause and in their need, will defend to the death their native soil, aiding each other like good comrades to the utmost of their strength. Even though large tracts of Europe and many old and famous States have

fallen or may fall into the grip of the Gestapo and all the odious apparatus of Nazi rule, we shall not flag or fail. We shall go on to the end, we shall fight in France, we shall fight on the seas and oceans, we shall fight with growing confidence and growing strength in the air, we shall defend our Island, whatever the cost may be, we shall fight on the beaches, we shall fight on the landing grounds, we shall fight in the fields and in the streets, we shall fight in the hills; we shall never surrender, and even if, which I do not for a moment believe, the Island or a large part of it were subjugated and starving, then our Empire beyond the seas, armed and guarded by the British Fleet, would carry on the struggle, until, in God's good time, the New World, with all its power and might, steps forth to the rescue and the liberation of the old. (Winston Churchill, June 4, 1940)

20. Only on account of its internal dissension was Germany defeated in 1918. The consequences were terrible.

After hypocritical declarations that the fight was solely against the Kaiser and his regime, the annihilation of the German Reich began according to plan after the German Army had laid down its arms.

While the prophecies of the French statement, that there were 20,000,000 Germans too many—in other words, that this number would have to be exterminated by murder, disease, or emigration—were apparently being fulfilled to the letter, the National Socialist movement began its work of unifying the German people and thereby initiating resurgence of the Reich. This rise of our people from distress, misery and shameful disregard bore all the signs of a purely internal renaissance. Britain especially was not in any way affected or threatened thereby.

Nevertheless, a new policy of encirclement against Germany, born as it was of hatred, recommenced immediately. Internally and externally there resulted that plot familiar to us all between Jews and democrats, Bolshevists and reactionaries, with the sole aim of inhibiting the establishment of the new German people's State, and of plunging the Reich anew into impotence and misery. (Adolf Hitler, June 22, 1941)

 ## 2.2   GOALS OF DEFINITION

Many disputes arise from unclear uses of language. It is important, therefore, to make sure that speaker and audience share an understanding of what the terms of a discussion mean. Definitions are the most common and effective way to clarify meaning.

Plato's early dialogues typically begin with a moral problem: an issue about courage, righteousness, or justice, for example. Socrates suggests that to know whether something is courageous or righteous or just, one must know what courage, righteousness, or justice are. This leads him to seek a definition of *courage, righteousness,* or *justice.* Usually, the search fails. But only if we have a clear definition, Socrates implies, can we hope to make progress.

In the *Euthyphro,* for instance, Socrates encounters Euthyphro at court. Euthyphro is there to testify against his father, whom he has accused of murder. His father suspected one of his slaves of killing another and threw the slave in a ditch to prevent his escape while he went to get the authorities. While he was gone, the slave died. Euthyphro feels certain that it is right for him to turn in his father and to testify against him. Socrates, dubious, asks Euthyphro how he knows. What is righteousness? What is Euthyphro's criterion for distinguishing right from wrong? Euthyphro has no answer and flees in discomfort.

Defining terms has often supplied a key to scientific progress. Imre Lakatos's book *Proofs and Refutations* shows how developing a general definition of *polygon*—something that took more than a century—enabled geometers to describe those figures and understand their properties. Similarly, some of the greatest achievements of nineteenth-century mathematics were definitions of *function, limit, continuity,* and *convergence.* Isaac Newton's mechanics rests on precise definitions of *kinetic energy, momentum,* and *force.* Important advances in physics over the past two centuries have relied on definitions of *magnetism, temperature, inertial mass,* and subatomic particles.[1]

There are various kinds of definitions. We may classify definitions according to their goal or purpose and also according to how they try to achieve it.

First, consider the goals a definition might have:

1. **Description.** Most definitions try to describe the meaning of a term as it is commonly used. Descriptive definitions are sometimes called **lexical,** because they are the kind of definition given in dictionaries or lexicons. Such definitions attempt to characterize the actual usage of an expression.

2. **Stipulation.** Some definitions introduce new terms into discourse. **Stipulative** definitions assign meanings to terms without regard to their ordinary use. The baseball rule book, for example, defines

---

[1] Some authors treat theoretical definitions such as these as falling in a separate category, with "theory" as their own distinctive goal. Admittedly, some theoretical definitions serve to reduce one theory to another. But that is orthogonal to the goals we discuss in the text. Some theoretical definitions are descriptive: The definition of *heat* as "mean kinetic energy of molecules" in statistical mechanics, for example, means to capture our concept of heat, already established from ordinary usage and from classical thermodynamics. Some, in contrast, are precising: The mathematical definitions of *function, limit, continuity,* and *convergence* were devised because confusion over these basic notions was obstructing mathematical progress. Some theoretical definitions are also stipulative: Consider, for example, the definition of π as the ratio of the circumference of a circle to its diameter.

out, hit, home run, sacrifice, and so on for the purposes of the game. It makes no attempt to define these terms as they are used outside baseball.

3. **Precision.** Some definitions try to remain close to the meaning of a term in natural language but also to make that meaning more precise or more useful for a particular purpose. Suppose that you hire someone to remove rocks from your yard. You need to agree on what the job entails. So, you may suggest that a rock, for this purpose, be considered any stone weighing more than a pound, anything less being a pebble. This is a "precising," or clarifying, definition of *rock* and *pebble.* Our ordinary concepts of rock and pebble are not very precise. To make sure you both have in mind the same task, you may need to sharpen those terms.

These definitions are helpful in resolving two kinds of problems resulting from two kinds of imprecision. The first, which the rock and pebble definition solves, is a problem of **vagueness.** Many terms in natural language have fuzzy boundaries. How many hairs can a man have on his head, for example, before he stops being bald? How high an IQ score must someone have to count as intelligent? How hard does it have to rain for a drizzle to become a rain and a rain to become a storm? These questions have no definite answers. In most circumstances, that is not a problem. Setting up the terms of a contract, however, or developing a scientific theory may demand more precision. Then, a clarifying definition can help.

The second sort of problem a clarifying definition can solve is **ambiguity.** Most words have more than one meaning. The context of an utterance usually makes clear which meaning the speaker intends. Sometimes context fails to disambiguate. Moreover, in special scientific, legal, or philosophical circumstances, ambiguity can cause so much trouble that precautions must be taken against it. Clarifying definitions can help. A definition may make it clear that, for the purposes of a contract, article, or discussion, a term is to have only one particular meaning. A food processing company that buys chickens, for example, should probably specify in a clarifying definition exactly what it expects to receive. An actual court case arose out of a company's failure to do just that; it expected fryers but received old hens and lost when it sued the supplier.

4. **Persuasion. Persuasive** definitions explain the meaning of a term, but contentiously—they express not only meaning but attitude as well. These definitions can be extremely effective in persuading and in exposing another's misuse of a term. Consider Alben Barkley's definition of *bureaucrat:*

(7) A bureaucrat is a Democrat who holds some office that a Republican wants.

Or, Frank Vanderlip's definition of *conservative:*

(8) A conservative is a man who does not believe that anything should be done for the first time.

On the other side of the political spectrum, Thomas Sowell offers a glossary of some political terms:

(9) Equal opportunity: Preferential treatment.

Compassion: The use of tax money to buy votes.

Insensitivity: Objection to the use of tax money to buy votes.

Private greed: Making money selling people what they want.

Public service: Gaining power to make people do what you want them to.

Simplistic: Argument you disagree with but cannot answer.

A matter of principle: A political controversy involving the convictions of liberals.

An emotional issue: A political controversy involving the convictions of conservatives.

Constitutional interpretation: Judges reading their own political views into the Constitution.

Politicizing the courts: Criticizing judges for reading their own political views into the Constitution.

Public-interest group: Politically organized liberals.

Special-interest lobby: Politically organized conservatives.

Persuasive definitions can provide important insights into the way words are used. Nevertheless, it is important to recognize them as persuasive and to distinguish the information they convey from the attitude they try to communicate. Persuasive definitions can expose appeals to the people by exposing slanted uses of language, but they can also perpetrate such appeals by slanting language themselves.

P R O B L E M S

Identify the goal of each definition.

1. A joint is a place where two things are joined.

2. Say that an argument is exact if it is valid and the omission of any premise would make it invalid.

3. A statement is mendacious if it is deceitful or false.

4. The building, for purposes of this contract, is the structure at 1 Gold Street, together with attached plaza and underground garage.

5. To prognosticate is to predict.

6. A miracle is a violation of the laws of nature. (David Hume)

7. Education is what survives when what has been learned has been forgotten. (B. F. Skinner)

8. Education is a state-controlled manufactory of echoes. (Norman Douglas)

9. Education: that which discloses to the wise and disguises from the foolish their lack of understanding. (Ambrose Bierce)

10. Education is the process of casting false pearls before real swine. (Irwin Edman)

11. Education is the process of driving a set of prejudices down your throat. (Martin Fischer)

12. Thrift is care and scruple in the spending of one's means. (Immanuel Kant)

13. A tout is a guy who goes around a race track giving out tips on the races, if he can find anybody who will listen to his tips, especially suckers, and a tout is nearly always broke. If he is not broke, he is by no means a tout, but a handicapper, and is respected by one and all. (Damon Runyan)

14. By Matter, therefore, we are to understand an inert, senseless substance, in which extension, figure, and motion do actually subsist. (George Berkeley)

15. Cynic, n.: a blackguard whose faulty vision sees things as they are, not as they ought to be. (Ambrose Bierce)

16. Cynicism is an unpleasant way of telling the truth. (Lillian Hellman)

17. Interest is the birth of money from money. (Aristotle)

18. Optimism is the content of small men in high places. (F. Scott Fitzgerald)

19. The optimist proclaims that we live in the best of all possible worlds; and the pessimist fears this is true. (James Branch Cabell)

20. A pessimist is a man who has been compelled to live with an optimist. (Elbert Hubbard)

21. Democracy demands that all of its citizens begin the race even. Egalitarianism insists that they all finish even. (Roger Price)

22. A liberal is a man too broadminded to take his own side in a quarrel. (Robert Frost)

23. Conservatism is the policy of make no change and consult your grandmother when in doubt. (Woodrow Wilson)

24. For war consisteth not in battle only, or the act of fighting, but in a tract of time, wherein the will to contend by battle is sufficiently known: and therefore the notion of time is to be considered in the nature of war, as it is in the nature of weather. For as the nature of foul weather lieth not in a shower or two of rain, but in an inclination thereto of many days together: so the nature of war consisteth not in actual fighting, but in the known disposition thereto during all the time there is no assurance to the contrary. (Thomas Hobbes)

25. Here therefore we may divide all the perceptions of the mind into two classes or species, which are distinguished by their different degrees of force and vivacity. The less forcible and lively are commonly denominated Thoughts or Ideas. The other species want a name in our language, and in most others; I suppose, because it was not requisite for any, but philosophical purposes, to rank them under a general term or appellation. Let us, therefore, use a little freedom, and call them Impressions; employing that word in a sense somewhat different from the usual. By the term impression, then, I mean all our more lively perceptions, when we hear, or see, or feel, or love, or hate, or desire, or will. (David Hume)

26. Devise a stipulative definition for a word you are introducing for the first time.

27. Devise a descriptive (lexical) definition for *album* and *pest*. Be careful to distinguish different meanings.

28. Devise precising definitions for *student* and *middle class* for purposes of a study concerning income levels and educational opportunities for college-age people.

29. Devise both descriptive and precising definitions for *livestock* and *rock star*.

30. Devise opposing pairs of persuasive definitions for *wealthy* and *kindhearted*.

2.3 MEANS OF DEFINITION

How do definitions achieve their goals?

1. **Ostension. Ostensive** definitions clarify the meaning of a term by pointing to examples of things to which it applies. In most circumstances, the easiest and most useful way to define *mango*, for

example, is to display a mango. The simplest way to explain the meaning of *laugh* is to point to people laughing.

2. **Listing.** Some definitions specify the meanings of words by listing things or kinds of things to which the word applies. The list may be complete or incomplete. To define *House of Congress,* for example, we might say "Senate and House of Representatives," a complete list; to define *fruit,* we might say "things like apples, bananas, oranges, and pears." Definitions by ostension and by listing are **denotative** or **extensional** definitions: They explain the meanings of a term by listing or pointing to part or all of its *denotation* or *extension,* the set of things of which it is true.

3. **Synonymy.** Some definitions simply give a synonym for the word being defined: *detective* for *gumshoe,* for example, or *freedom* for *liberty.* These are sometimes called **synonymous** definitions.

4. **Analysis.** Other definitions try to explain the meaning of a term by indicating what the things to which it applies have in common. Such definitions are often called **analytical.** Plato's dialogues revolve around attempts to devise analytical definitions of such terms as *courage, piety, friendship, knowledge,* and *justice.*

   Aristotelian logicians refer to such definitions as definitions *per genus et differentiae,* definitions "by genus and difference." The idea behind this terminology is that we can define a species or kind of thing by indicating *(a)* what genus, or more general classification, it falls under and *(b)* what distinguishes the species from others of the same genus. Thus, we might define a bachelor as a male adult who has never been married; *adult* here specifies the genus, while *male* and *who has never been married* specify the differences, or distinguishing features, of bachelors.

Definitions, we have seen, vary in both goal and method. The type of a definition depends on both its goal and its method. We can summarize the possible kinds of definition in this table:

KINDS OF DEFINITION

		METHODS			
		OSTENSION	LISTING	SYNONYMY	ANALYSIS
GOALS	Description				
	Stipulation				
	Precision			xxxxxxxxx	
	Persuasion				

We can classify any definition according to its goal and the method it uses to achieve it. There seem, therefore, to be sixteen possible kinds of definition.[2]

Are all these kinds really possible, however? We have blocked out one combination, for a definition by synonymy cannot meet a goal of precision: If the defining term is more precise than the term being defined, it cannot be synonymous with it. So, at least one of the sixteen pigeonholes in the preceding table has no pigeons.

Some authors argue that other pigeonholes are empty as well. Some contend, for example, that definitions by synonymy cannot be stipulative, for the same definition cannot both establish the meaning of a new term and identify two terms that already have the same meaning. But nothing in our characterization of synonymous definition requires that the meaning of the term being defined already be established. An example of a stipulative, synonymous definition is the common practice in logic of writing *iff* for *if and only if*.

Some authors also hold that synonymous definitions cannot be persuasive. It is easy to see why: A definition can persuade only if the defining expression has an emotional charge or other significance that the expression being defined lacks. In that case, however, how could the two be synonyms? Still, we reject this reasoning. Carried further, it would suggest that there are few if any synonymous definitions, for few or no terms are not only logically but also emotionally equivalent. Consider these synonymous terms: *lie, fib,* and *prevaricate*. They are true of precisely the same things. But they do not have the same emotional power. The differences are subtle, but, roughly, fibs are more forgiveable, and prevarications more technical, than lies. For another example, consider *dung, feces, excrement,* and their more colloquial synonyms. They are equivalent logically but not emotionally.

Finally, some contend that extensional definitions—definitions by ostension and listing—cannot meet the goal of precision, for giving examples or even a complete list of things to which a word applies does not determine its meaning. But this seems wrong. First, sometimes an example or list of examples *does* determine meaning: Consider the listing definition of *House of Congress* as "Senate and House of Representatives," a listing of children in a will, or an ostensive definition of *Newt Gingrich*. Second, precising definitions make terms more precise and useful; they need not determine meaning completely. An ostensive or listing definition can make terms less vague: You might specify the meanings

---

[2]Some writers list other kinds of definitions, for example, operational (specifying experimental procedures that determine the application of a term) or etymological (specifying meaning by giving a history of a term). We are disinclined to view these, in themselves, as definitions.

of *rock* and *pebble* for your contractor, for example, by holding up examples: "This is a rock. Remove things like this. *That* is a pebble. Leave things like that." Such a definition can also reduce or eliminate ambiguity: "When I told you to put your money in the bank, I meant *that* [pointing at the Signet Bank building], not *that* [pointing at a piggy bank]."

Distinguishing definitions by their goals and their means of achieving them thus allows us to distinguish fifteen kinds of definition.

P R O B L E M S

Identify the goal and kind of each of the following definitions.

1. A period is a punctuation mark ending a declarative sentence.

2. Period: .

3. A punctuation mark is a period, comma, question mark, exclamation point, colon, or semicolon.

4. A punctuation mark is a , ! ? ; .

5. A punctuation mark is a mark used to separate sentences or their elements.

6. A sphere is a ball.

7. A sphere is a round object.

8. A sphere is a set of points equidistant from a single point in three-dimensional space.

9. A continent is a large land mass.

10. Continent: Africa, Antarctica, Asia, Australia, Europe, North America, South America

11. Common sense is genius dressed in its working clothes. (Ralph Waldo Emerson)

12. Positive means being mistaken at the top of one's voice. (Ambrose Bierce)

13. Maturity of mind is the capacity to endure uncertainty. (John Finley)

14. A lie is an excuse guarded. (Jonathan Swift)

15. By happiness is intended pleasure, and the absence of pain; by unhappiness, pain, and the privation of pleasure. (John Stuart Mill)

16. Pleasure therefore, or delight, is the appearance or sense of good; and molestation or displeasure, the appearance or sense of evil. (Thomas Hobbes)

17. Happiness is the satisfaction of all our desires. . . . (Immanuel Kant)

18. Courage is a special kind of knowledge: the knowledge of how to fear what ought to be feared and how not to fear what ought not to be feared. (David Ben-Gurion)

19. The obvious is that which is never seen until someone expresses it simply. (Kahlil Gibran)

20. Politics is the art of looking for trouble, finding it everywhere, diagnosing it incorrectly, and applying the wrong remedies. (Groucho Marx)

21. Politics is the gentle art of getting votes from the poor and campaign funds from the rich, by promising to protect each from the other. (Oscar Ameringer)

22. Politicians are people who resolve through linguistic processes conflicts that would otherwise have to be solved by force. (S. I. Hayakawa)

23. The liberty of men in groups is power. (Edmund Burke)

24. Political power, properly so-called, is merely the organized power of one class for oppressing another. (Marx and Engels)

25. Charm is a way of getting the answer yes without having asked any clear question. (Albert Camus)

26. Brain: the apparatus with which we think we think. (Ambrose Bierce)

27. Prejudice, n. A vagrant opinion without visible means of support. (Ambrose Bierce)

28. A fanatic is a man that does what he thinks th' Lord wud do if He knew th' facts in th' case. (Finley Peter Dunne)

29. The greatest fool is he who thinks he is not one and all others are. (Baltasar Gracian)

30. Confidence is that feeling by which the mind embarks on great and honorable courses with a sure hope and trust in itself. (Cicero)

31. The original members of the League of Nations shall be those of the Signatories which are named in the Annex to this Covenant and also such of those other States named in the Annex as shall accede without reservation to this Covenant. (Covenant of the League of Nations, 1919)

32. The Assembly shall consist of Representatives of the Members of the League. (Covenant of the League of Nations, 1919)

33. By a 'denoting phrase' I mean a phrase such as any one of the following: a man, some man, any man, every man, all men, the present King of England, the presenting King of France, the center of mass of the solar system at the first instant of the twentieth century, the revolution of the earth around the sun, the revolution of the sun around the earth. (Bertrand Russell)

34. In the first place, it is mostly considered unjust to deprive any one of his personal liberty, his property, or any other thing which belongs to him by law. Here, therefore, is one instance of the application of the terms just and unjust in a perfectly definite sense, namely, that it is just to respect, unjust to violate, the legal rights of any one. (John Stuart Mill)

35. Now it is known that ethical writers divide moral duties into two classes, denoted by the ill-chosen expressions, duties of perfect and of imperfect obligation; the latter being those in which, though the act is obligatory, the particular occasions of performing it are left to our choice, as in the case of charity or beneficence, which we are indeed bound to practise, but not towards any definite person, nor at any prescribed time. In the more precise language of philosophic jurists, duties of perfect obligation are those duties in virtue of which a correlative right resides in some person or persons; duties of imperfect obligation are those moral obligations which do not give birth to any right. (John Stuart Mill)

For items 36–60, give an example, not already given in the text, of each kind of definition.

36. description, ostension

37. description, listing

38. description, synonymy

39. description, analysis

40. stipulation, ostension

41. stipulation, listing

42. stipulation, synonymy

43. stipulation, analysis

44. precision, ostension

45. precision, listing

46. precision, analysis

47. persuasion, ostension

48. persuasion, listing

49. persuasion, synonymy

50. persuasion, analysis

51. Devise a stipulative definition for a word you are introducing for the first time and then give equivalent definitions by the other three methods.

52. Devise analytical definitions for *album* and *pest*. Be careful to distinguish different meanings.

53. Devise precising definitions for *employed* and *low-income* for purposes of a study concerning income levels and educational opportunities for college-age people.

54. Devise ostensive or listing definitions for *livestock* and *rock star*.

55. Devise analytical and listing definitions for *wealthy* and *kindhearted*.

56. Devise a listing definition for *New England state*.

57. Devise a descriptive (lexical) definition for *lunatic* and *psychologist*. Are your definitions synonymous or analytic?

58. Devise both synonymous and analytic definitions of *odor* and *calamity*.

59. Devise ostensive or listing definitions for *religion* and *president*.

60. Devise analytic definitions of *acorn, sow, grandfather, professor,* and *church*.

■ 2.4 CRITERIA FOR DEFINITIONS

Plato and Aristotle held that analytical definitions are better than other definitions, for they can explain not only *what* falls under a term but also *why* it does so. Logicians have traditionally used four rules to evaluate analytical definitions:

1. A definition must state the essential attributes of the kind.
2. A definition must not be circular.
3. A definition must not be negative when it can be positive.
4. A definition must not use ambiguous, obscure, or figurative language.

We examine these rules in turn.

1. *A definition must state the essential attributes of the kind.* This rule requires that a definition be neither too broad nor too narrow. The term

being defined, the **definiendum,** should apply to exactly those things to which the defining expression, the **definiens,** applies. A definition is *too narrow* if the defining expression applies to too little and *too broad* if it applies to too much. Consider, for example, the standard dictionary definition of *chair:* "a piece of furniture for one person to sit on, having a back and, usually, four legs."[3] To determine whether the definition is too narrow, we can ask, could there be a chair that was not a piece of furniture for one person to sit on, having a back? Arguably, the answer is yes: A beanbag chair has no back. This definition, therefore, may be too narrow. Is it also too broad? Could there be a piece of furniture for one person to sit on, having a back, that was not a chair? Probably not. So, the definition is too narrow but not too broad.

Plato defined *human being* as "featherless biped." This is too broad; a skeptic is said to have refuted Plato by throwing a plucked chicken over the walls of the Academy! The Chinese philosopher Hsün Tzu (310–212 B.C.), interestingly, rejects the same definition for a similar reason: "The yellow-haired ape also has two feet and no feathers."[4] But Hsün Tzu's chief complaint is that the definition does not explain what makes someone human. Aristotle tried to rectify this, defining *human being* as "rational animal." Rationality does seem more central to humanity than being featherless. But this definition is probably both too broad and too narrow. It is too broad because extraterrestrial beings could be rational animals without being human. It is too narrow because there are irrational humans.

Rule 1 requires that the definiens and definiendum be equivalent. The sentence asserting that something satisfies the definiens must always agree in truth value with the sentence asserting that it satisfies the definiendum.

2. *A definition must not be circular.* An expression should not be defined in terms of itself. If the definiendum appears anywhere in the definiens, the definition cannot fulfill its purpose of clarifying meaning. Nobody who did not understand the term already could understand the definition. Gertrude Stein's "A rose is a rose is a rose" is not a definition of *rose.* (She did not intend it to be.)

Rarely is a proposed definition directly circular. But, in systems of definitions, indirect circularity is not unusual. Indeed, dictionaries are condemned to it. A dictionary tries to give a definition of every word in a language. But, as Aristotle first realized, trying to define everything leads to a dilemma. Either the chain of definitions must be infinite or it must loop back on itself. Dictionaries have a finite number of pages, so they

---

[3]*Webster's New World Dictionary* (New York, 1970), p. 235.

[4]Hsün Tzu (in the Pinyin transliteration, Xunzi), "Against Physiognomy," in *The Works of Hsüntze,* translated by Homer H. Dubs (London: Arthur Probsthain, 1928), reprinted in *Understanding Non-Western Philosophy,* Daniel Bonevac and Stephen Phillips, eds. (Mountain View: Mayfield, 1993), p. 255.

must choose circularity. Usually a dictionary's circles are so large that few users ever recognize them.

Consider a definition of *friendly:* "like, characteristic of, or suitable for a friend."[5] Is this circular? Certainly it explains the meaning of an expression in terms of another very closely related to it. But whether it is circular depends on the system of definitions of which this forms a part. How is *friend* defined? If the answer is "person who acts friendly," then the system of definitions is circular. If the definition of *friend* is independent, however—like the dictionary definition of *friend* as "a person whom one knows well and is fond of"—then there is no circle.

Sometimes this rule is taken to ban from the defining expression not only the term being defined but also any of its synonyms. Defining *fib* as *lie* is not directly or indirectly circular, so long as there is an independent definition of *lie*, but it is not analytical either. Synonymous definitions, in general, may clarify meaning but do little to explain the meaning of a term; they are not definitions by genus and difference. Whether rule 2 prohibits the use of synonyms, therefore, depends on whether the synonymous expressions are defined in a circle—*fib* as *lie*, *lie* as *prevaricate*, *prevaricate* as *dissimulate*, and *dissimulate* as *fib*, for example—or whether some link is explained analytically.

Some good definitions may appear circular at first glance. Aristotle defined *citizen* as the son or daughter of a citizen. By itself, this is circular. But it is part of a more adequate definition of *citizen*. We might specify an initial class of citizens—the residents at some time, for example—and then say that any child of a citizen is also a citizen. If we add that there are no other citizens, we have defined the class of citizens: the class of residents, children of residents, children of children of residents, and so on.

This is the form of an **inductive** or **recursive** definition. All inductive definitions have three parts. The first, the **basis** of the definition, states that the term being defined applies to certain objects or categories of objects. The basis of Aristotle's definition says that residents at a certain time are citizens. **Inductive clauses** constitute the second part of an inductive definition. They have a conditional form, specifying that if certain objects satisfy the term being defined, then so do certain others. These are the clauses—children of citizens are citizens, for example—that appear circular in isolation. Finally, every inductive definition has a third part, its **closure condition,** saying that the term being defined applies to nothing else: The term applies only to the objects to which the basis and inductive clauses of the definition force it to apply.

3. *A definition must not be negative when it can be positive.* We could try to define *white* as a color that is neither black, red, orange, yellow,

---

[5]*Webster's New World Dictionary,* p. 559.

green, blue, or purple, but that is not very informative; it is a poor definition of *white*. Similarly, defining *claret* as wine that is not sweet and is neither white nor rosé seems perverse, even though it is equivalent to the acceptable definition *dry red wine*. In general, we should prefer positive to negative definitions.

Sometimes, however, negative definitions are unavoidable. *Dark* may be defined as "wholly or partly without light"; *bald* as "without hair"; *poor* as "lacking material possessions"; and *amorphous* as "without shape or form." These definitions have no obvious positive equivalents.

4. *A definition must not use ambiguous, obscure, or figurative language.* Definitions are tools for clarifying meaning. A definition using ambiguous, obscure, or figurative language is flawed; it cannot fulfill its chief purpose, for it tries to clarify the meaning of an expression by equating it with another expression whose meaning is unclear. Khalil Gibran gives the figurative definition "Beauty is eternity gazing at itself in a mirror"; this hardly helps to explain the meaning of the term *beauty*. To explain the meaning of a term in terms more obscure than the term itself is, usually, to explain nothing.

This is not to say, however, that all definitions using figurative or obscure language are worthless. Samuel Johnson defined *language* as the dress of thought. That is metaphorical but also rich in insight. Bertrand Russell called a series *continuous* when it is Dedekindian and contains a median class having $\aleph_0$ terms.[6] To most of us, this definition is obscure. But, to anyone who has worked through Russell's discussion of continuity, it is clear. Obscurity is, to some extent, in the eye of the beholder. That is, obscurity is relative to the context. One audience may find obscure what another finds intelligible. Whether a definition satisfies rule 4, consequently, may depend upon the audience to whom it is addressed and the conversation of which it forms a part.

P R O B L E M S

For each of the following definitions, *(a)* identify its goal, *(b)* identify the method it uses to achieve that goal, and *(c)* evaluate it according to the four rules of definition.

1. Originality is the art of concealing your source. (Franklin P. Jones)

2. Wisdom denotes the pursuing of the best ends by the best means. (Frances Hutcheson)

3. An intellectual is someone whose mind watches itself. (Albert Camus)

---

[6]*Introduction to Mathematical Philosophy* (London, 1919), p. 104.

4. Anger is momentary madness. (Horace)

5. Character is that which reveals moral purpose, exposing the class of things a man chooses or avoids. (Aristotle)

6. What is virtue? To hold yourself to your fullest development as a person and as a responsible member of the human community. (Arthur Dobrin)

7. Love is liking someone better than you like yourself. (Frank Tyger)

8. Political power is the power to oppress others. (Lin Piao)

9. Charity is the sterilized milk of human kindness. (Oliver Herford)

10. Philanthropist: A rich (and usually bald) old gentleman who has trained himself to grin while his conscience is picking his pocket. (Ambrose Bierce)

11. Courage is almost a contradiction in terms. It means a strong desire to live taking the form of a readiness to die. (G. K. Chesterton)

12. Democracy is a process by which the people are free to choose the man who will get the blame. (Laurence J. Peter)

13. A man is called selfish, not for pursuing his own good, but for neglecting his neighbors. (Richard Whately)

14. Patience is the art of hoping. (Vauvenargues)

15. Once an editor explained to me that a journalist was just an out-of-work reporter. (Linda Ellerbe)

16. Altruism is the art of using others with the air of loving them. (René Dubreuil)

17. Action is but coarsened thought—thought become concrete, obscure, and unconscious. (Frédéric Amiel)

18. Acquaintance, n. A person whom we know well enough to borrow from, but not well enough to lend to. (Ambrose Bierce)

19. Gossip is when you hear something you like about someone you do not. (Earl Wilson)

20. Conscience: A small, still voice that makes minority reports. (Franklin P. Adams)

The following are definitions from Samuel Johnson's groundbreaking dictionary, first published in 1755. Identify the goal and method of each. Have the meanings of these terms changed since the eighteenth century? To answer this question, say whether these definitions are now too broad or too narrow, and, if so, explain why.

21. anecdote: Something yet unpublished; secret history.

22. to belch: To throw out from the stomach; to eject from any hollow place. It is a word implying coarseness; hatefulness; or horrour.

23. blockhead: A stupid fellow; a dolt; a man without parts.

24. bug: A stinking insect bred in old household stuff.

25. chitchat: Prattle; idle prate; idle talk. A word used only in ludicrous conversation.

26. compliment: An act, or expression of civility, usually understood to include some hypocrisy, and to mean less than it declares.

27. diploma: A letter or writing conferring some privilege, so called because they used formerly to be written on waxed tables, and folded together.

28. excise: A hateful tax levied upon commodities, and adjudged not by the common judges of property, but wretches hired by those to whom excise is paid.

29. fastidious: Disdainful; sqeamish; delicate to a vice; insolently nice.

30. funk: A stink. A low word.

31. glee: Joy; merriment; gayety. It anciently signified musick played at feasts. It is not now used, except in ludicrous writing, or with some mixture of irony or contempt.

32. horseplay: Coarse, rough, rugged play.

33. informal: Offering an information; accusing. A word not used.

34. jargon: Unintelligible talk; gabble; gibberish.

35. jogger: One who moves heavily and dully.

36. lizard: An animal resembling a serpent, with legs added to it. . . . In America they eat *lizards*. . . .

37. lunch, luncheon: As much food as one's hand can hold.

38. to marvel: To wonder; to be astonished. Disused.

39. nincompoop: A fool; a trifler.

40. occult: Secret; hidden; unknown; undiscoverable.

41. to pamper: To glut; to fill with food; to saginate; to feed luxuriously.

42. rug: (1) A coarse, nappy, woollen cloath. (2) A coarse nappy coverlet used for mean beds. (3) A rough woolly dog.

43. ruse: Cunning; artifice; little stratagem; trick; wile; fraud; deceit.

44. scalp: The scull; the cranium; the bone that encloses the brain.

45. tarantula: An insect whose bite is only cured by musick.

46. unkempt: Not combed. Obsolete.

47. vegetable: Anything that has grown without sensation, as plants.

48. vermicelli: A paste rolled and broken in the form of worms.

49. warren: A kind of park for rabits.

50. zealous: Ardently passionate in any cause.

Many philosophical debates can be construed as contests over definitions. Consider the following definitions of key philosophical terms advanced by various characters in Plato's dialogues. Do they meet the criteria of definition? Explain.

51. Justice is the advantage of the stronger.

52. Justice is giving each person his/her due.

53. Justice is doing one's own [task].

54. Knowledge is perception.

55. Knowledge is true belief.

56. Knowledge is true belief with an account (i.e., an explanation or justification).

57. Courage is standing and fighting, and not running away.

58. Courage is endurance of the soul.

59. Courage is wise endurance of the soul.

60. Courage is the knowledge of that which inspires fear or confidence.

61. Courage is knowing what ought to be feared.

62. A is a friend of B if A likes B.

63. A is a friend of B if A and B like each other.

64. David Hume defines *cause* in three different ways. Do these definitions meet our criteria? Are they equivalent?

    Similar objects are always conjoined with similar. Of this we have experience. Suitably to this experience, therefore, we may define a cause to be (1) an object, followed by another, and where all the objects similar to the first are followed by objects similar to the second. Or in other words (2) where, if the first object had not been, the second never had existed. The appearance of a cause always conveys the mind, by a customary transition, to the idea of the effect. Of this also we have experience. We may, therefore, suitably to this experience, form another definition of cause, and call it, (3) an object followed by another, and whose appearance always conveys the thought to that other.

CHAPTER 3

INFORMAL FALLACIES

A good argument generally presents premises to provide evidence for a conclusion that follows from those premises and to which the premises are relevant in context. The definitions of validity and soundness, however, do not mention evidence, relevance, or context. Even sound arguments can be unsuccessful in a particular context.

In general, a **fallacy** is a kind of bad argument. Arguments committing **formal fallacies** have bad forms; they are invalid. Arguments committing **informal fallacies** make other kinds of errors, typically violating considerations of evidence, relevance, or clarity in particular contexts. The most interesting and dangerous fallacies are those that look like good arguments. This chapter discusses many of the most common and dangerous informal fallacies.

3.1 FALLACIES OF EVIDENCE

The premises of a good argument offer evidence for its conclusion. Aristotle spoke of some sentences as coming before others "in the order of knowledge." If we can think of some sentences being better known or more evident than others, then we can say that the premises of a successful argument should be more evident than the conclusion. Aristotle pointed out that "demonstration is from things more creditable and prior" (*Prior Analytics* ii, 16). Or, in the words of the influential seventeenth-century *Port Royal Logic,* "what serves as proof must be clearer and better known than what we seek to prove."[1] The premises of a good argument should be justifiable independently of the conclusion.

---

[1]Antoine Arnauld (1611–1694), *Logic, or, The Art of Thinking* (Indianapolis: Bobbs-Merrill, 1964; originally published in 1662), p. 247.

What fallacies result from violating this? Consider this argument:

(1) The earth will be invaded by purple people eaters next year.

∴ The earth will be invaded by purple people eaters next year.

The premise does not provide evidence for the conclusion; it *is* the conclusion. Unless we already accept the conclusion, we have no reason to accept the premise. The premise is not independently justifiable: The premise and conclusion, being the same sentence, are evident to the same degree in any context.

1. BEGGING THE QUESTION *(PETITIO PRINCIPII).* Argument (1) commits a fallacy called **begging the question:**

> An argument **begs the question** if and only if the premises include or presuppose the conclusion.[2]

Various sorts of argument exemplify this fallacy. For example, the conclusion may appear among the premises, not *verbatim,* but substantially rephrased. The conclusion still depends on a premise no more evident than the conclusion itself.

(2) Next year, violet, human-devouring creatures will invade this planet.

∴ The earth will be invaded by purple people eaters next year.

A more interesting kind of question begging occurs when the conclusion does not appear in the premises directly but is **presupposed** by the premises—when, that is, no one could assert the premises without already taking the conclusion for granted. The presuppositions of a statement are more evident than it; a statement is thus no more evident than its presuppositions. If the premises presuppose the argument's conclusion, then the argument fails to establish its conclusion. The premises cannot provide evidence for the conclusion, because they already assume it. A classic instance of an argument whose premises presuppose its conclusion is this:

---

[2]This definition is Richard Whately's. See his *Elements of Logic* (New York: Sheldon and Company, 1869), p. 179: An argument begs the question "when one of the Premisses (whether true or false) is either plainly equivalent to the conclusion, or depends on that for its own reception." See also p. 220: In a question-begging argument, "one of the Premisses either is manifestly the same in sense with the Conclusion, or is actually proved from it or is such as the persons you are addressing are not likely to know, or admit, except as an inference from the Conclusion. . . ." Compare Aristotle's definition, which relates even more directly to evidence: "Whether a person does not conclude at all, or whether he does so through things more unknown, or equally unknown, or whether he concludes what is prior through what is posterior" (*Prior Analytics* ii, 16).

(3) The Bible says that God exists.

The Bible was inspired by God.

Any writing inspired by God is true.

∴ God exists.

The second premise, *The Bible was inspired by God,* presupposes the conclusion. One cannot assert it without assuming that God exists. Consequently, any evidence for the second premise would automatically be evidence for the conclusion, and any evidence against the conclusion would tell against that premise. The premise cannot be any more evident than the conclusion. The argument, therefore, is fallacious.

2. COMPLEX QUESTION. Sometimes questions have presuppositions: They cannot be asked without taking certain things for granted. They cannot be answered unless what they presuppose is true. The classic case is

(4) Have you stopped beating your wife?

You cannot answer yes or no without, in effect, admitting your former wife-beating. If you respond, "Yes, I have," your answer presupposes that you have beaten your wife in the past. If you respond, "No, I haven't," it is even worse: Not only have you beaten your wife in the past, but also you are still doing it! The only way to respond without admitting prior wife-beating is to challenge the presupposition: "What? I've never beaten my wife." This amounts to rejecting the question rather than answering it.

If we have established that you have beaten your wife, of course, (4) is acceptable. In that case, the presupposition is already known. Virtually all questions have presuppositions; the danger of fallacy arises when the presuppositions are no more evident than the conclusion being sought.

An excellent example of a complex question occurs at the beginning of Cicero's most famous oration against Cataline:

How much longer will you abuse us, O Cataline? How much longer will your unbridled boldness thrash itself about?

These questions presuppose that Cataline has been abusing us with reckless audacity.

Does Cicero commit a fallacy? Not necessarily. A question gets a reasoner into trouble only when it presupposes the conclusion of the argument.

An argument commits the fallacy of **complex question** if and only if its premises include a question that presupposes the conclusion.

Cicero is in danger, therefore, of committing the fallacy of complex question. His oration avoids it by going on to present evidence of Cataline's abuses. This suggests an important way of repairing complex-question or begging-the-question fallacies: *Supply independent evidence for the conclusion.*

Other cases of complex question involve compound questions that, in effect, pose two questions and presuppose that a single answer suffices for both. "Will you be kind and let me have an extension on that paper?" suggests that kindness and agreeing to the extension go together. "Are we going to sign and be cheated by them again?" presupposes that they have cheated us in the past and that our signing will let them cheat us again. To respond to these questions without assenting to their presuppositions, we can split them into their components and insist on giving different answers to each. For example, we might respond, "I'll be kind, but you cannot have the extension," or, "We're going to sign, but the contract is solid; they won't be able to cheat us again." In parliamentary procedure, a motion to divide the question—to split a question into two or more component questions—takes precedence over other motions precisely because of the danger of fallacy that complex questions involve.

3. INCOMPLETE ENUMERATION (FALSE DICHOTOMY). Sometimes we seem to face dilemmas: your money or your life, truth or consequences, trick or treat. These dilemmas may be real. At times, however, they are only apparent; there are other possibilities. Whenever an argument passes over possibilities it ought to recognize, it is guilty of the fallacy of incomplete enumeration.

> An argument commits the fallacy of **incomplete enumeration** if and only if it presupposes a disjunction that does not include all available possibilities.

An argument may commit this fallacy by "failing to make an exhaustive enumeration of alternatives, that is, not sufficiently considering all the ways in which something may exist or happen."[3]

Historian David Hackett Fischer has compiled a list of titles of actual works by reputable historians, published by reputable presses, that exhibit this fallacy. Among them are the following:

Napoleon III: Enlightened Statesman or Proto-Fascist?

The Causes of the War of 1812: National Honor or National Interest?

The Abolitionists: Reformers or Fanatics?

Plato: Totalitarian or Democrat?

---

[3]Arnauld, *Logic*, p. 255.

John D. Rockefeller—Robber Baron or Industrial Statesman?
The Robber Barons—Pirates or Pioneers?
The New Deal—Revolution or Evolution?
Renaissance Man—Medieval or Modern?
What Is History—Fact or Fancy?

It should be obvious that something is wrong with these titles. Perhaps Napoleon III was neither an enlightened statesman nor a proto-fascist. Perhaps the War of 1812 was fought for national honor and national interest and for other reasons as well. Perhaps some abolitionists were both reformers and fanatics, some were one and not the other, and some were neither. Every title presents an incomplete enumeration of the possibilities.

P R O B L E M S

Discuss the fallacies (or dangers thereof) in the following examples.

1. How is it that little children are so intelligent and men so stupid? (Alexandre Dumas, fils)

2. Certainly there is life after death. The proof may be found in the Ouija board, since the messages the board transmits are from departed souls.

3. Terrorism is bad, because it encourages further acts of terrorism.

4. Why was American slavery the most awful the world has ever known? (Nathan Glazer)

5. Morality is just a matter of taste. Why? It's just like arguing about whether chocolate ice cream is better than vanilla; there's no fact of the matter.

6. Adultery is immoral, since sexual relations outside marriage violate ethical principles, and the alternative of unrestrained sexual relations would destroy the institution of marriage.

7. Word 7.0, the fancy Windows '95 version of Microsoft's popular word processor, has a partisan spellchecker. It recognizes the names Kennedy, Johnson, Carter and Clinton, but not Nixon (for which it offers to substitute "Nikon") or Reagan ("regain," "reign"). Either Bill Gates favors Democrats, or he appreciates that conservatives don't need spellcheckers. *(National Review)*

Identify the presuppositions of the following questions.

8. a. Why aren't American goods competitive abroad?
   b. Why do things break?

  c. Why don't they work anymore?
  d. Whatever happened to good old American workmanship and quality? (James Brady)

9. a. Was Reconstruction shamefully harsh or surprisingly lenient?
  b. Could Lincoln have succeeded where Johnson failed?
  c. Was the latter a miserable bungler or a heroic victim?
  d. How well did the freedman meet his new responsibilities?
  e. When did racial segregation harden into its elaborate mold? (Don E. Fehrenbacher)

10. There are some questions that make a ranger's day. "What state is Utah in?" is one of them. (A very good question, when you think about it.) Following are some challengers from a past park season, based on a spot survey of national parks.

"How many years does it take for a deer to turn into an elk?"

"Are there any trails that go up, but don't come down?"

"What does that rock taste like?"

"Are those baby bats?" (The answer was, "No, those are flies.")

"How long did it take Walt Disney to build this?" ("This" was Carlsbad Caverns.)

"Is the world cave underground?"

"Are all the islands surrounded by water?" (Stu Stuller)

11. Begging the question has sometimes been called the most serious fallacy, on the grounds that an argument committing it has a fatal flaw and cannot easily be repaired. Do you agree? Why, or why not?

12. For each of the book titles listed on pages 90–91, give at least three alternatives not enumerated in the title.

3.2   FALLACIES OF RELEVANCE: CREDIBILITY

Relevance requires that an argument be directed at the point at issue. Sometimes a person advances an irrelevant argument as an intentional bit of sophistry to avoid a tough issue. Sometimes the parties to the conversation are confused about what the issue really is; sometimes they are arguing at cross purposes. And sometimes people agree about what the issue is at the outset but change the issue during the course of the conversation.

  An argument that violates relevance is called an **ignoratio elenchi**—an argument ignorant of its own goal or purpose—and is therefore fallacious. It misses the point, in effect slipping one conclusion in place of another. The next three sections discuss particular kinds of relevance fallacies. But they do not exhaust such fallacies. So, we will use *ignoratio elenchi* to refer to fallacies of relevance that do not fall under one of the more specific kinds discussed below.

In this section we consider some fallacies resulting from focusing, not on the issue, but on the arguer. **Ad hominem** (or **ad personam**) arguments, arguments "to the man (or person)," attempt to refute positions by attacking those who hold or argue for them. There are three kinds of *ad hominem* argument: (1) The attack may consist of an assault on a person's integrity, moral character, psychological health, or intellectual ability. If so, the *ad hominem* is termed **abusive**. (2) The attack may consist of a charge of unreliability due to a person's special circumstances. In such cases, the *ad hominem* is termed **circumstantial**. (3) The attack may accuse someone of inconsistency or hypocrisy. It is then called a **tu quoque**. All three kinds of argument may violate the condition of relevance, for they are directed, not at the issue at hand, but at the people holding some view of that issue.

4. *AD HOMINEM* ABUSIVE.   Abusive *ad hominem* arguments involve insults:

> An argument is an **abusive ad hominem** argument if and only if it purports to discredit a position by insulting those who hold it.
>
> Form: A is bad. (Insult)
> ∴ What A says is false.

The insult may assail a person's integrity or moral character. The political opponents of Edward Kennedy, for example, have occasionally raised the issue of his accident at Chappaquidick. This may be relevant to the senator's character, which in turn may be relevant to his qualifications for office. When used to argue against a position he holds, however, the charge is *ad hominem*. Whether or not the senator is personally irresponsible has no bearing on the truth of his beliefs on whether the government should provide universal health insurance or raise the minimum wage or have an activist judiciary, for example.

The insult may strike at a person's mental health. Franklin Roosevelt, while in office, was derided as insane, suffering from an Oedipus complex and a "Silver Cord complex"; his next-door neighbor termed him "a swollen-headed nitwit with a Messiah complex and the brain of a boy scout."[4] These insults are not arguments, let alone fallacious arguments.

---

[4]See George Wolfskill and John Hudson, *All But the People: Franklin D. Roosevelt and His Critics* (New York: Macmillan, 1969), pp. 5–16.

To use them against Roosevelt's positions or policies, however, is fallacious. Some writers have argued that Bishop Berkeley's philosophy, according to which all "material" objects are really just complexes of ideas, stems from an obsessive-compulsive neurosis (in Freudian theory, bad toilet training). This psychological hypothesis is not itself a fallacy. But to use it to discredit Berkeley's philosophy, to argue that it is false, is to commit an *ad hominem* fallacy. Other writers have condemned the German philosopher Friedrich Nietzsche's works as the ravings of a lunatic. Nietzsche did spend the last years of his life in a mental institution; this does not show, however, that what he wrote is false. Neurotics and even madmen may argue validly and speak truly.

Finally, an *ad hominem* argument may insult the intellectual capacities of an opponent. American journalist H. L. Mencken abused American presidents fiercely for intellectual as well as other failings. He referred to Warren Harding as a "stonehead," but most of his attacks concerned moral character. He called Theodore Roosevelt "blatant, crude, overly confidential, devious, tyrannical, vainglorious and sometimes quite childish." Coolidge was "petty, sordid and dull," "a cheap and trashy fellow . . . almost devoid of any notion of honour . . . a dreadful little cad." Mencken attributed to Herbert Hoover "a natural instinct for low, disingenuous, fraudulent manipulators" but saved his best abuse for Franklin Roosevelt, calling him "the Führer" and "the quack" and terming his administration "an astounding rabble of impudent nobodies," "a gang of half-educated pedagogues, non-constitutional lawyers, starry-eyed uplifters and other such sorry wizards."[5]

Abraham Lincoln was one of the United States' greatest presidents. But Lincoln was widely abused in both the Northern and Southern press while in office. In the South, he was called "a drunk" and "a cross between a sandhill crane and an Andalusian jackass," among other things. Northern newspapers, between 1861 and 1865, called Lincoln "an awful, woeful ass," a "baboon," the "craftiest and most dishonest politician that ever disgraced an office in America," "empty skull," a "mole-eyed monster with a soul of leather," and an "obscene clown."[6] Again, these are not themselves fallacies, for they are not arguments. In the context of attacking one of Lincoln's positions, however, they easily become fallacies.

5. *AD HOMINEM* CIRCUMSTANTIAL. Circumstantial *ad hominem* arguments try to discredit an argument by appealing to the situation, motives, beliefs, or other characteristics of the arguer.

---

[5]Paul Johnson, *Modern Times* (New York: Harper and Row, 1983), p. 258.

[6]Thomas Keiser, "'The Illinois Beast': One of Our Greatest Presidents," *Wall Street Journal*, February 11, 1988.

An argument is a **circumstantial ad hominem** argument if and only if it purports to discredit a position by appealing to the circumstances or characteristics of those who hold it.

Form: A has self-interested motives for saying p.
∴ p is false.

Usually, circumstantial *ad hominem* arguments charge a person or group with holding a position solely because it serves their own interests. People tend to advance arguments of this sort against lobbying groups. The Tobacco Institute, for example, frequently releases reports raising questions about the link between smoking and disease and routinely denounces reports claiming to establish such links. Critics of the Institute often dismiss its statements on the ground that the tobacco industry funds its research. This is a circumstantial *ad hominem*: The critics charge that the motives of those who pay for the Institute's work suffice to discredit it.

Of course, to say that this is a fallacy is not to recommend the Institute's research or its conclusions. The point is rather that, to determine the health effects of smoking, we must evaluate the content of research on both sides of the question. To reject a position because of the motives or sources of financial support behind it is to advance a circumstantial *ad hominem* argument.

Evaluating the reliability of an argument is very different from evaluating the reliability of a person. We decide whether to accept what someone else says, in part, on the basis of how reliable we judge that person to be. This is as it should be. If we find out that someone is biased, we may question that person's reliability. When someone advances an argument, however, giving reasons for a conclusion, then reliability is no longer so important. The reliability of others bears on whether we should accept any premises they introduce but not on whether their arguments are good and deserve to convince us. In the words of Samuel Johnson,

> Nay, Sir, argument is argument. You cannot help paying regard to their arguments if they are good. If it were testimony, you might disregard it. . . . Testimony is like an arrow shot from a long bow; the force of it depends on the strength of the hand that draws it. Argument is like an arrow shot from a cross bow, which has equal force though shot by a child.[7]

---

[7]Samuel Johnson, quoted in David Hackett Fischer, *Historians' Fallacies* (New York: Harper and Row, 1970), p. 282.

Circumstantial *ad hominem* arguments may establish something about the reliability of an opponent but do not establish the incorrectness of that person's position.

Of course, sometimes the reliability of a person is the issue at hand. Suppose that Jones is testifying in court about Hanlon's racketeering. The opposing lawyer may introduce evidence to show that Jones himself is an ex-convict who still associates with shady characters; that his initial statement to the police conflicts with his current testimony; that he is a known liar and confidence artist; and that he owed Hanlon money, giving him a motive to help send Hanlon to prison. Some of these charges may seem to be abusive *ad hominem* arguments, others circumstantial. They do not, however, make the lawyer's argument fallacious. Jones's reliability is precisely the issue. Therefore, his character and situation are relevant. Similar points apply to political campaigns.

6. TU QUOQUE.   A special sort of circumstantial *ad hominem* argument appeals to the situation, motives, beliefs, or other characteristics of the audience to charge them with inconsistency or hypocrisy unless they accept the conclusion. Unlike the other forms of *ad hominem* argument, it is used primarily for defense.

> An argument is **tu quoque** if and only if it purports to discredit a position by charging those who hold it with inconsistency or hypocrisy.
>
> Form: A does not really believe (or does not act as if he or she believes) p.
>   ∴ p is false.

Tu quoque (pronounced too-KWOH-kway), literally, means "you, too." When charging someone with inconsistency, a *tu quoque* accuses an opponent of not believing his or her own argument: If the opponent does not believe it, then surely the rest of us do not need to take it seriously. When charging someone with hypocrisy, a *tu quoque* accuses an opponent of failing to practice what he or she is preaching: If the opponent does not act on the basis of his or her own argument, then the rest of us may ignore it also.

A classic case of this fallacy involves a confrontation between a hunter and a person who eats meat but objects to hunting. The latter denounces shooting animals; the sportsman responds by saying, "If you think killing animals is wrong, why do you eat meat?" The hunter is charging the meat eater with inconsistency or hypocrisy, depending on whether that person believes that his or her own meat eating is justifiable or not.

Inconsistent or hypocritical people may nevertheless advance successful arguments and utter true sentences. The fallacy, then, consists in inferring from someone's inconsistency or hypocrisy the falsity of one of his or her beliefs—without considering the merits of those beliefs themselves.

Like other forms of circumstantial *ad hominem* arguments, *tu quoque* arguments often persuade. Nobody likes to be inconsistent or hypocritical. If the argument convinces the audience members that they exhibit one of these failings, they will naturally try to remove the inconsistency or hypocrisy by revising their beliefs or practices. For this reason, *tu quoque* arguments may succeed in persuading despite their irrelevance.

The problem with *tu quoque* arguments is that "the conclusion which is actually established is not the *absolute* and *general* one in question, but *relative* and particular; viz., not that 'such and such is the fact,' but that '*this man* is bound to admit it.' . . ." on pain of inconsistency or hypocrisy.[8] Consequently, *tu quoque* arguments, while they fail to establish the desired conclusion, can be effective. The opponent who really is inconsistent or hypocritical has a problem. A *tu quoque* argument thus forces the opponent onto the defensive. But the opponent's problem—inconsistency or hypocrisy—is a problem for the *person*, not necessarily for the position being advanced. That is why even a persuasive *tu quoque* counts as a fallacy.

Charges of inconsistency are more effective than allegations of hypocrisy. The opponent may admit to hypocrisy or weakness of will but cannot blunt the attack by similarly admitting to inconsistency. To return to the dispute between the hunter and the critic, the meat eater could respond to a charge of hypocrisy by granting his or her hypocrisy. Meat eaters could admit that eating meat is wrong but say they cannot give it up. There is no analogue with inconsistency. The meat eaters could admit that they do not have the strength to *act* as they believe but not that they do not have the strength to *believe* as they believe.

P R O B L E M S

Identify the fallacies of credibility, if any, in the following passages.

1. People who smoke can hardly tell others not to.

2. Nijinsky's theories about dance are probably not very worthwhile, since, after all, he spent the last twenty-five years of his life in an asylum.

3. How can the wealthy bureaucrats and technocrats on Capitol Hill be expected to create good laws for the common people?

---

[8]Whately, *Elements of Logic*, p. 237. (Emphasis in original.)

4. A: I think you ought to tell your girlfriend that you're having second thoughts about your relationship. You shouldn't just start seeing other people without telling her.
B: I don't remember you being a paragon of honesty when you and Rachel started having some trouble.

5. Mr. Fast . . . says, "I have been active in one part and another of the peace movement over the past 40 years." 1989 minus 40 equals 1949. In 1949, Howard Fast was everyday defending Josef Stalin. Fast was the true Stalinist Stakhanovite. . . . (William F. Buckley, Jr.)

6. Dear Madam: You should know that some idiot in your town is sending me a series of crazed letters—and signing your name. (Senator Stephen Young to a constituent)

7. Striped bass may be endangered by the Westway's dredging and landfill. Rita Thompson, a Westway supporter, questioned both the moral character of the bass and the wisdom of those who eat them. "Bass eat sewage and garbage; I wouldn't think of eating bass," she said. "A study should be made of people who eat bass." *(New York Times)*

8. Chelsea Clinton will attend the private Sidwell Friends School. Sidwell is a haven for children of the capital's rich and famous (tuition is about $10,800) and a world away from the metal detectors and gangs in the city's public schools. We understand why the Clintons would exercise this private-school choice, but why don't they want poor parents and children to have the same opportunity? *(Wall Street Journal)*

9. Nick thanked Chairman Finisterre for the opportunity to present his views before such a distinguished committee of senators: over two thousand bounced checks between them, a seducer of underage Senate pages, three DUIs, one income-tax evader, a wife beater, and a case of plagiarism, from, of all sources, a campaign speech of Benito Mussolini. (Christopher Buckley)

10. . . . Robert Ball—who incidentally started working for the government in 1939—argued against privitization on the following grounds: "Social Security is perhaps our strongest expression of community solidarity. Social Security is based on the premise that we're all in this together, with everyone sharing responsibility not only for contributing to their own and their family's security, but also to the security of everyone else, present and future." . . .

    Leave aside for a moment the vast empirical data demonstrating that privitization would improve retirement security, fuel economic growth, and make the system more fair. When opponents attack privitization because they fear it would weaken the "community," what they really fear is that it would take government out of the picture. Their point, it would seem, is not to expand the pie of benefits for each individual retiree; the point is to keep the public pie cutters employed. (Ted Forstmann)

Identify the fallacies, if any, in the following passages. (Fallacies from any of the previous sections may be included.)

11. Either you complete all your courses this semester or you might as well drop out. Since it would be a terrible waste for you to drop out, you should complete all your courses.

12. Of course, I don't believe that grades are all that important. Only a nerd like you would think they are.

13. Surely this proposal isn't in the best interests of the students. It was passed by an all-faculty committee.

14. When did you tell Jones that you suspected that what you were doing was against the law? The conclusion is clear: Either you were lying then or you were willing to break the law.

15. You can trust Joan. She's told me that she's always been known for her honesty.

16. Evaluate the following attempt to avoid fallacy:
    If the Senate is entitled to pass judgment on John Tower's personal history, then the public is also entitled to know salient facts about those who judge Mr. Tower. . . . The point here is not merely "you're another," *[tu quoque]* but that the essence of the Tower debate cannot be understood without first understanding its total hypocrisy. If the Senate wanted to improve ethical standards in Washington it could start in its own backyard, particularly by outlawing honorariums. But it has no such intention; about drinking and conflicts of interest it is utterly cynical. These are not ethical concerns, but weapons in a political battle. *(Wall Street Journal)*

17. Discuss the following, which seems to presuppose that at least some circumstantial *ad hominem* arguments are good:
    If a stockbroker were so smart, he would not be making his riches by selling stock tips to widows and orphans. In the style of the chain letter, the tipster divulges inside information for his gain and your loss. The rhetorical pose of stockbrokers and racetrack tipsters to be offering prudent advice is contradicted by their circumstances, a contradiction catalogued in

rhetoric as the "circumstantial *ad hominem.*" That is to say, "Being so smart, why don't you do it yourself, if it's such good advice?" (Donald N. McCloskey)

■ 3.3   FALLACIES OF RELEVANCE: CONFUSION

Often, arguments miss the mark because they confuse the issue.

7. RED HERRING.   A **red herring** is an irrelevant point introduced into an argument or debate. It is so called because hunters sometimes drag a red herring, a particularly smelly fish, to obscure their own scent. Such a point proves nothing concerning the issue. But a red herring is usually introduced as if it were pertinent and may succeed in convincing an audience in spite of its irrelevance. It may show something resembling the desired conclusion, or it may wield great emotional power. In either case, however, it confuses the issue. What it proves is not the conclusion or even a weaker version of it but something quite beside the point.

> An argument is a *red herring* if and only if it tries to undermine an opponent's argument by introducing a point irrelevant to the issue at hand.

Consider, for example, a debate on the morality of extravagant expenditure. Don says that buying a Mercedes is immoral, because many people are homeless and hungry; the money could and should be put to better use helping them. Michelle disagrees. Among other things, she argues that spending money on expensive toys is perfectly legal. She asks, "What would you do? Throw Mercedes owners in jail? Execute them? Reeducate them?" But the legality of extravagance is a red herring. Punishment is also a red herring. The debate is about whether extravagant expenditure is moral or immoral, not about whether it is legal or illegal or whether it ought to be punished.

Whether something is truly a red herring may be controversial. Occasionally the parties to a dispute disagree about what the issue is. One side may regard as a red herring what the other treats as central to the problem. In 1988, the U.S. Senate debated the Intermediate Nuclear Forces treaty between the United States and the Soviet Union. Senator Jesse Helms argued against ratification of the treaty, pointing out that although it required dismantling European intermediate-range missiles, the treaty did not mandate the destruction of any nuclear warheads—the part of the missiles designed to explode. Senator Dan Evans responded by calling Helms's charge "more than a red herring; I'd call it a crimson whale."

Is Helms's argument a red herring? Is an arms control treaty's failure to eliminate warheads irrelevant to the decision whether to ratify it? The answer depends on what the treaty is supposed to accomplish. If the two sides in the debate cannot agree about the proper purposes of the treaty and of arms control in general, then they are unlikely to agree about what is relevant to the issue of ratification. What is or is not a red herring can thus itself become a matter of dispute.

8. STRAW MAN.   The **straw man** fallacy occurs when someone attacks an opponent's position by exaggerating it to make it seem ridiculous. It is named for the practice of training soldiers by having them attack straw men. Defeating an opponent made of straw is obviously not the same as defeating a real enemy.

An opponent's stance is easier to refute, of course, if you attack an exaggerated, distorted, misrepresented, or particularly problem-riddled form of it. People therefore sometimes set up a "straw man"—a feeble version of the opponents' convictions—and knock it down. This can succeed in demonstrating the inadequacy of the tottery view set up for assault but fails to show the inadequacy of the opponents' actual beliefs.

> An argument erects a *straw man* if and only if it tries to justify rejecting a position by attacking a different, and usually weaker, position.

Opponents of abortion, for example, are often attacked for rejecting abortion in all cases, including even such unusual cases as severe deformity, rape, incest, or threats to life. Although some abortion opponents do believe that *all* abortions are immoral, the attack is irrelevant to the more common moderate position that allows for such exceptions. Attacking moderates in this way constitutes a straw man.

PROBLEMS

Identify the fallacies of confusion, if any, in the following passages.

1. If children's writing skills are as bad as everyone says, we ought to take another look at standardized tests. If these tests don't measure writing ability, it's no wonder that students never develop it.

2. It's just wrong to change the rules, raising the academic standards for athletic scholarships. It's going to keep these kids out

of the college of their choice, and that's not the American way. That's not what the American dream is all about.

3. Penny: It's important to work for equal rights and opportunities for the disabled. Too often, they've been excluded, from prejudice or just thoughtlessness.
   Enid: Don't you think we're all disabled in some way or another? My uncle, for example, walks with a limp. My grandfather has a bad back. My grandmother has arthritis. But your programs aren't set up to help them. In fact, they don't need help.

4. There is an incredible amount of empty space in the universe. The distance from the sun to the nearest star is about 4.2 light years, or 25 followed by 12 noughts miles. . . . And as to mass: the sun weighs about 2 followed by 27 noughts tons, the Milky Way weighs about 160,000 times as much as the sun and is one of a collection of galaxies of which, as I said before, about 30 million are known. It is not very easy to retain a belief in one's own cosmic importance in view of such overwhelming statistics. (Bertrand Russell)

5. A Canadian judge threw a case out of court because a witness was too boring. The case was originally reported in the *Forensic Accountant Newsletter*. The judge said the man was "beyond doubt the dullest witness I've ever had in court . . . [He] speaks in a monotonal voice . . . and uses language so drab and convoluted that even the court reporter cannot stay conscious." The judge said: "I've had it. Three solid days of this steady drone is enough. I cannot face the prospect of another 14 indictments. It's probably unethical but I don't care." (Nury Vittachi)

6. The isolationist believes that while international trade may be desirable, it is not necessary. He believes that we can build a wall around America and that democracy can live behind that wall. But the internationalist declares that, to remain free, men must trade with one another—must trade freely in goods, in ideas, in customs and traditions and values of all sorts. (Wendell Wilkie)

7. The Center for Science in the Public Interest [recently] declared that chocolate croissants, butter-laden pastries and other toothsome sweets contain heart-clogging amounts of fat. Spokesbeings insinuated that vendors might want to attach warning labels to their baguettes and assorted delicacies. What began as a polite lecture on digestion quickly turned into a covert bid for expanded federal power.

The Lifestyle Police want to turn us into veggie-munching, meathating, teetotaling, public-TV watching, power-walking, euphemism-talking zombies. (Tony Snow)

Identify the fallacies, if any, in the following passages. (Fallacies from any of the previous sections may be included.)

8. Why are the nations with the most developed systems of professional management education, the United States and Great Britain, performing so poorly, when two nations that provide almost no professional management training, Germany and Japan, have been the outstanding successes of the postwar period? (David Vogel)

9. It's important that we place restrictions on foreign investment in this country. Almost 75 percent of the commercial real estate in Los Angeles is owned by the Japanese. More and more American companies are being bought by foreign firms. Some people want to put a big "For Sale" sign on this country.

10. To a great extent, right-wing ideologies are modes of thought designed not to spur action or to enhance collective responsibility, but to provide a moral justification for egocentricity. . . . Insofar as political commitment requires qualities of self-sacrifice or a capacity to find personal fulfillment through social participation, such personal qualities do not resonate well with right-wing ideological perspectives. (Richard Flacks)

11. If we think to regulate printing, thereby to rectify manners, we must regulate all recreations and pastimes, all that is delightful to man. No music must be heard, no song be set or sung, but what is grave and Doric. There must be licensing dancers, that no gesture, motion, or deportment be taught our youth but what by their allowance shall be thought honest. . . . (John Milton)

12. Howard Fast [arguing that antiabortion advocates cannot be sincere, because their concern for life stops at birth]: I have never heard a right-to-life voice raised in protest against the 60,000 innocents murdered by the death squads of El Salvador.
   William F. Buckley, Jr.: The lifers are, by Mr. Fast and others who think as he does, encumbered by the responsibility for everything that happens to the fetus after it materializes into a human being in the eyes of the law. And if you aren't around to see to it that at age 14 the kid is receiving the right education, ingesting the right foods, leading a happy, prosperous life, why,

you had no business bringing him into this world. You are a hypocrite to the extent that you support life for everyone who suffers in life. It is only left for Mr. Fast to close the logic of his own argument, which would involve him in a syllogistic attempt along the lines of:

Everyone alive suffers.

No one not living suffers.

Therefore, no one should live.

13. In the following passage, accounting firms claim that their critics are committing a fallacy. Do you think their claim is correct? Do you find any fallacies in the accounting firms' response?

> Some critics aren't surprised at the recent strains at major accounting firms because they have long believed that consulting and auditing, like oil and water, don't mix.
>
> That's because accounting firms serving as outside auditors are supposed to judge objectively whether their client companies are making proper business decisions. But auditors are likely to be less critical of decisions in which they themselves have been deeply involved, critics say.
>
> Some consultants and legislators call this a potential conflict of interest. And some independent consulting firms complain that consultants at accounting firms gain an unfair competitive advantage by pushing their services to clients that their firms may also audit.
>
> In response, accounting firms call the problem of a seeming conflict of interest a red herring. They contend that the experience gained as auditors makes them better consultants and that criticism of their gains in the consulting field comes simply from competitors irked by their own loss of business. *(Wall Street Journal)*

14. Suppose you are arguing about changing the grading system at your school from letter grades to pass/fail. *(a)* Pick a side of the issue, and construct an argument for your position. *(b)* Evaluate your argument critically. Does it contain any red herrings or straw men? *(c)* If so, can you add something to your argument to relate them to the issue? *(d)* If not, devise a red herring, and explain why it is one.

15. Suppose you wish to argue that television has an extremely negative effect on its viewers but don't quite know how to proceed. What are some red herrings or straw claims you might argue instead? Make a list and, for each, explain why it does not address the original issue.

■ 3.4 FALLACIES OF RELEVANCE: MANIPULATION

Sometimes arguments direct themselves not to the issue at hand but to manipulating the audience. They use primarily nonrational means to persuade people of their conclusions.

9. APPEAL TO THE PEOPLE (OR GALLERY; *AD POPULUM*). Some politicians have made reputations for using nonrational, emotional appeals to win support for their positions. Any argument based not on rational considerations but on emotional appeals counts as an appeal to the people.[9] (Such arguments are sometimes called "appeals to the gallery" to allude to politicians' practice of directing arguments at the observers in the gallery rather than at fellow legislators.)

> An argument is an **appeal to the people** (or **to the gallery**) if and only if it tries to justify its conclusion by appealing to the audience's emotions.

Cicero employed appeals to the people brilliantly. Consider this portion of his argument to the Roman Senate in favor of the death penalty for Cataline, with its emotional description of the terror wrought by Cataline's plots:

> Yet I am not a man so iron-hearted to be unaffected by the grief of my most dear and loving brother, present here, nor by the tears of all these friends you see seated around me. Nor can I prevent my thoughts being often recalled to my own home by my despondent wife, my terrified daughter and my infant son (whom I think the state is cherishing as a sort of pledge for my loyalty as consul), or by my son-in-law who stands within my view awaiting anxiously the result of this day.[10]

Appeals to the people are not confined to the ancients. Almost every modern politician uses them, often with the aid of a team of media consultants or spin doctors to maximize their effect. Here, for example, is a classic from Richard Nixon's "Checkers" speech:

> My family was one of modest circumstances, and most of my early life was spent in a store out in East Whittier. It was a grocery store—one

---

[9]Some writers use this term differently, for what appears in the following section as "appeal to common practice."

[10]Cicero, fourth oration against Cataline, *The Speeches*, trans. Louis E. Lord (London: W. Heineman, 1937), p. 1.

of those family enterprises. I worked my way through college and, to a great extent, through law school.

The only reason we were able to make it go was because my mother and dad have five boys and we all worked in the store.

Why do I feel so deeply? Why do I feel that, in spite of the smears, the misunderstanding, the necessity for a man to come up here and bare his soul as I have? And I want to tell you why. Because, you see, I love my country.[11]

Often appeals to the people take the form of using slanted language to convey a certain attitude toward what is being said, without doing anything to support that language. In 1977, Leonid Brezhnev defended the persecution of political protestors in the Soviet Union:

> In our country it is not forbidden "to think differently" from the majority. . . . It is quite another matter if a few individuals who have . . . actively come out against the socialist system, embark on the road of anti-Soviet activity, violate laws and, finding no support inside their own country, turn for support abroad, to imperialist subversive centres. . . . Our people demand that such . . . activists be treated as opponents of socialism, as persons acting against their own motherland, as accomplices if not actual agents of imperialism. . . . We have taken and will continue to take against them measures envisaged by our law.[12]

Emotional or slanted language may lead the audience to overestimate the evidence in favor of certain conclusions or to overemphasize its importance. Notice the emotionally packed words and phrases Brezhnev used: *come out against the socialist system, anti-Soviet, imperialist subversive centres, motherland, accomplices,* and *agents of imperialism.* Such terms try to lead the listener to think of protestors as enemies rather than fellow citizens.

Nevertheless, the use of slanted terms does not always mark an argument as unacceptable. What matters is whether the argument relies on slanting to lead the audience toward its conclusion. Political magazines with a definite and strongly expressed perspective tend to use much language that would strike an uncommitted observer as slanted. A magazine on the right of the political spectrum may speak of a conflict between Communists and anti-Communists as a fight between Marxist-Leninist thugs and freedom fighters, while one on the left speaks of it as a battle between progressives and reactionary totalitarians. Is a fallacy lurking here? Perhaps, but not necessarily. We need to answer two questions: (1) Are the

---

[11]Richard Nixon, "Checkers" speech.

[12]*Reprints from the Soviet Press,* April 30, 1977, pp. 22–23.

slanted terms essential to deriving the conclusion of the argument? and (2) What is the intended audience? Many publications are mostly preaching to the converted. Their use of slanted language, therefore, is not fallacious but does have a negative consequence: Their arguments, to the extent that they rely on slanting, can succeed only with a limited audience.

10. APPEAL TO COMMON PRACTICE. "When in Rome, do as the Romans do." This might be the slogan of appeals to common practice. Virtually everyone can recall, as a child, using an argument of this kind. You and your friends, perhaps, were playing outside on a warm summer evening. Someone said, "Let's go to the park!" Some of the children readily assented; you and a few others decided to ask permission. Your mother, it turned out, did not like the idea at all. "But everyone else is going!" you cried. If you were particularly sophisticated, you bent the truth a bit and added, "All the other mothers said it was OK!"

This kind of argument is called an **appeal to common practice** (or **steamrolling**). The conclusion asserts that a certain kind of action is acceptable, permissible, or even obligatory. The premises indicate that the kind of action is common practice: that everyone, or most people, or at least many people are doing it. The argument urges the listener to jump on the bandwagon and agree with the conclusion.

> An argument *appeals to common practice* if and only if it tries to justify a kind of action by appealing to the common practice of the community.
>
> Form: x is common practice. (All, or most, or many people do x.)
> ∴ x is acceptable (or obligatory).

What is wrong with such arguments? Sometimes, nothing. Suppose you travel to Europe and observe people eating with their forks in their left hands. You may infer, quite properly, that holding your fork in your left hand is acceptable behavior there. Or, suppose you visit Central America and note that when introduced, people say *Mucho gusto* to each other. You may legitimately conclude that this is not only acceptable but also obligatory in that part of the world when meeting someone. In general, what people do is a good guide to what a community counts as acceptable. So, there is nothing wrong with appealing to common practice to support a conclusion about what a certain community's standards allow.

Appeals to common practice are fallacious, however, when the conclusion involves a stronger and more serious sense of acceptability or obligation. Consider your argument, directed to your mother, that you should be allowed to go to the park with your friends. That everyone else's mother has granted permission may show that community standards allow her to grant permission. She will not be a social pariah if she says you may go. But this, of course, is not what you want to establish. You are trying to show that she *ought* to let you go. She might indicate the problem with appeals to common practice with a question: "Would you jump off a bridge if all your friends were doing it?"

The unstated assumption behind an appeal to common practice is that anything commonly done is right. Occasionally advertisers make this assumption explicit: "50 million Americans can't be wrong!" But, of course, this is not true: 50 million Americans *can* be wrong. Common practices in many societies, at many different times, appear to later observers wrongheaded and even barbarous.

11. APPEAL TO FORCE (*AD BACULUM*).   A rather different sort of fallacy occurs when someone uses a threat. The threat may be physical: A mobster may make you an offer you cannot refuse. (As Sherlock Holmes tells Watson, "I should be very much obliged if you would slip your revolver into your pocket. An Eley's No. 2 is an excellent argument with gentlemen who can twist steel pokers into knots."[13]) Or, the threat may be financial: An employer may tell workers that they will be fired if they reveal secrets; a law may penalize speeding or littering with a hefty fine. Or, the threat may involve embarrassment: A lobbyist may intimidate a legislator by threatening to expose sordid details of past business dealings.

These are not arguments. They are nonrational attempts at persuasion or, more accurately, coercion. But arguments may also try to engender feelings of fear or insecurity. An appeal to force is an argument relying on a threat:

> An argument **appeals to force** if and only if it tries to justify a kind of action by threatening the audience.
>
> Form: If A does x, A will suffer.
> ∴ A should not do x.

Although the appeal to force has been counted among logical fallacies since ancient times, it is not very clear when threats should count as arguments rather than as coercion. Consider this case: A lobbyist threatens a Congressman with defeat in the next election if he does not support subsidies for the lobbyist's industry. The lobbyist may have large sums to spend on political activity, and the Congressman's margin of victory in the

---

[13]Sir Arthur Conan Doyle, "The Adventure of the Speckled Band," *The Adventures of Sherlock Holmes* (Secaucus, NJ: Castle Books, 1980; originally published in *The Strand*, 1891), p. 115.

last election may have been slim. So, suppose that the lobbyist can make good on the threat. In effect, the lobbyist argues:

(5) If you do not vote for subsidies, you will not be reelected.

∴ You should vote for subsidies.

The lobbyist argues that it will be best for the Congressman if he votes in favor of subsidies. If subsidies are a bad idea, however—if they impose significant costs on consumers, say, or if they harm our allies economically—then those costs almost certainly outweigh the harm to the Congressman of not being reelected. Appeals to force point to a possible harm or frustrated desire and treat it as overriding all other considerations. If, in a particular circumstance, that harm is not overriding, the appeal is fallacious.

12. APPEAL TO PITY *(AD MISERICORDIAM)*.

An appeal to pity tries to arouse pity for someone and exploit it to persuade the listener to act for that person's benefit. In general, an appeal to pity is an argument relying on arousing feelings of sympathy or pity among the listeners:

> An argument **appeals to pity** if and only if it tries to justify a kind of action by arousing sympathy or pity in the audience over the consequences of the action.
>
> Forms: A's doing x will harm B.
> ∴ A should not do x.
>
> A's doing y will help B.
> ∴ A should do y.

A classic example, close to college professors' hearts, is an argument used by a student seeking a higher grade. "I'm pre-med. I really want to go to medical school and become a doctor. But my grades, so far, are just borderline. If this C stays on my record, I'll never make it. All these years of hard work will be wasted."

Appeals to pity try to play on our altruism. Some such arguments succeed. If doing Y will help someone and there is no reason not to do it, then you probably should do Y. If by giving to charity you can help save lives without producing any harm to yourself or others, then you should.

But the condition "there is no reason not to do it" is very important. The rule "Help others" holds only when other things are equal. Suppose that giving the pre-med student a C really will harm that student. Before concluding anything about the grade, it is necessary to consider the effect of raising the grade on others, as well as any general principles of fairness involved. If changing the grade to a B would help the student get into a medical school, the change would help that particular student but would cost some other student a place. So, the consequences are not unambiguously positive; one student would benefit, and another would suffer. Further, another C student in the course getting the same or perhaps even better grades than the pre-med student would end up with a lower grade than the pre-med student's B. This seems unfair: It violates a general principle of justice, that similar cases should be judged similarly.

P R O B L E M S

Identify the fallacies of manipulation, if any, in the following passages.

1. Gentlemen: You have undertaken to cheat me. I won't sue you, for the law is too slow. I'll ruin you. Yours truly, Cornelius Vanderbilt

2. No [college football] team that's ever appeared in your Top Twenty over the past 100 years is guiltless of cheating in one way or another. (Dan Jenkins)

3. No man will take counsel, but every man will take money; therefore money is better than counsel. (Jonathan Swift)

4. "You tell me. Your kids go to bed crying at night because they're hungry. Is 'off the books' going to bother you?" asks a former steelworker. (Clare Ansberry)

5. I cannot stand this proliferation of paperwork. It's useless to fight the forms. You've got to kill the people producing them. (Vladimir P. Kubaidze)

6. William Jennings Bryan, a 36-year-old Congressman from Nebraska, gave the silver forces their war cry at the Democratic convention in July [1896]. "You shall not crucify mankind upon a cross of gold!" were the words that made Bryan the Democratic nominee for President. *(Wall Street Journal)*

7. Good sense is, of all things among men, the most evenly distributed; for everyone thinks himself so abundantly provided with it, that those even who are the most difficult to satisfy in everything else, do not usually desire a larger measure of this quality than they already possess. (René Descartes)

8. The bourgeoisie, wherever it has got the upper hand, has put an end to all feudal, patriarchal, idyllic relations. It has pitilessly torn asunder the motley feudal ties that bound man to his "natural superiors," and has left remaining no other nexus between man and man than naked self-interest, than callous "cash payment." (Karl Marx and Friedrich Engels)

9. Plea 5. *Some Good men think it is lawful! Ans:* We are not to walk by the opinion of this or that good Man, but by the Scriptures. *To the Law and to the Testimony, if they speak not according to that there is no light in them.* Fearful Judgments have befallen a Professing People for doing such things as some good Men through error of judgment have approved of. (Increase Mather)

10. Mrs. Harris has serious heart problems, and clearly she has suffered over the last eight years. Just as it is unlikely that she poses a threat to society, so it is certain that she could do more good as a sensible prison reformer on the outside than as a lifetime prisoner on the inside. According to Mrs. Harris, Bedford's correctional officers always refer to the inmates as "ladies." It is time to release the lady who wrote this book. (John J. DiIulio, Jr.)

11. It cannot be said more plainly: Parental choice in education is the American way. Parents choose where to live, what to feed their children, what church to belong to, what careers to follow, whom to call their friends. A system of financing that doesn't allow them to choose schools for their children is so out of kilter that one wonders how it ever came to exist in this free nation. (Gary L. Bauer)

12. The statist businessman is, by definition, a lobbyist. Having made his peace with 20th century collectivism, he is fundamentally concerned with "who gets what" from government's redistributive powers. He seeks subsidies for himself and penalties and regulations for his competitors. He is the miserable figure Ronald Reagan described as the fellow who hoped the crocodile would eat him last. (Theodore J. Forstmann)

Identify the fallacies, if any, in the following passages. (Fallacies from any of the previous sections may be included.)

13. The judiciary has thus reached into the Constitution's spirit and structure, and has elaborated from the spare text an idea of the "human" and a conception of "being" not merely contemplated but required. (Laurence Tribe)

14. Jane: We shouldn't be sending aid to that country. I've seen several anti-American demonstrations from their capital on the news. There are allegations of human rights abuses.
Ned: All foreign capitals have anti-American demonstrations from time to time. I see them on the news all the time from all parts of the world. I guess people like to blame problems on some powerful factor they cannot control.

15. Why does not the pope, whose riches are at this day more ample than those of Croesus, build the basilica of St. Peter with his own money rather than with that of poor believers? (Martin Luther, 1517)

16. . . . in last spring's House debate over a bill to send . . . aid to the Nicaraguan rebels . . . a Democratic congressman rose, said that he was for the aid, and urged his fellow Democrats to vote for it too. Otherwise, he said, they would be vulnerable to attack in the next election for being "soft on communism." Reporters later cited the argument as swaying many other Democrats to support the bill, which passed 248-184. . . . [This] motive was a base one. It is plainly contemptible to vote for the killing of men, women and children so that your political career may be prolonged. (Jack Beatty)

17. Son: What is a traitor?
    Wife: Why, one that swears, and lies.
    Son: And be all traitors that do so?
    Wife: Everyone that does so is a traitor and must be hanged.
    Son: And must they all be hanged that swear and lie?
    Wife: Every one.
    Son: Who must hang them?
    Wife: Why, the honest men.
    Son: Then the liars and swearers are fools; for there are liars and swearers enow to beat the honest men and hang them up. (William Shakespeare, *Macbeth* IV, ii)

18. a. Do you believe that you should keep more of your hard-earned money or that the government should get more of your money for increasing bureaucratic government programs?
    b. Do you support a large and bureaucratic government-run health care system or a private-sector solution to America's health care problems?
    c. Which tax measure should the Senate consider this legislative session? (Please check only one) [] Flat Tax [] Abolish the IRS [] National Sales Tax [] Super Majority Amendment: A ⅗ vote required to pass any future tax increases. (National Republican Senatorial Committee)

19. Make a list of appeals to force that parents typically use to make their children do things. Make another list of appeals to pity that children typically use to avoid doing those things. In each case, explain why the appeal does or does not constitute a good argument.

20. Devise an argument justifying some potentially controversial everyday action or habit—for example, eating meat, using disposable plastic utensils, driving a car, buying record albums when there are starving people, etc.—without appealing to common practice. *(a)* Why is this difficult? Explain. *(b)* What must be done to avoid appealing to common practice in these cases?

21. List some appeals to pity that students might make when asking for a grade change. In each case, do such appeals constitute good arguments? Why, or why not?

22. In 1984, a bumper sticker favoring the Democratic candidate for president over then-President Ronald Reagan read, "No Mo' Ron for President." What argument is implicit in the bumper sticker's message? Does this commit an *ad hominem* fallacy? Why, or why not?

23. Faced with an audience of college students and the issue of whether the following should be legal, what mistakes could an arguer make in appealing to the emotions of the audience? Make a list, and explain why each emotional appeal is logically irrelevant.
    a. sport hunting
    b. animal trapping
    c. boxing
    d. marijuana use
    e. marijuana dealing
    f. cocaine use
    g. cocaine dealing
    h. cigarette smoking
    i. cigarette smoking in public places
    j. prostitution
    k. pornography involving consenting adults only
    l. child pornography
    m. prayer in schools
    n. Christmas or Hanukkah displays on public property
    o. use of experimental drugs by terminally ill patients
    p. allowing young workers to labor for less than minimum wage
    q. clubs admitting members of only one sex

24. You are a political consultant charged with convincing voters in your state or county *(a)* to raise taxes or *(b)* not to raise taxes. What emotional appeals can you use in your campaign? Write a TV commercial using them, and explain whether, and why, they are or aren't logically relevant.

■ 3.5 INDUCTIVE FALLACIES

Recall that some good arguments are inductively strong rather than deductively valid. If their premises are true, their conclusions are probable but not guaranteed. Some fallacies are common inductive errors.

13. APPEAL TO IGNORANCE. Most arguments try to establish a conclusion by relying on what is known. Some, however, rely on what is unknown. This is not always fallacious. A famous example of a good inference from lack of evidence is in a Sherlock Holmes story:

> "Is there any other point to which you would wish to draw my attention?"
> "To the curious incident of the dog in the night-time."
> "The dog did nothing in the night-time."
> "That is the curious incident," remarked Sherlock Holmes.[14]

Similarly, if a yearlong investigation into Ben's activities has produced no evidence that he has committed any crime, we might argue from the lack of evidence of wrongdoing for Ben's innocence. The argument, furthermore, may be good. The lack of evidence of criminal activity may itself be evidence of innocence.

Nevertheless, arguments on the basis of a lack of knowledge easily go awry. Consider what we need to know about the investigation before we can justifiably conclude that Ben is innocent. First, the investigation must have been extensive and thorough. Second, the investigators must have been competent and unbiased. Third, the investigation must have been in a position to turn up evidence, if there were any. If the investigation received inadequate funding or could not examine relevant documents or if several key associates of Ben met sudden, unnatural deaths just before being questioned, a conclusion of innocence based on the results of the investigation would be hasty.

Appeals to ignorance rely on the unknown to support their conclusions:

An argument **appeals to ignorance** if and only if it tries to justify its conclusion by appealing to what is not known.

Form: There is no evidence against p.
∴ p.

---

[14]Doyle, "The Adventure of Silber Blaze," *The Adventures of Sherlock Holmes*, pp. 196–197.

Such appeals succeed only under certain conditions. A lack of knowledge shows something only if there has been a competent, thorough, unfrustrated attempt to secure that knowledge. Otherwise, pointing toward a lack of knowledge is an unfair attempt to shift the burden of proof onto one's opponents.

Some dramatic examples of this kind of fallacy arose in the context of Senator Joseph McCarthy's crusade against Communists in government. Presenting a list of alleged Communists in the State Department, he said about one official, "I do not have much information on this except the general statement of the agency that there is nothing in the files to disprove his Communist connections."[15] Lack of disproof, clearly, is not tantamount to proof.

Other examples concern parapsychic phenomena (such as ESP) or controversial but unproven hypotheses (such as the existence of UFOs). On one side, people tend to argue that nobody has proven that ESP does not exist or that all unidentified flying objects are merely natural phenomena. Sometimes, they conclude that ESP or UFOs are real. This is an appeal to ignorance. On the other side, people tend to argue that there is no scientific evidence that anyone has extrasensory perception or that UFOs are anything but natural phenomena. Sometimes, they conclude that there are no such things. Whether this is also an appeal to ignorance depends on how extensively such matters have been investigated. We can draw conclusions from a lack of evidence only if we know a great deal about attempts to find such evidence.

14. APPEAL TO AUTHORITY.   Appealing to authorities to establish facts or conclusions is not in itself a fallacy. Modern life is so complex that we all rely extensively on other people for information. You probably know that Bismarck is the capital of North Dakota. But how do you know? Have you been there, seen the legislature meet there, and seen the official documents declaring it the state capital? Most people have not. They have obtained their information from geography books, maps, or teachers. Those sources, in turn, have usually derived their knowledge from other geography books, maps, or teachers. Between you and Bismarck may be a long chain of people, each relying on the authority of another.

This is fine, provided that the people relied on are trustworthy and knowledgeable about the matter at hand. But we must be careful to assess the reliability of our sources. The views of a rightful authority on a particular subject are relevant evidence concerning that subject. Opinions of authorities count, however, only in their areas of expertise. And, even at best, expert testimony is only part of the story.

---

[15]Richard H. Rovere, *Senator Joe McCarthy* (New York: Harcourt Brace, 1959), p. 132.

For an argument relying on authority to succeed, the opinions it cites must fall within the authority's expertise. To quote Albert Einstein on physics, Henry Kissinger on foreign policy, and Bill James on baseball is perfectly acceptable. (It is acceptable, that is, unless the arguer is also an expert in the same field. Physicists do not quote other physicists at each other; they appeal to facts, experiments, and theoretical results.) To quote Einstein on foreign policy, Kissinger on baseball, and James on physics, however, yields no evidence concerning those areas.

> An argument **appeals to authority** if and only if it tries to justify its conclusion by citing the opinions of authorities.
>
> Form: A says p.
> ∴ p.

When are these appeals fallacious? When is it appropriate to base a conclusion on the opinions of authorities? (1) The subject must be one on which there can be authorities. Some philosophers have held that there are no authorities on moral questions. Very likely nobody can be an authority on whether God exists or on what the one true religion is. For very different reasons, probably nobody can be an authority on what the stock market will do next week or next year. (2) The opinions cited in the argument must fall within the authority's area of expertise. (3) The authority must be trustworthy; there must be no additional factors that might make the expert misrepresent matters.[16] If the authority is a pathological liar or a comedian, has strong political reasons for being on a particular side of an issue, or is in danger of persecution from the Inquisition or a congressional committee, the testimony may not be very reliable.

A special problem arises when legitimate experts disagree. Frequently, the testimony of one authority conflicts with that of another. At times, the preponderance of expert opinion is on one side of a question; in the mid-1980s, labor union leaders were able to find only one economist in the United States willing to speak out in favor of protectionism. Even such an overwhelming majority could be wrong, although the weight of such opinion provides substantial evidence in favor of the majority's view. In many circumstances, however, opinion is more evenly divided. This makes it much more difficult for a nonexpert to judge who is correct. Whenever legitimate experts disagree, appeals to authority lose much of their force. It becomes necessary to look at the arguments in detail.

---

[16]Compare the definition in Vatsyayana's Commentary on the *Nyaya-sutra*, translated by Mrinalkanti Gangopaghyay (Calcutta: Indian Studies, 1982): "A trustworthy person is the speaker who has the direct knowledge of an object and is motivated by the desire of communicating the object as directly known by him."

15. ACCIDENT.   Writers on logic have described the fallacy of *accident* in two quite different ways. First, and as we will use the term, the fallacy of accident has been described as making "an unqualified judgment of a thing on the basis of an accidental characteristic."[17] Philosophers have traditionally held that some attributes of a thing are essential to it in the sense that without them the thing would not be what it is, while others are accidental in the sense that the thing could retain its identity without them. Your having a mind, for example, is essential to you; without a mind, you would not be the person you are. But reading this book is accidental to you; you could stop reading it without ceasing to be you. Arguments that mistake accidental for essential attributes commit the fallacy of accident.

A famous and humorous example occurs in Shakespeare's *Henry V,* where Fluellen argues that King Henry of Monmouth is the equal of Alexander the Great of Macedon:

> If you look at the maps of the 'orld, I warrant you sall find, in the comparisons between Macedon and Monmouth, that the situations, look you, is both alike. There is a river in Macedon, and there is also moreover a river at Monmouth. It is called Wye at Monmouth; but it is out of my prains what is the name of the other river. But 'tis all one; 'tis alike as my fingers is to my fingers, and there is salmons in both.

Ruling a kingdom with a river containing salmon is accidental to being a great king; Fluellen's argument mistakes it for an essential feature of greatness. Literary critic Richard Levin has correspondingly called the logic of arguments committing the fallacy of accident "Fluellenism."[18]

> An argument commits the fallacy of **accident** if and only if it tries to justify its conclusion by treating an accidental feature of something as essential.

A grimmer example: In 1959, in Cuba, Castro tried to justify throwing out "innocent" verdicts and sentencing forty-four political opponents to long prison terms by declaring, "Revolutionary Justice is based not on legal precepts but on moral conviction."[19] But someone else's moral indignation hardly justifies punishment.

---

[17]Arnauld, *Logic*, p. 259.

[18]See Richard Levin, *New Readings v. Old Plays* (Chicago: University of Chicago Press, 1979).

[19]Quoted in Hugh Thomas, *Cuba, or the Pursuit of Freedom* (New York: Harper and Row, 1971), pp. 1202–1203.

16. MISAPPLICATION.   The fallacy of accident has also been characterized as "applying a general rule to a particular case whose 'accidental' circumstances render the rule inapplicable."[20] To avoid confusion, we will call this fallacy *misapplication.* Suppose, for example, that you have promised to meet a friend. On your way there, you are involved in a collision; the other driver is injured. You can meet your friend only if you leave the scene before the police arrive. There is a general rule that you should keep your promises. Nevertheless, to leave the scene because a rule requires you to keep your promise is foolish. To apply the rule in such a contrary case is to commit a fallacy of misapplication.

> An argument commits a fallacy of **misapplication** if and only if it tries to justify its conclusion about a particular case by appealing to a rule that is generally sound but inapplicable or outweighed by other considerations in that case.

Much of what we ordinarily call "common sense" consists of rules about what to do in various situations. The rules, however, are fairly rough guidelines; they do not hold in every possible circumstance. They are *default* rules: They say what to do under normal circumstances, if no other relevant circumstances arise. They specify a default course of action. But, in unusual cases, the default can be overridden. Rules such as "Follow your emotions," "Do what is best for you," "Help others," and "Listen to the experts" are valuable, but they have exceptions. Failing to recognize exceptions involves a fallacy of misapplication.

Philosophers have tended to think of these rules as containing "other things being equal" clauses. We should think of them as saying, for example, "Follow your emotions, other things being equal," and so on. In effect, these rules say what to do if nothing else indicates otherwise. Ignoring the cautionary "other things being equal" clauses leads to misapplication fallacies.

Plato discusses a classic example of misapplication in the *Republic:*

> . . . are we to say that justice or right is simply to speak the truth and to pay back any debt one may have contracted? Or are these same actions sometimes right and sometimes wrong? I mean this sort of thing, for example: everyone would surely agree that if a friend has deposited weapons with you when he was sane, and he asks for them when he is out of his mind, you should not return them. The man who returns them is not doing right, nor is one who is willing to tell the whole truth to a man in such a state.[21]

---

[20]Irving M. Copi, *Introduction to Logic* (New York: Macmillan, 1978), p. 95.

[21]Plato, *Republic,* 331c, translated by G. M. A. Grube (Indianapolis: Hackett Publishing Company, 1974), pp. 5–6.

To apply the rules "Tell the truth" and "Keep your promises" would be, in this case, to misapply them. Other things are not equal. Your obligation to prevent harm to your friend and others outweighs your obligation to tell the truth and keep your promise.

For other examples of misapplication fallacies, we can turn to the history of the United States Supreme Court, which has not only declared many lower-court decisions to be misapplications of law but has also reversed its own earlier decisions on a number of occasions. The *Dred Scott* decision (1857), for example, held that Congress did not have the authority to exclude slavery from the territories. After the Civil War, the Court held that this had been a misapplication of Constitutional principles. *Plessy v. Ferguson* (1895) held that segregation did not violate the Fourteenth Amendment's guarantee of equal protection of the laws; in *Brown v. Board of Education* (1954) the Court held that to be a misapplication of the Constitution, finding the "separate but equal" doctrine of the earlier decision oxymoronic.

17. HASTY GENERALIZATION. To draw conclusions about a kind or large class of things, we typically examine only some of them. If we choose carefully which and how many things to sample, our conclusions may be highly reliable. (For a detailed discussion of generalization and sampling, see Chapter 10.) If we do not, however, our conclusions may be unreliable. We may sample too few objects. Or, the objects we examine may be atypical or all of one particular sort. In such cases, we commit the fallacy of hasty generalization. To conclude that welfare fraud is common from seeing a flashy dresser driving a Lexus use food stamps at the grocery store is to generalize far too quickly. But to reach the same conclusion from government studies estimating fraud rates in certain programs at over 30 percent may not be fallacious at all.

18. FALSE CAUSE. Fallacies of causal reasoning occur frequently enough to be mentioned here, though we will not consider causal reasoning in detail until Chapter 11. One form of causal fallacy has a Latin name, *post hoc, ergo propter hoc*—"after this, therefore because of this." As the name suggests, this fallacy involves drawing a causal conclusion simply from the temporal ordering of events. So, the form of the argument is

> E preceded F.
> ∴ E caused F.

or, in terms of causal generalizations,

> Events of kind E precede events of kind F.
> ∴ Events of kind E cause events of kind F.

Clearly, this is a poor and unreliable form of argument. The Hundred Years' War occurred before the discovery of uranium, but it would be preposterous to conclude that the Hundred Years' War caused uranium to be discoved. Similarly, the moon landing occurred before the Mets won the 1969 World Series, but the moon landing did not cause the Mets to win. John Allen Paulos tells a joke that makes the point:

> Two Australian aborigines were brought to this country and see for the first time a waterskier winding and cavorting his way around a lake. "Why is the boat going so fast?" asked one of the aborigines. The second answered, "Because it is being chased by a madman on a string."[22]

Nevertheless, arguments of this form are tempting. Political candidates rely on them constantly, proclaiming proudly that the economy did well during their tenure, concluding that they were responsible for its excellent performance, or decrying the unfortunate events occurring during their opponent's tenure, concluding that their opponent was the cause of all that misfortune. Political commentators also tend to be glib about extending temporal relations to causal relations. Large U.S. government budget deficits were a consistent feature of the world economy of the 1980s. At various times, commentators seemed to reach a consensus that the deficits caused an expansion, a recession, increases in interest rates, decreases in interest rates, increases in inflation rates, decreases in inflation rates, rises in the value of the dollar, declines in the value of the dollar, and almost every other economic occurrence of the decade. Sometimes these hypotheses gained some support from economic theory. Often, however, they were little more than *post hoc, ergo propter hoc* arguments.

Another unreliable form of causal inference concerns statistical correlation. As we shall see in Chapter 11, a correlation between two kinds of events allows inferring the existence of some causal connection but yields no information about its exact form or nature. If one quantity varies with another, changes in the former may cause changes in the latter, or changes in the latter may cause changes in the former, or changes in both quantities may result from some other causal mechanism. Finally, of course, the correlation may be just a coincidence. But it can be tempting to draw a straightforward causal conclusion from a correlation:

E correlates with F.

∴ E causes F.

That also constitutes a fallacy of false cause.

Sports figures and fans sometimes fall prey to such arguments. The Dallas Cowboys, for example, declined to wear their blue home uniforms for years because they believed them to be unlucky; they seemed to lose more often while wearing those uniforms. Almost certainly, however,

---

[22]John Allen Paulos, *I Think, Therefore I Laugh* (New York: Columbia University Press, 1985), p. 67.

there was no causal relationship. Similarly, the career of Lyman Bostock, a Detroit outfielder during the 1970s, tracked closely the career of Austin McHenry, a St. Louis outfielder who played during the first decades of the twentieth century. Their achievements correlate, but almost certainly by coincidence. For a while after the assassination of John F. Kennedy, a sheet circulated describing various parallels between the assassinations of Kennedy and Abraham Lincoln. The sheet observed, for example, that Lincoln's assassin shot him in a theatre and fled to a warehouse, while Kennedy's shot him from a warehouse and fled to a theatre. Lincoln's secretary, named Kennedy, urged him not to go to the theatre; Kennedy's secretary, named Lincoln, urged him not to go to Dallas. The coincidences, if true, are remarkable, but they point to no causal link.

Bad causal arguments are more tempting, and more deceptive, when the correlation probably is not coincidental. As most barometer faces indicate explicitly, low barometric pressure correlates with rain. It can be tempting to conclude that low atmospheric pressure causes rain or that rain causes low pressure. In fact, however, both tend to be effects of a common cause, namely, the movements of an air mass. During the 1970s, medical researchers established a correlation between heart disease and what they called "Type A" behavior—aggressiveness, hard work, ambition, and so on. Some people were tempted to conclude that Type A behavior causes heart disease. That may be true, but it may also be true that some common genetic or environmentally produced characteristic produces both Type A behavior and a tendency toward heart disease.

The moral, then, is that correlations, even when not coincidental, signal the existence of a causal link but reveal nothing about its nature.

P R O B L E M S

Identify the inductive fallacies, if any, in the following passages.

1. REV. KEY RESIGNS; ATTENDANCE DOUBLES. *(LaGrange [Ga.] News)*

2. The thing that appalls me about the newspaper business is the number of trees it consumes. (Prince Charles)

3. I will make no concessions . . . my father made concessions and he was beheaded. (King James II)

4. Most of the people who are supposed to be powerful really aren't. They may be powerful in their company—but they still cannot get a cab in the rain. (Ed Kosner)

5. . . . the satellites are invisible to the naked eye and therefore can have no influence on the earth and therefore would be useless and therefore do not exist. (Francesco Sizi, arguing, against Galileo, that Jupiter has no moons)

6. How can anyone govern a nation that has 246 different kinds of cheese? (Charles de Gaulle)

7. Aunt Margaret said to little Alvin, "Would you like some more stuffing?" Alvin said, "Nope. And I don't know why the turkey eats it either!" (Milton Berle)

8. "This then is not a definition of right or justice, namely to tell the truth and pay one's debts."
   "It certainly is," said Polemarchus interrrupting, "if we are to put any trust in Simonides." (Plato)

9. A Saudi bank took a customer to court over 10 million riyals. The guy refused to pay interest because he said it was against his religion. The judge asked, "Didn't you know that before?" And the guy said, "Yes, but at that time I was not religious." (Abdulaziz O'Hali)

10. In Plymouth, Mass., Anthony Varrasso was called for jury duty. No big deal—except that Anthony [was] just 3 years old. His mother [told] officials that Anthony was a bit young to be deciding cases, but they insisted that he was 18. That's what it said on a census form. Plymouth has called dead people, pets, even a building for jury duty because of incorrect census information. (Charles Oliver)

11. Attorney General Earl Warren testified at the congressional Tolan Committee hearings in San Francisco that "because we have had no sabotage and no fifth column activities in this State," therefore "the sabotage we are to get" has been carefully concealed and precisely timed, as at Pearl Harbor and as in occupied Europe. (Eric J. Sundquist)

12. Inmate sues chairman of [Mississippi] parole board for not receiving a scheduled parole hearing, although said inmate was on the lam when the hearing was to have been held. (*New York Times Magazine*)

13. A little arson joke: These two fellas are walking down the beach in Miami. "My factory burned down," says the first, "and I retired on the insurance."
    "*That's* funny," says the second. "*My* factory got washed away in a flood, and I retired on the insurance, too!"
    "Really?" asks the first, incredulous—"How do you start a flood?" (Andrew Tobias)

14. A raspberry to the U.S. Customs Service, for seizing two antique British telephone booths. The colorful red cast-iron and glass boxes are being sold as novelty antiques by a British firm. Cus-

toms agents insist that the 1,500-pound booths should be classified as fabricated steel and they've demanded that the British government issue a certificate that counts the booths against the steel import quotas. (Citizens for a Sound Economy Foundation)

15. Tony Maggerise went to traffic court in Tacoma, Wash., to protest a speeding ticket. The officer didn't show, so the judge had the defendant flip a coin. Maggerise objected—even though the fine was reduced to $20—but the judge ruled his complaint ex post facto: "I bet he wouldn't have been griping if he had called it right." (*Life*)

16. Watch how the New York Yankees perform in the baseball playoffs. Since 1945, they have appeared in five election-year World Series. When they lose (1960, 1964, 1976), a Democrat wins the White House; when they win (1952, 1956), the Republican triumphs. *(The Economist)*

17. . . . an historian's professed inability to discern any plot, rhythm, or predetermined pattern is no evidence that blind Samson has actually won his boasted freedom from the bondage of "Laws of Nature." The presumption is indeed the opposite; for, when bonds are imperceptible to the wearer of them, they are likely to prove more difficult to shake off than when they betray their presence and reveal something of their shape and texture by clanking and galling. (Arnold Toynbee)

18. Talking of snakes, Mrs. Montgomery told me that once she nearly trod upon a krait—one of the most venomous snakes in India. "I was going back in the evening to my bungalow, preceded by a servant who was carrying a lamp. Suddenly he stopped and said, 'Krait, Mem-sahib!'—but I was far too ill to notice what he was saying, and went straight on. Then the servant did a thing absolutely without precedent in India—he touched me!—he put his hand on my shoulder and pulled me back. Of course if he hadn't done that I should undoubtedly have been killed; but I didn't like it all the same, and got rid of him soon after." (J. R. Ackerley)

19. The photo editor of the *Herald Tribune* sent me out on an assignment. I'd been too embarrassed to bring my camera to the [job] interview. So he lent me one—loaded, thank goodness, because the camera he gave me I had no idea how to open or close. So I took off to this luncheon and shot my roll of film, and the man standing behind me in the lunch line said, "Oh, I see you're using the new Pentax. I just wrote a brochure about that camera." And I said, "Oh, that's wonderful. Would you mind unloading

it?" Because I really didn't want to go back to the *Tribune* and hand in my loaded camera; I thought it would look more sophisticated to be able to give them the roll of film. So he unloaded the camera and he said, "And what brings you to this lunch?" I told him I was covering it for the *Herald Tribune*—and I'm sure to this very day that man has no question in his mind as to why the paper folded. (Jill Krementz)

20. Has the world made it too easy for places to become countries? I have no wish to undo what is done; places that have gone the route of grim self-determination should be allowed to live with the consequences. But it may be useful to adopt criteria that prospective new countries must hereinafter meet before being upgraded from provisional status to true sovereignty.

   The country must have won an Olympic medal in an event that is not the bobsled or the luge. It must at some point in its history have produced an epic poem that can now be bought in a Penguin edition. Its army must have served with distinction as a UN peacekeeping force—outside the country itself. A restaurant featuring the country's cuisine must exist somewhere in the United States outside of New York City.

   Upon receipt of a hard-currency bond to ensure payment of its diplomats' parking tickets abroad, we may safely say to such a place, Welcome to the club. (Cullen Murphy)

Identify the fallacies, if any, in the following passages. (Fallacies from any of the previous sections may be included.)

21. Man walks because things in nature move. . . . Similarly, man digests because nature is chemical. . . . Man predicts because nature is mechanical. . . . Once more man is free because nature is contingent. (Sterling P. Lamprecht)

22. We repudiate all morality taken apart from human society and classes. We say that it is a deception, a fraud, a befogging of the minds of the workers and peasants by the landlords and capitalists. (Nicolai Lenin)

23. We are wrong if we suppose that man alone is gifted with esthetic feeling. Many animals are more beautiful than the featherless biped that transiently rules the earth. . . . (Will Durant)

24. Because all—and I underline the word all—taxes have a repressive effect on economic activity, that means they are costly on the economy to impose. (Alan Greenspan)

25. On Wednesday 22 March 1933, the first concentration camp will be opened near Dachau. It will accommodate 5,000 prisoners.

Planning on such a scale, we refuse to be influenced by any petty objection, since we are convinced this will reassure all those who have regard for the nation and serve their interests. (Heinrich Himmler)

26. The finest historians will not be those who succumb to the dehumanizing methods of social science, whatever their uses and values, which I hasten to acknowledge. Nor will the historian worship at the shrine of the Bitch-Goddess QUANTIFICATION. (Carl Bridenbaugh)

27. While the role of dope in damping social unrest in early industrial England has not been extensively investigated, every historian of the period knows that it was common practice at the time for working mothers to start the habit in the cradle by dosing their hungry babies on laudanum ("mother's blessing," it was called). (Theodore Roszak)

28. Mr. Muravchik is troubled by my characterization of IPS angel Samuel Rubin as a Communist party member, when the evidence suggests only that Rubin was registered as a Communist party voter and that therefore he was not under party discipline. Let me just say that no one on the Left has denied Rubin's CP membership, and I am hardly the first to have made the charge. . . . (Scott Steven Powell)

29. This is a sad day for all of us, and to none is it sadder than to me. Everything that I have worked for, everything that I have hoped for, everything that I have believed in during my public life, has crashed into ruins. There is only one thing left for me to do; that is, to devote what strength and powers I have to forwarding the victory of the cause for which we have to sacrifice so much. (Neville Chamberlin, 1940)

30. They [U.S. soldiers] were totally healthy when they went over to the Persian Gulf. No problems whatsoever. They come back and all of the sudden their children are [born] defective, they can't have children, or they [the children] die. [Something in the Gulf made them sick.] (Senator John D. Rockefeller)

31. Discuss the allegation of fallacy in the following:
    "Shoes! Frivolity! Marie Antoinette!" In the Philippines I built houses for 30,000 slum dwellers. I planted 80 million trees around Manila . . . and they talk only about shoes! (Imelda Marcos)

32. Using the front page of a recent newspaper, identify five authorities used by newswriters in their stories. Given the information you have, is the appeal to authority justified? Explain.

33. Following are some standard proverbs. For each, create two arguments, one that represents a proper application of the proverb and one that represents a misapplication of the proverb.
    a. A stitch in time saves nine.
    b. Haste makes waste.
    c. A penny saved is a penny earned.
    d. Look before you leap.
    e. He who hesitates is lost.
    f. Absence makes the heart grow fonder.
    g. Out of sight, out of mind.
    h. Less is more.
    i. Honey catches more flies than vinegar.
    j. The apple never falls far from the tree.
    k. A bird in the hand is worth two in the bush.
    l. There are two sides to every question.
    m. The early bird gets the worm.

34. In each of the following pairs, both sentences could be argued by appeals to ignorance. Construct such an appeal for each sentence.

    a. God exists.
    b. God does not exist.

    c. There is intelligent life elsewhere in the universe.
    d. There's no intelligent life elsewhere in the universe.

    e. Humans are smarter than dolphins.
    f. Dolphins are smarter than humans.

35. Following are several fundamental rights or freedoms. Consider the example cases and explain why each is a correct application, or misapplication, of the listed right or freedom. Some are controversial: In those cases, choose one side, and give reasons.

    Freedom of speech:
    a. Publishing a newspaper critical of the government.
    b. Publishing an article claiming, without evidence, that a public figure is a criminal.
    c. Publishing an article claiming, without evidence, that an ordinary citizen is a criminal.
    d. Producing a highly pornographic movie.
    e. Showing a highly pornographic movie in your theater.
    f. Shouting "Fire!" in a crowded theater.
    g. Burning the American flag in public.

    Freedom of religion:
    a. Holding a worship service in a private building.
    b. Holding a worship service in a public park.

c. Holding a worship service in the middle of a busy public roadway.

d. Privately displaying religious articles or symbols.

e. Publicly displaying religious articles or symbols.

f. Refusing medical help on religious grounds.

g. Refusing medical help for one's children on religious grounds.

h. Using illegal drugs as part of a religious ceremony.

i. Conducting a religious ceremony involving the handling of poisonous snakes.

j. Handing out religious pamphlets in a public airport.

Freedom to keep and bear arms:

a. Keeping sharp knives in the kitchen.

b. Keeping a baseball bat under the bed.

c. Keeping a pistol in a desk drawer.

d. Keeping a hunting rifle in the basement.

e. Keeping a semiautomatic rifle in the closet.

f. Keeping a submachine gun in the nightstand.

g. Carrying a stickpin in a purse.

h. Carrying a switchblade in a pocket.

i. Carrying a pistol in a shoulder holster under a jacket.

j. Carrying a rifle over a shoulder.

■ 3.6 FALLACIES OF CLARITY

Sometimes arguments fail by being unclear. A premise understood in one way might be true but fail to support the conclusion; read in another way, it might support the conclusion but turn out to be false. Several fallacies arise from unclarity.

19. EQUIVOCATION.   Perhaps the simplest mistake about meaning involves misconstruing a word or phrase. Most words have more than one meaning. Sometimes context does not determine which meaning is intended. If the listener attaches a meaning to a word that the speaker did not intend, then miscommunication takes place; the information the listener draws from the conversation is not what the speaker meant to convey. Evelyn Waugh, in East Africa in 1959, wrote of a tragic misunderstanding based on an ambiguous word:

> I spent one day with the Masai. . . . They had a lovely time during the Mau Mau rising. They were enlisted and told to bring in all the Kikuyu's arms. Back they proudly came with baskets of severed limbs.[23]

---

[23]Mark Amory, ed., *Letters of Evelyn Waugh* (London: Weiderfeld & Nicolson, 1980), p. 517.

The fallacy of equivocation is more than miscommunication. It involves deriving a conclusion by relying on differing meanings of a word or phrase:

> An argument is guilty of **equivocation** if and only if it tries to justify its conclusion by relying on an ambiguity in a word or phrase.

Equivocation is easiest to commit, and hardest to detect, in arguments relying on confusions between the meanings of terms with several closely related meanings. Consider, for example, the political terms *liberal* and *conservative*. Each has several different meanings relevant to politics:

Liberal:

a. giving freely; generous

b. not restricted to the literal meaning; not strict

c. tolerant of views different from one's own

d. of democratic or republican forms of government

e. favoring reform

Conservative:

a. conserving or tending to conserve

b. tending to preserve established traditions

c. moderate; cautious; safe

The English philosophers John Locke (1632–1704) and John Stuart Mill (1806–1873) are often called liberals; they advocated tolerance and democracy and, so, are liberals in senses c and d. In their own time, they also favored reform and, so, were liberals in sense e as well. Today, however, they would be described as conservatives or libertarians. Franklin Delano Roosevelt (1882–1945) was called a liberal for instituting the New Deal; that collection of policies involved government spending on programs directed at helping people and, thus, was liberal in sense a, an economic rather than social sense of *liberal*. Most people who now consider themselves liberals mean the term in sense e; they favor reform and social change.

An argument can easily go astray by confusing these senses of the word. Thus, someone who deduces that a candidate will try to help the poor from the fact that the candidate is a liberal, where that term is used in any of senses b–e, equivocates. Similarly, someone who argues that a candidate would promote tolerance simply because that candidate is a liberal in the sense of favoring social change commits a fallacy of equivocation.

The ambiguity of the term *conservative* can also promote equivocation. A voter who chooses the conservative as a safe or moderate choice may be making a mistake, if that candidate is conservative in the sense of favoring established traditions. After all, one might use extreme measures to defend established tradition. Conversely, a candidate who favors moderation does not necessarily seek to preserve the establishment.

Equivocation may involve phrases containing more than one word as well as individual words. Consider these exchanges (the ambiguous phrases are in italics):

(6) A: The painter got so angry he *kicked the bucket.*
   B: You mean he died right there?

(7) A: Do you *believe in* infant baptism?
   B: Yes—I've seen it done!

A particularly interesting case of an inference based on equivocation concerns J. Edgar Hoover, former director of the Federal Bureau of Investigation.

> One of [J. Edgar] Hoover's many quirks was his demand that no memo should exceed one page, with wide margins. An agent ran into trouble getting his reports onto one page, so he encroached on the prescribed margin width. Hoover wrote back, "Good analysis, but watch the borders." Since no subordinate was willing to question "The Chief," the FBI dispatched agents to the Canadian and Mexican borders, to "watch." Nobody knew what they were looking for, but for a while, these borders were watched as never before.[24]

20. AMPHIBOLY. Some sentences are ambiguous, not because they contain ambiguous words or phrases, but because of their grammatical construction. Such sentences are **syntactically ambiguous** or **amphibolous**—they can be read in two different ways, because they have two different syntactic or grammatical structures.

This item from the Bangkok *Post* illustrates amphiboly:

(8) As in previous years, the evening concluded with a toast to the new president on champagne provided by the retiring president, drunk as usual at midnight.

One might conclude that the retiring president has an alcohol problem. Usually, amphiboly results from misplacing a clause or phrase in a sentence. We can rewrite the amphibolous item above as the unambiguous

(9) As in previous years, the evening concluded with a toast, drunk as usual at midnight, to the new president on champagne provided by the retiring president.

---

[24]Mortimer R. Feinberg, "When to Engender Fear . . . or at Least a High Degree of Anxiety," *Wall Street Journal*, October 24, 1988, p. A14.

Sometimes amphiboly results from an elliptical construction. The classic example is a wartime poster urging citizens to "Save soap and waste paper." Another is the topic of a debate between some well-known conservatives and liberals in 1988: "Resolved, that the Right is better able to deal with the Soviets than the Left." Finally, a joke: A bear hunter was driving north to go hunting when he saw a sign that said "Bear Left." He went home.

In general, an argument is guilty of the fallacy of amphiboly if it confuses sentence structure in a way that affects the sentence's meaning:

> An argument is guilty of **amphiboly** if and only if it tries to justify its conclusion by relying on an ambiguity in sentence structure.

It might seem from the preceding that amphiboly is more a source of jokes than of fallacious inferences. But there are serious examples. The German philosopher Immanuel Kant (1724–1804) titled his greatest work *Kritik der Reinen Vernunft (Critique of Pure Reason)*. This is ambiguous in both English and German. Is pure reason the agent or the object of the critique? In other words, is the title the nominalized form of *x critiques pure reason* (as with *the defeat of the Spanish Armada*) or of *pure reason critiques x* (as with *the ride of Paul Revere*)? Probably, Kant intended both.

Another serious example concerns versions of the cosmological argument for God's existence offered by Thomas Aquinas (1225–1274) and John Locke (1632–1704). At a crucial point in Aquinas's third proof, for example, he argues:

(10) We observe in things something that can be, and can not be, for we observe them springing up and dying away, and consequently being and not being. Now not everything can be like this, for whatever can not be, once was not. If all things could not be, therefore, at one time there was nothing.

But *All things could not be* is ambiguous. His argument for it interprets it as meaning *Everything is such that it is possible for it not to exist*. But his conclusion, "at one time there was nothing," interprets it as meaning *It would be possible for everything not to exist* (all at once, that is). The proof seems to depend on an amphiboly.

As G. W. F. Leibniz (1646–1716) pointed out, something similar happens in Locke's proof. He maintains that something has existed from all eternity. In support of this, he argues that if, at any time, nothing had existed, nothing could exist now. He concludes from it that there is an eternal being. But this move relies on two different readings of *something has existed from all eternity*. Locke's premises support the assertion that at any time something or other existed at that time but not the stronger claim that some one thing—God—has existed at every point in time.

An example that is both serious and humorous is a law that the Missouri legislature passed on August 28, 1994, which seems to outlaw sex:

> (11) A person commits the crime of sexual misconduct in the first degree if he has deviate sexual intercourse with another person of the same sex, or he purposely subjects another person to sexual contact or engages in conduct which would constitute sexual contact except that the touching occurs through the clothing without that person's consent.[25]

Some have argued that this law bans all gay sex; others, that it bans all nonconsensual sex (but permits, for example, consensual gay sex); others, that it bans all sex; and yet others, that it is so unclear it cannot be construed as banning anything!

A special kind of ambiguity results from the indeterminacy of the reference of pronouns or other anaphoric devices. A pronoun, such as *I, me, my, you, your, he, him, his, she, her, it, they, them,* or *their,* derives its reference from context. First- and second-person pronouns derive their reference from extralinguistic features of the context: primarily, from who is speaking or writing to whom. Third-person pronouns may derive their reference in this way but may also derive their reference from linguistic features of the context. Often, the context does not determine a single referent; the pronoun may refer to any of several objects. This, too, can produce ambiguity.

A good example is from Shakespeare:

> (12) The Duke yet lives that Henry shall depose
> But him outlive, and die a violent death.

Who is "him"? Henry or the Duke? Sometimes, as in this example, the ambiguity is intentional. Most often, however, it is a sign of bad writing.

A funny and unintended example of this kind of ambiguity occurs in a translation of the Old Testament:

> (13) "Saddle me the ass," he said unto his sons. And they saddled him. (II Samuel)

An example from Yankee sportscaster Jerry Coleman:

> (14) The ball's hit deep to center. Winfield's going back, back. He slams his head into the wall! It's rolling toward second base!

How can referential ambiguity be avoided? Sentences in which a pronoun may plausibly be interpreted as referring to several different objects

---

[25]Kate Bailey, "New Law Bans Sex in Missouri, Some Say," *Austin American-Statesman,* November 6, 1994, p. A4.

should be rewritten to remove the ambiguity. Rephrasing sometimes is enough. Sometimes, a definite description—*my donkey, the Duke*, etc.—clarifies the reference without introducing awkwardness. Replacing *him* with *the Duke* or *the King* resolves the ambiguity of who will outlive whom.

21. ACCENT.   Sentences often have presuppositions. Accenting or stressing a word in a sentence can change those presuppositions. Most frequently, emphasis on a particular word adds presuppositions to those required by the unstressed sentence. Consider this fairly straightforward, unambiguous sentence:

(15) John shouldn't berate his employees in public.

This presupposes only that John has employees. But, if we stress certain words in the sentence, additional presuppositions arise. Stressing the subject of the sentence, for example, suggests that (15) is true because of something special about John:

(16) **John** shouldn't berate his employees in public.

This presupposes that it may be acceptable for others to berate their employees publicly. Stressing the verb suggests a contrast between berating and some other activity:

(17) John shouldn't **berate** his employees in public.

This presupposes that he should instead do something else.

(18) John shouldn't berate **his** employees in public.

This seems to presuppose that it's acceptable to berate other people's employees.

(19) John shouldn't berate his **employees** in public.

This suggests that it is alright for him to berate others. Finally,

(20) John shouldn't berate his employees in **public.**

presupposes that it is acceptable for him to berate them in private.

In each case, emphasis on a particular word or phrase suggests that there is an important contrast between what that word or phrase conveys and some alternative. To put this in the context of reasoning, suppose that someone is quoted saying the above sentence in a newspaper story about John. What may we infer about that person's views? There are many possibilities. To jump to a conclusion, in the absence of additional information, is to commit the fallacy of accent.

An old joke embodies the fallacy of accent. The first mate on a sailing ship had been doing a good job but went ashore one day and got rip-roaring drunk. The captain wrote in the log, "The first mate was drunk today." The next day, when the first mate read the log, he became angry and began thinking of ways to take revenge. That evening, he came up with an idea. He wrote in the log, "The captain was sober today."

In general, an argument is guilty of the fallacy of accent if, by misplacing emphasis, it relies on presuppositions not justified originally by the premises:

> An argument is guilty of **accent** if and only if it tries to justify its conclusion by relying on presuppositions arising from a change in stress in a premise.

Guarding against this fallacy requires making emphasis clear in one's own writing and reporting fairly the use of emphasis in the writings of others. Quotations with italics or other emphatic devices should be accompanied by a note saying whether the emphasis is in the original.

A striking example of the accent fallacy occurred during Richard Nixon's presidency. Shortly after the highly publicized Tate-LaBianca murders, for which Charles Manson and his followers were arrested, Nixon gave a speech. In an aside, he mentioned Manson's crime. Apparently by oversight, Nixon said, "Manson murdered," rather than "Manson allegedly murdered." Though the reference to Manson was a minor illustration of a point, at least one big-city newspaper the next day carried the headline MANSON GUILTY, NIXON DECLARES. The headline plainly encouraged readers to think that Nixon had tried to prejudice the outcome of a criminal trial. To draw such an inference, however, would be to commit the fallacy of accent.

22. COMPOSITION AND DIVISION. The fallacy of **composition** consists in attributing something to a whole or group because it can be attributed to the parts or members. An individual pin dropping makes no perceptible sound, but it is not reasonable to conclude that dropping a boxful of pins would make no perceptible sound. Similarly, each book in a library collection may be good without the collection as a whole being good.

We have already seen a case of the fallacy of composition. Appeals to pity may involve an assertion that someone can be helped without significant harm to anyone else. But a large number of acts, each of which causes only tiny, virtually imperceptible harms, may collectively cause great harm. Raising one student's grade harms others only a small amount, but a general policy of grade inflation harms students, employers, and universities substantially. To argue that the overall effects will be minor because the effect in each case is minor is to commit the fallacy of composition.

A classic case of the fallacy of composition occurs in a proof of the principle of utility given by British philosopher John Stuart Mill (1806–1873).

The principle of utility says that each of us ought to maximize happiness—not just our own happiness but everyone's happiness ("the general happiness"). In a key passage, Mill's argument moves from individual happiness to everyone's happiness:

> No reason can be given why the general happiness is desirable, except that each person, so far as he believes it to be attainable, desires his own happiness. This, however, being a fact, we have not only all the proof which the case admits of, but all which it is possible to require, that happiness is a good, that each person's happiness is a good to that person, and the general happiness, therefore, a good to all aggregate of all persons.[26]

Having shown that "each person's happiness is a good to that person," Mill concludes that "the general happiness" is "a good to the aggregate of all persons." But this does not follow. Consider his premise: "each person . . . desires his own happiness." This does not imply that everyone desires everyone's happiness, which is what his conclusion seems to require.

The fallacy of **division** is the inverse of the fallacy of composition. It involves arguing from the properties of a group or whole to the properties of members or parts. That a forest is verdant and dense does not imply that all, or even most, trees in it are verdant and dense. That a group tends toward violent behavior does not establish that the individual members tend toward violent behavior. A comical example of this fallacy is the argument

(21)   Whales are in danger of extinction.

This is a whale.

∴ This is in danger of extinction.

The fallacies of composition and division are more common than these examples may suggest. Academics, for example, frequently rate academic departments by rating the individual faculty members in those departments. In turn, they rate universities on the basis of the quality of their departments. Performed carelessly, this involves two fallacies of composition. That the members of a department are excellent does not entail that the department itself is excellent; it may be sharply divided, or extremely narrow, or unbalanced, or simply not organized into an effective unit. Similarly, that most or all departments of a university are good does not establish that the university is good in the sense of educating its students effectively. A university may be more or less than the sum of its parts. The parts may function well or poorly together, and that has a large effect on the quality of the university as a whole. Conversely,

---

[26]John Stuart Mill, *Utilitarianism* (Indianapolis: Hackett, 1979; originally published, 1861), p. 34.

the excellence of a college or university does not entail the excellence of its departments taken individually, much less the excellence of its individual faculty members. Obviously, the quality of the whole and the quality of the parts are not unrelated. But to argue, without further support, from the quality of one to the quality of the other is to commit a fallacy of composition or division.

P R O B L E M S

Identify the fallacies of clarity, if any, in the following arguments.

1. All that glitters is not gold. Gold glitters. So, gold is not gold.

2. India's citizens are mostly poor. Therefore, India is a poor country.

3. Lawrence is a member of a wealthy family. Lawrence, therefore, is wealthy.

4. No cat has nine tails. Every cat has one more tail than no cat. So, every cat has ten tails.

5. Food is necessary to life. Sauerkraut is food. So, sauerkraut is necessary to life.

6. Nothing is better than liberty. Prison life is better than nothing. So, prison life is better than liberty.

7. Humans are the only animals that laugh. Brenda is human. So, Brenda is the only animal that laughs.

8. Not to give to charity is selfish. To pay the rent is not to give to charity. So, to pay the rent is selfish.

9. Each part of Socrates is smaller than Socrates. All of Socrates, therefore, is smaller than Socrates.

10. The meat I bought yesterday was raw. We're eating the meat I bought yesterday. So, we're eating raw meat.

11. The company has performed badly this year. We can only conclude that the employees have performed badly this year.

12. A: We must distribute the money fairly, so we'll give everyone equal shares.
    B: That means I get more, since I'm starting out with less.

Identify fallacies, if any, in the following. Fallacies may be any from this chapter.

13. Then a whole city which is established according to nature would be wise because of the smallest group or part of itself, the commanding or ruling group. This group seems to be the smallest by

nature and to it belongs a share in that knowledge which, alone of them all, must be called wisdom. (Plato)

14. *. . . primary matter is nothing if considered at rest. . . .* For whatever is unthinking is nothing, and whatever lacks variety is unthinking. Now if primary matter simply moves in one direction in parallel lines, *it is uniformly at rest,* and consequently is nothing. But *all things are full* in so far as primary matter is identical with space. Hence, *all motion is circular,* or composed of circular motions, or those returning into themselves. (G. W. F. Leibniz)

15. *Objection.* Whoever produces all that is real in a thing, is its cause. God produces all that is real in sin. Hence, God is the cause of sin.

   Answer. . . . Real signifies either that which is positive only, or, it includes also privative beings: in the first case, I deny the major and admit the minor; in the second case, I do the contrary. (G. W. F. Leibniz)

16. The danger of [political and media] consultants is not the money they spend but the power they assume. Recently, *The Wall Street Journal* exposed the fact that the Republican team of Black, Manafort and Stone and their Democratic associate, Peter Kelly, had raised money for both Democratic Senate candidate John Breaux and his Republican opponent, Henson Moore of Louisiana. When one firm works for both candidates, it's not hard to guess who wins on election night. (Raymond Struther)

17. Henry, king not by usurpation, but by the holy ordination of God, to Hildebrand, not pope, but false monk.

   This is the salutation which you deserve, for you have never held any office in the Church without making it a source of confusion and a curse to Christian men, instead of an honor and a blessing. To mention only the most obvious cases out of many, you have not only dared to lay hands on the Lord's anointed, the archbishops, bishops, and priests, but you have scorned them and abused them, as if they were ignorant servants not fit to know what their master was doing. This you have done to gain favor with the vulgar crowd. (Henry IV, 1076)

18. A young woman, a smartly outfitted executive type, was talking about the counterman in the deli in her office building. He had presented her with a stuffed lion and a stocking filled with chocolates. In a little more than a year, he had proposed to her four times. The woman insisted that she had no romantic interest in the man.

   "So what do I do?" she asked her companion-possibly-mother.

There was a thoughtful pause. Finally, the older woman spoke. "I have one question. Does he own the deli?" (Dale Boorstein)

19. The distress and misery that oppress all the Christian estates, more especially in Germany, have led not only myself, but every one else, to cry aloud and to ask for help, and have now forced me, too, to cry out and to ask if God would give His Spirit to any one to reach a hand to His wretched people. Councils have often put forward some remedy, but it has adroitly been frustrated, and the evils have become worse, through the cunning of certain men. Their malice and wickedness I will now, by the help of God, expose. . . . (Martin Luther)

20. Few issues this year have generated more hysteria than the problems of American agriculture. You can be sure that common sense has departed when a congressional panel reverentially takes testimony on the farm crisis from actresses who recently played farm wives in the movies. Tears flowed copiously from the lovely eyes of Jessica Lange, Sissy Spacek, and Jane Fonda as they described their feelings for beleaguered farmers. Miss Fonda's testimony was especially credible—after all, she is the second generation of Fonda thespians to play a farmer. (Henry Hyde)

21. Plea 4. *Such dancing is now become customary among Christians. Ans:* Which cannot be thought on without horror. A great and Learned Divine takes notice of it as a very sad thing, that all the profane Dances in use amongst the Lascivious Greeks of Old, have of late years been revived in the Christian World; yea, and in places where the Reformed Religion has taught men better. But shall [a] Christian follow the course of the World? They ought to swim against the stream, and to keep themselves pure from the sins of the Times of which this of *mixed dancing* is none the least. (Increase Mather)

22. In Saudi Arabia, yes, they cut off the hands of a thief. So, every year, they cut off—what, six, seven people's hands? And what is the population of Saudi Arabia? Seven million? In Saudi Arabia, people do not steal. Is that worse than in America, where there are not these "brutal" punishments, and 15 people get killed or banged on the head or something every week on Park Avenue? Which is the more important? The lives of the people who get murdered there? Or the hands of the people they punish in Saudi Arabia? The one is the price you pay to avoid the other. I do not know. Perhaps you prefer to see the dead men on the streets in New York? Perhaps that is not so offensive to you? (Mohammed Mannei)

23. If we consider, by whom this practice of Promiscuous Dancing was first invented, by whom patronized, and by whom witnessed against, we may well conclude that the admitting of it, in such a place as *New-England,* will be a thing highly pleasing to the Devil, but provoking to the Holy God. Who were the inventors of Petulant Dancings? They had not their origin amongst the people of God, but amongst the heathen. By whom have the Promiscuous Dances been patronized? Truly, by the worst of the heathen. *Caligula, Nero,* and such like atheists and Epicures were delighted in them. *Lucius* (that infamous apostate) hath written an oration in defence of profane and promiscuous dancings. (Increase Mather, in a Puritan tract against dancing between men and women)

24. The current campaign introduces a new variation on an old theme: If women buy designer clothes and play along, they can "go places" in a man's world. The will to succeed is presented as the ticket to power, though in reality, it is the system itself that locks women out.

    Consider the retail industry. In 1983, it employed more than fifteen million workers, three-quarters of them women. But the decision-makers are mostly men. Though the work force at J. C. Penney's is 77 per cent female, just three women serve among the corporation's top sixty-one officers. . . . Behind the glitter and hype for professional women, real wages for retail workers have declined since 1966, and the giant department stores are eager to roll back wages even further. (Richard Moore and Elizabeth Marsis)

25. "Why should freedom of speech and freedom of the press be allowed? Why should a government which is doing what it believes to be right allow itself to be criticized? It would not allow opposition by lethal weapons. Ideas are much more fatal things than guns. Why should any man be allowed to buy a printing press and disseminate pernicious opinions calculated to embarrass the government?" asked Nikolai Lenin.

    Oh, him. (Liz Smith)

26. Our Dec. 29 cover story, "Are we spending too much on education?" simply suggested that spending more money wasn't necessarily the best way to deal with the decline in educational standards, but from some of the furious reactions you would have thought we were defending ignorance and illiteracy. I mean some of those educational types really got nasty in criticizing our article. One college president went so far as to make ethnic

slurs against the writer. Can it be that a certain kind of educator has a guilty conscience about the quality of the product his kind is delivering? (James W. Michaels, *Forbes*)

27. Indeed, of all the millions of square miles of territory conquered through aggression by various nations since 1945 alone, only those taken by Israel in a war of self-defense were expected to be returned. "It is natural enough," said J. William Fulbright, who was Chairman of the Senate Foreign Relations Committee at the time of the Six-Day War, "for Israel to resist the honor of being the first modern military victor to be obliged by the principles of the United Nations Charter, especially when the greater powers who dominate the Security Council have set such a wretched example. Be that as it may, the principle is too important to be cast away because of the hypocrisy or self-interest of its proponents." In other words, as one commentator sardonically remarked, "all the self-interested hypocrites have a right to ask of Israel what they would not dream of doing themselves." (Norman Podhoretz)

28. The Reverend Jesse Jackson has been rotating around the world, broadcasting over Radio Havana to add his bit to the South African mess and comparing Botha to Adolf Hitler, among other insults to human intelligence.

    Americans have indulged this demagogue too long. Official Democrats do not dare anathematize him because he allegedly speaks for blacks, who vote Democratic. If that is so, at least non-Democrats can make the point that needs making, namely that if Mr. Jackson were white, he'd be run out of town. (William F. Buckley, Jr.)

29. To argue, as Marxists do, that the primacy of the material factor in history follows logically from the metaphysical primacy of matter is to be guilty of a fallacy based on nothing more complicated than a pun. "Material" in dialectical materialism refers to matter in a neutral technical sense in which the material is simply that which has mass and position and is in motion. "Material" in historical materialism refers to material things in an economic sense. (Historical materialism is, in fact, often called the economic interpretation of history.) . . . If "matter" in this [neutral] sense is primary, why should we suppose it to follow that matter in the evaluative [economic] sense is also primary in the scheme of human valuations? (Kwasi Wiredu)

30. *If it is said:* This leads to the conclusion that a necessary being can be made of possible things. But the conclusion is absurd.

we will answer: . . . each cause has a cause, but the aggregate of these causes has no cause. For all that can be truly said of the individuals cannot similarly be said of their aggregate. (al-Ghazali)

31. Hitherto I had stuck to my resolution of not eating animal food, and on this occasion, I consider'd, with my master Tryon, the taking every fish as a kind of unprovoked murder, since none of them had, or ever could do us any injury that might justify the slaughter. All this seemed very reasonable. But I had formerly been a great lover of fish, and, when this came hot out of the frying-pan, it smelt admirably well. I balanc'd some time between principle and inclination, till I recollected that, when the fish were opened, I saw smaller fish taken out of their stomachs; then thought I, "If you eat one another, I don't see why we mayn't eat you." So I din'd upon cod very heartily, and continued to eat with other people, returning only now and then occasionally to a vegetable diet. (Benjamin Franklin)

32. Two battleships assigned to the training squadron had been at sea on maneuvers in heavy weather for several days. I was serving on the lead battleship and was on watch on the bridge as night fell. The visibility was poor with patchy fog, so the captain remained on the bridge keeping an eye on all activities.

     Shortly after dark, the lookout on the wing of the bridge reported, "Light, bearing on the starboard bow."

     "Is it steady or moving astern?" the captain called out.

     Lookout replied, "Steady, captain," which meant we were on a dangerous collision course with that ship.

     The captain then called to the signalman, "Signal that ship: We are on a collision course, advise you to change course 20°."

     Back came a signal, "Advisable for you to change course 20°."

     In reply, the captain said, "Send: I'm a captain, change course 20°."

     "I am a seaman second class," came the reply. "You had better change course 20°."

     By that time, the captain was furious. He spit out, "Send: I'm a battleship, change course 20°."

     Back came the flashing light, "I'm a lighthouse!"

     We changed course. (Frank Koch)

Give at least two meanings for each of the following words.

33. art

34. bass

35. carriage

36. funny

37. go

38. interest

39. jog

40. key

41. lump

42. mind

43. out

44. pin

45. run

46. sign

47. tip

48. well

49. Can you think of any words that are not ambiguous? List three.

Discuss the confusions in meaning, intentional or unintentional, involved in the following passages.

50. In Massachusetts, "Slow Children" street signs have been judged demeaning and at $100 a pop are being replaced by "Watch Children" signs. (Wladyslaw Pleszczynski)

51. I am a marvelous housekeeper. Every time I leave a man I keep his house. (Zsa Zsa Gabor)

52. "The Louis XV bed is too small," a client once told the legendary decorator, Ruby Ross Wood. "I think I'd better get a Louis XVI." (Gerald Clarke)

53. If you aren't fired with enthusiasm, you will be fired with enthusiasm. (Vince Lombardi)

54. Chicago Cub outfielder Andre Dawson recently paid a $1,000 fine for disputing a strike call by umpire Joe West. On the memo line of his check Dawson wrote, "Donation for the blind." (*Sports Illustrated*)

55. President Clinton modestly [announced] that we get but one chance in a hundred years to elect someone of his caliber. Just as well, considering the country's chance of surviving two of him. (*National Review*)

56. At a time of grave crisis during the Civil War, Abe Lincoln was awakened by an opportunist who reported that the head of customs had just died. "Mr. President, would it be all right if I took his place?" "Well," said Lincoln, "if it's all right with the undertaker, it's all right with me." (Morris Udall)

57. Abraham Lincoln attended his first ball in Springfield because he wanted to see Mary Todd. "Miss Todd," he said, "I should like to dance with you in the worst way." Afterward Mary told a friend: "He certainly did!" (Paul F. Boller, Jr.)

58. There was a time when you were an imperialist if you invaded an alien territory and imposed on independent peoples an authority they rejected. Today, you are an imperialist if you oppose such aggression. (Jean-Francis Revel)

59. T. J. was one of those linebackers who didn't need pharmaceuticals to get ready to play. One day a sportswriter asked him how he always managed to get "up" for the games. T. J. said, "Aw, Coach just comes by and knocks on the door." (Dan Jenkins)

60. [I remember] the 1930s story about a policeman in New York's Union Square wielding a billy club to break up a Communist rally. "But I'm an *anti*-Communist," one demonstrator protested, to which the policeman—stepping up his blows—replied, "I don't care what kind of a Communist you are!" (Allen Weinstein)

61. I nodded to New York's most famous autograph hunter, a pleasant-faced, stout young man who always recognized me as someone he'd seen before and whose perennial question—"Are you anybody?"—doubtless seemed less philosophical to him than to me. (Steven Bach)

62. Randolph Churchill, the journalist son of Winston Churchill and not remarkable for the sweetness of his character, went into the hospital to have his lung removed. It was announced in the press that the trouble was not cancer. Novelist Evelyn Waugh commented: "A typical triumph of modern science to find the only part of Randolph that was not malignant and remove it." *(The Little, Brown Book of Anecdotes)*

63. Adele F. Weiss telephones the Whitney Museum of American Art to find out if her husband should bring along his camera.
   "Hello," she says. "Is picture-taking permitted at the Sargent exhibit?"
   "Would you repeat that, please?"
   Mrs. Weiss asks the question a second time.

"No. You have to look at the pictures on the wall. You cannot take them off the wall." (Ron Alexander)

Explain the equivocation or amphiboly, if any, that creates confusion in the following passages.

64. At Wednesday's meeting both mothers and fathers of twins will meet for the first time. *(Bloomington [Ill.] Pantagraph)*

65. Mrs. Samuel Calcotte entertained at luncheon Monday, honoring Mrs. Dale McIntyre, who was again celebrating her 29th birthday. *(Drayton Plains [Mich.] Lakeland Tribune)*

66. Workmen swept the snow from the seats of 20,000 spectators. *(Cincinnati Post and Times-Star)*

67. Strong westerly winds blew down the Rockies in Wyoming. *(Kansas City Times)*

68. Mrs. Pike C. Ross left today for Le Harpe and the Brookfield Zoo in Chicago to visit relatives. *(Lewistown [Ill.] Evening Record)*

69. No parking signs often indicate "Violators will be towed," yet I've never yet seen a tow truck dragging anyone down the street. (John Allen Paulos)

70. Columbia, Tenn., which calls itself the largest outdoor mule market in the world, held a mule parade yesterday, headed by the governor. (Jefferson City, Missouri, paper)

71. Local police are puzzled over the finding of a car parked outside the Methodist Church containing a full case of Scotch whisky. So far they have found no trace of the owner, but Captain Casey is diligently working on the case. *(Muscatine [Iowa] Journal and News-Tribune)*

72. Mr. Okun lives with his wife, his high-school sweetheart, and three sons. (Leo Rosten, quoting Okun)

73. Class Q allotments are based upon the number of dependents up to a maximum of three, so, if the birth of a child will mean your husband is entitled to more quarters allowance, notify him to take the necessary action. (Form enclosed with soldiers' allotment checks, 1955)

74. Otto von Hapsburg, heir apparent to the defunct Austro-Hungarian throne, in 1988 was asked if he was going to see the Austria-Hungary soccer match. "No," replied the head Hapsburg, "But how interesting. Tell me, whom are we playing?" (M. S. Forbes)

75. When Sargent's painting of Belle Stewart Gardner in a low-cut dress was exhibited at a Boston club, a member was heard to remark crudely (in a reference understood by anyone familiar with the White Mountains) that she was naked "all the way down to Crawford's Notch." Belle was unperturbed, although she did remark, when asked for a contribution to the Charitable Eye and Ear Infirmary, that she did not know there was a charitable eye or ear in Boston. (Joseph J. Thorndike, Jr.)

76. Two clergymen were discussing the present sad state of sexual morality. "I didn't sleep with my wife before we were married," one clergyman stated self-righteously. "Did you?"

"I'm not sure," said the other. "What was her maiden name?" (John Allen Paulos)

Explain and correct the equivocation or amphiboly in the following newspaper headlines.

77. MOMS MAY LIKE 'HARRIET' MORE THAN THEIR DAUGHTERS. *(Austin American-Statesman)*

78. RUBBERIZED ROADS SPRINGING UP, SAYS GOODYEAR EXECUTIVE. *(Los Angeles Times)*

79. BRITISH VIRGINS GET SET FOR TOURIST BOOM. *(Charlotte Amalie [Virgin Islands] News)*

80. PARKS WILL NOT ISSUE PARKING PERMITS TO FISH. *(Highland Park [Ill.] Advertiser)*

81. JURY GETS DRUNK DRIVING CASE HERE. (Austin, Texas)

82. NIGHT SCHOOL TO HEAR PEST TALK. (Oakland, California)

83. COUNTY OFFICIALS TO TALK RUBBISH. (Los Angeles, California)

84. POCATELLO MATTRESS FACTORY PLAYS IMPORTANT ROLE IN CITY'S GROWTH. (Pocatello, Idaho)

85. YOUNG DEMOCRATS ELECT BONE HEAD. *(Selma [Ala.] Times-Journal)*

86. FATHER OF 11 FINED $200 FOR FAILING TO STOP. *(Lancaster [Pa.] New Era)*

87. ANTIQUE STRIPPER TO DEMONSTRATE WARES AT STORE. *(Hartford Courant)*

88. CONTINUING EDUCATION FOR WOMEN MUSHROOMS. *(Washington Star)*

89. ZOO: OPEN UNLESS WILDCAT STRIKE RESUMES. *(Philadelphia Inquirer)*

Explain these examples of equivocation or amphiboly in advertisements.

90. Hear . . . the Weatherman. The complete dope on the weather! *(Rochester Times-Union)*

91. Wanted—Old piano for child in fairly good condition. *(Westfield [Mass.] Wallace Pennysaver)*

92. Now is your chance to have your ears pierced and get an extra pair to take home too! *(Auburn [N.Y.] Citizen-Advertiser)*

93. EVERYBODY'S buying our turkeys, geese and fowl because they know no better. *(Telegraph [Belfast, Ireland])*

94. Wildlife Museum and Taxidermy Studio. Come and see—
    Mounted Wildlife in Natural Habitat
    Fish Mounted in Under Water Scenes
    Special Mountings for Children. *(Eagle River [Wis.] Vilas County News-Review)*

Explain the differences among the following by indicating what each stressed sentence presupposes.

95. a. **I** never see any roaches around here.
    b. I never **see** any roaches around here.
    c. I never see any **roaches** around here.
    d. I never see any roaches around **here.**

96. a. **Larry** bet on the Seahawks this time.
    b. Larry **bet** on the Seahawks this time.
    c. Larry bet on the **Seahawks** this time.
    d. Larry bet on the Seahawks **this time.**

Some sentences are very difficult to decipher for various reasons. What do the following sentences mean? What did their authors probably intend them to mean? Why are they hard to understand?

97. Live within your income, even if you have to borrow to do so. (Josh Billings)

98. a. If I could drop dead right now, I'd be the happiest man alive.
    b. I never liked you, and I always will.
    c. We're overpaying him, but he's worth it.
    d. I may not always be right, but I'm never wrong.
    e. Anybody who goes to a psychiatrist ought to have his head examined.
    f. Going to call him William? What kind of a name is that? Every Tom, Dick, and Harry is called William. (Sam Goldwyn)

99. Nobody said it would be easy, and nobody was right. (George Bush)

100. a. Line up alphabetically according to height.
   b. Nobody goes there anymore; it's too crowded.
   c. When you come to a fork in the road, take it.
   d. If you don't know where you're going, you'll end up somewhere else.
   e. [Asked whether he wanted the pizza he ordered cut into four pieces or eight:] Better make it four. I don't think I can eat eight. (Yogi Berra)

PART II

ARISTOTELIAN LOGIC

CHAPTER 4

CATEGORICAL PROPOSITIONS

Aristotle invented the discipline of logic. He also developed a logical theory that survived as the discipline's core for more than two thousand years. In this chapter we will begin to develop his theory of **categorical syllogisms.**[1] Other kinds of syllogisms fall under what is now called **propositional logic,** to be treated in Chapters 6–8. Because all syllogisms we consider in this chapter and the next are categorical, we will drop the adjective *categorical* and speak simply of syllogisms.

The key concept of the theory is that of a *term*.

> A **term** is an expression that applies to objects taken individually.

All terms are true or false of individual objects. Pick any object: It will either be a cat or not be a cat. It will either make money or not make money. *Cat* and *makes money*, then, are terms. They apply to cats and to things that make money, respectively. Alone, terms are not statements; they are not true or false. But they can be spoken of as true or false of objects. Objects of which a term is true *satisfy* it.

---

[1]Aristotle's most important logical works are the *Categories, On Interpretation,* the *Prior Analytics,* and the *Posterior Analytics.* The theory of the syllogism occupies Book I, Chapters 1–7 of the *Prior Analytics.*

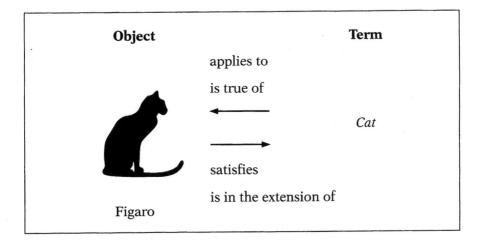

A term sorts objects into two groups: those of which it is true and those of which it is false. The set of objects of which a term is true is the **extension** of the term.[2] *Cat*, for example, applies to all cats and is false of everything else. The set of cats, then, is the extension of *cat*.

Terms occupy different grammatical categories:

TERMS

Adjectives: *red, noble, clever, logical, friendly, fair*

Adjective phrases: *somewhat strange, extremely nice, very friendly*

Common nouns: *man, woman, cat, dog, chair, molecule, number*

Modified nouns: *tall tree, very affectionate kitten, well-written answer, man who left, child I saw last Thursday*

Participles: *excited, exciting, made, understanding, disturbed*

Participial phrases: *made in Costa Rica, troubled by his friend's comment, interested in pursuing the deal*

Verb phrases: *eats, makes money, gave Fred the key, will see Kim on Tuesday, knows everybody who is anybody*

Expressions that are not terms either do not apply to objects (e.g., articles, determiners, connectives, noun phrases, and sentences) or apply to

---

[2]Antoine Arnauld and Pierre Nicole introduced this concept in *Logic, or the Art of Thinking* (the *Port Royal Logic*) in 1662. (See Chapter 3.)

them in pairs, triples, and so on rather than individually (e.g., prepositions and transitive verbs).

NONTERMS

Articles: *a, an, the*

Determiners (Quantifiers): *some, no, all, every, most, one, two, few, several*

Connectives: *and, or, if, not, but, however, although*

Adverbs: *quickly, somewhere, hardly, unintentionally, cleverly*

Adverbial phrases: *with a smile, to Pittsburgh, in five minutes, for an hour*

Prepositions: *up, in, under, for, to, of, from, with*

Noun phrases: *all people, some friends, no cities above 7,000 feet*

Transitive verbs: *make, believe, give, hit, thank*

Sentences: *Some friends gave me a watch, John quickly thanked Sally*

We can ask of any object whether it is a cat or whether it makes money, but not whether it some, or hardly, or with a smile, or of, or makes.

P R O B L E M S

Which of the following are terms?

1. animal
2. admire
3. carpenter
4. friendly
5. wisely
6. kennel
7. many
8. but
9. ridiculous
10. sleep
11. admit
12. the
13. people

14. most people

15. people who watch *Geraldo*

■ 4.1 Kinds of Categorical Proposition

The theory of syllogisms treats a limited class of statements called *categorical propositions*. They fall into four groups. This table summarizes the forms of categorical propositions, together with the letters *A, E, I,* and *O*, which medieval logicians used to abbreviate them. (*A* and *I* are the first two vowels of *affirmo*, Latin for "I affirm"; *E* and *O*, of *nego*, "I deny.") The other uppercase letters abbreviate terms.

		QUALITY	
		AFFIRMATIVE	NEGATIVE
	UNIVERSAL	A: All F are G	E: No F are G
QUANTITY			
	PARTICULAR	I: Some F are G	O: Some F are not G

When we abbreviate a categorical proposition by writing, All F are G, Some F are G, No F are G, Some F are not G, and so on, *All, Some,* or *No* represents the proposition's main quantifier; F and G represent the terms it relates—its grammatical subject and predicate, that is—expressed in a noun phrase (for example, as *cat lovers* or *people who love cats* rather than *love cats*).

UNIVERSAL AFFIRMATIVE PROPOSITIONS

These categorical propositions are both universal and affirmative. They affirm that all the things they talk about satisfy some term. Each of these propositions is universal affirmative:

(1) a. All men are mortal.
   b. Each student is scheduled for an appointment.
   c. Every stock with an initially low P/E ratio is a good buy.
   d. Anybody who would believe that story is a fool.

UNIVERSAL NEGATIVE PROPOSITIONS

These propositions are universal, but they deny that some term applies to certain objects. Typically, their main quantifiers are negative. These propositions are universal negative:

(2) a. Nobody is angry with me.
    b. No Philadelphia team was a pennant winner between 1950 and 1980.
    c. No even prime numbers are greater than 2.

PARTICULAR AFFIRMATIVE PROPOSITIONS

These propositions have an existential main quantifier and affirm that some term applies to certain objects. The following propositions are particular affirmative:

(3) a. Some quarterbacks are partial owners of their teams.
    b. A research group devoted to the issue is making progress.
    c. A good comedian is appearing at the Warehouse tonight.

To make the meanings of such propositions clear, we will assume that *some* is equivalent to *at least one*.

PARTICULAR NEGATIVE PROPOSITIONS

These propositions have an existential main quantifier but deny that some term applies to certain objects. Each of the following is particular negative:

(4) a. Some roaches aren't affected by common insecticides.
    b. A new approach Jane has developed is not a likely candidate.
    c. A painting recently acquired by the museum is not an original.

Again, we assume that *some* is synonymous with *at least one*. So, we take *Some roaches aren't affected by common insecticides* to be true if there is even one roach that is not so affected.

P R O B L E M S

Which of the following propositions are categorical? Which convert to categorical propositions if the subject or predicate is expressed in nominal form (*tree climber* vs. *climbs trees*, for instance)? Classify each categorical proposition as universal or particular and affirmative or negative.

1. Some cats are friendly.

2. Some cats are not tame.

3. Most cats chase mice.

4. All cats chase mice.

5. No students listen to graduation speeches.

6. Few students listen to graduation speeches.

7. John is a student.

8. John didn't listen to the speech at his graduation.

9. The king planned the feast carefully.

10. Every king plans feasts carefully.

11. The king plans every feast carefully.

12. A dean was late for the meeting.

13. The dean was late for the meeting.

14. At least two deans were late for the meeting.

15. Every one of my friends showed up.

■ 4.2 CATEGORICAL PROPOSITIONS IN NATURAL LANGUAGE

Aristotle's logic, understood as the theory of the categorical propositions *All F are G, Some F are G, No F are G,* and *Some F are not G,* appears quite limited. Most statements and arguments fall outside its scope. But the theory is not quite as restricted as it first appears. Many statements can be expressed in categorical form. We have already seen that terms easily convert into nominals—*climbs trees* into *tree climbers,* for example, or *swim* into *swimmers.* This section discusses other ways of rendering natural language statements as categorical propositions.

GENERIC PROPOSITIONS

The indefinite articles *a* and *an* often have particular force. We might paraphrase *I've consulted an attorney* as *Some attorney is a person I've consulted.* But indefinite articles sometimes have a different, **generic** force that approximates a universal quantifier. The same is true of the definite article *the* and **bare plurals,** plurals without accompanying quantifiers.

Consider the difference between the propositions in the following pairs. The first member of each pair seems to introduce a particular object. The second seems to speak about the typical instances of a kind.

Particular	Generic
A whale followed the boat into the harbor.	A whale is a mammal.
The whale appeared angry.	The whale is a mammal.
Whales have followed us before.	Whales are mammals.

The semantic rules governing the meaning of the articles and bare plurals are complex and involve verb tense. But there are two simple tests to determine whether such an expression has particular or generic force in a given statement. (1) Substitute *some* for the article (or add it to the bare plural). If the meaning of the sentence does not change, the expression is acting as an existential. If it does, it is generic. (2) Drop the article and make the noun plural. (Or, if testing a bare plural, add an article and make the noun singular.) If the meaning changes, the expression is acting as existential. If it does not, it is generic.

Traditionally, generics are interpreted as universals. The generic statements listed above can be treated as equivalent to *All whales are mammals*. But this is only approximate. *Birds fly* is true even though penguins do not fly, and *Dogs bark* is true even though basenjis do not bark. Some contemporary theories account for this by interpreting generics as universals limited to normal instances of a kind. In this view, *Birds fly* is equivalent, not to *All birds fly*, but to *All normal birds fly*.

Exclusive Propositions

The expressions *only* and *none but* are not really quantifiers. But they often act like quantifiers. They have the odd feature, however, that they tend to "reverse" a statement's terms. That is, the order of terms in the logical form of the statement is the opposite of their order in English. *Only female animals have babies* does not mean that every female animal has babies but, rather, that no animal that is not female has babies. This is equivalent to saying that every animal who has babies is female. *The only*, however, does not reverse the terms. *The only animals that have babies are females* is equivalent to *Only female animals have babies* and means that all animals that have babies are females, not that all female animals have babies.

Only F are G ⇔ All G are F

None but F are G ⇔ All G are F

The only F are G ⇔ All F are G

Exceptive Propositions

Statements containing *all but* or *all except* are sometimes called **exceptive** propositions. Usually, logic textbooks say that an exceptive proposition is equivalent to a conjunction of two categorical propositions:

All but F are G ⇔ All nonF are G AND No F are G (?)

This, however, seems too strong, as this example shows:

(5) All schoolchildren except those taught at home wake up early on weekday mornings.

This clearly implies that all schoolchildren who are not taught at home wake up early on weekday mornings. Does it also imply that no home-schooled children wake up early? Surely not. According to (5), home-schooled children are possible, not invariable, exceptions to the rule.[3]

A better method for rendering exceptive propositions as categoricals, then, is

All but F are G ⇔ All nonF are G

Proper Names and Pronouns

Statements with proper names and pronouns may be treated as categorical within syllogistic logic with little risk of distorting meanings. To do this, replace the name or pronoun that forms the grammatical subject with a term:

Amanda	thing (person) identical to Amanda
	thing (person) who is Amanda
He, she, it, I, you	thing (person) identical to him, her, it, me, you

Then, make the proposition universal:

(6) Amanda knows who committed the murder.

(7) All people who are Amanda know who committed the murder.

---

[3]Moreover, the usual account is inconsistent with the standard treatment of *none but*. *None but* has the effect of reversing terms because *but*, by indicating qualifying or contravening considerations, has a negative character: *None but* means roughly *none who are not*. Relations of contraposition and obversion, to be discussed in section 4.5, give us this chain of equivalences: *None but F are G* ⇔ *No nonF are G* ⇔ *All nonF are nonG* ⇔ *All G are F*. Applying the same reasoning to *all but* gives us one, not two, categorical propositions: *All but F are G* ⇔ *All nonF are G*. Thus, *All schoolchildren except those taught at home wake up early on weekday mornings* should translate simply as *All schoolchildren who are not taught at home wake up early on weekday mornings*. Anyone who insists that *All but F are G* implies *No F are G* should also, to be consistent, maintain that *None but F are G* implies *All F are G*. As we have seen, however, it does not, for *No animals but females have babies* does not imply *All female animals have babies*.

The general rule is thus:

> a is F ⇔ All things identical to a are F

Definite Descriptions

Whether to treat sentences with the definite article—*The president is angry,* for example—as expressing categorical propositions is controversial. Traditional logicians and contemporary linguists generally treat them as singular rather than categorical, for they purport to say something about an object—the president, in the example given. Bertrand Russell, in his influential 1905 paper "On Denoting," treats them as general. *The president is angry,* in his view, is equivalent to *There is one and only one president, and every president is angry.* So, for Russell, a statement of the form *The F is G* implies the categorical proposition *All F are G* but also implies that there is a unique F.

This is not the place to sort out this disagreement. It turns, however, on how to interpret definite phrases, such as *the present king of France* or *the golden fleece,* that denote nothing. In the traditional view, sentences containing these are usually neither true nor false, for they are infelicitous; they purport to say something about an object, but they fail. In Russell's view, they are false (except when they explicitly assert nonexistence), for they say partly that the described object exists, and it does not.

Some logic texts treat sentences containing definite descriptions as singular and then urge students, as in the last section, to translate them into universals (*All things identical to the president are angry,* for example). This is tantamount to adopting the equivalence

> The F is G ⇔ All F are G

In this approach, however, statements with nondenoting terms are true. As we shall see, Aristotle's interpretation of categorical propositions assumes that all terms have nonempty extensions—that is, that all terms apply to something. On that assumption, there is no problem with nonexistence. Nevertheless, the equivalence is not perfect: Definite descriptions seem to imply uniqueness (there is one and only one president of the United States, for example), but universal statements do not.

Adverbs and Pronouns

Certain adverbs are known as **adverbs of quantification** because they have built-in quantifiers over times, places, or cases. Statements containing them may not appear to be categorical, but they often translate into categorical propositions. Consider these statements:

(8) a. I always wear shorts to work in the summer.
   b. My friends never believe me when I tell them I'm broke.
   c. Wherever I go, my little sister goes too.

We can represent them as categorical propositions by universally quantifying over places or times:

(9) a. All times when I go to work in the summer are times when I wear shorts.
b. No times when I tell my friends I'm broke are times when they believe me.
c. All places I go are places my little sister goes.

A similar strategy works for pronouns with built-in quantifiers, as in these statements:

(10) a. Anything you can do, I can do better.
b. Nobody knows the trouble I've seen.

They, too, are equivalent to categorical propositions, formed by splitting the pronouns into quantifiers and the nouns *thing* or *person*:

(11) a. All things you can do are things I can do better.
b. No people are people who know the trouble I've seen.

CONNECTIVES AND QUANTIFIERS

Combinations of quantifiers and *not* deserve special attention. They are often equivalent to categorical propositions, but their grammatical form can be misleading. Rules governing the combinations are complex; the best technique is simply to ask whether a possible categorical rendering has the same meaning as the original. Consider these examples:

(12) a. All that glitters is not gold.
b. I don't have any quarters.
c. Not all our classmates are adventurous.

These are equivalent to the categorical propositions

(13) a. Some things that glitter are not gold. (NOT: Nothing that glitters is gold.)
b. No quarters are things that I have. (NOT: I do not have all quarters.)
c. Some of our classmates are not adventurous.

OTHER QUANTIFIERS

Many quantifiers, and adverbs and pronouns incorporating them, go beyond the resources of syllogistic logic. Many of them nevertheless form families that behave somewhat analogously to *some, all,* and *no,* having similar patterns of distribution and licensing many of the same inferences. This table summarizes a few of these relationships.

PARTICULAR	UNIVERSAL? AFFIRMATIVE	UNIVERSAL? NEGATIVE
Some	All	No
(At least) a few	Almost all	Almost no (or hardly any)
Many	Most	Few
At least n + 1	All but at most n	At most n
Often (frequently)	Usually (mostly)	Rarely

These families are structured similarly; the immediate inferences to be discussed in this chapter and the techniques for evaluating syllogisms to be developed in Chapter 5 can be adapted to work for all of them. But they are not equivalent. One cannot replace *many* with *some*, or *few* with *no*, without altering meaning. The families are nevertheless related to each other in strength: *Many* implies *at least a few*, which implies *some*. Similarly, *all* implies *almost all*, which implies *most*, and *no* implies *hardly any*, which implies *few*.

Some of these implications sound strange, because we assume that people make the strongest assertion the facts allow. So, it's strange to say *Most whales are mammals* when, in fact, all are. Ordinarily, we would infer from the assertion that most whales are mammals that not all are—if they were, why wouldn't the speaker have said so?—but this is not part of the meaning of *most*. Recall the criterion of Chapter 2: If something can be cancelled, it cannot be entailed by an expression and, so, cannot be part of its meaning. *Most cats can be tamed—in fact, all can* is not contradictory, but *Most cats can be tamed, although few can* is.

Some textbooks recommend translating *Few patients survived more than five years* as *Some patients survived more than five years, and some did not*. Obviously, the translation loses content. But it has other problems as well. The positive half of the translation can be cancelled; *Few, in fact, no patients survived more than five years* is not a contradiction. So, *Few patients survived more than five years* does not entail *Some patients survived more than five years*. The negative half gives *few* existential import. As we shall see in section 4.6, that is acceptable in Aristotle's interpretation of categorical propositions, but not in a modern interpretation.

PROBLEMS

Find a categorical equivalent of one of the four standard forms for each of the following, labeling terms.

1. Some attorneys are ruthless.

2. All my friends came to see me.

3. A dog is barking at my cat.

4. A dog is a good companion.

5. I have not been able to contact some people who are listed as references.

6. Only a vindictive person could have written that letter.

7. Potatoes contain vitamin C.

8. Samantha knows everyone here.

9. None but the brave deserve the fair.

10. Her colleagues are the only suspects.

11. Squirrels are foraging for the nuts we threw.

12. Nell is considering what to do.

13. Everyone but the professor dreads the exam.

14. I am not upset!

15. John always thinks he knows the answer.

16. Max does not have a clue.

17. I only read books with pictures.

18. I cannot find it anywhere.

19. All but the best students need to study.

20. Not everywhere can you see that!

■ 4.3 DIAGRAMMING CATEGORICAL PROPOSITIONS

The Swiss mathematician Leonhard Euler (1707–1783) and the British mathematician John Venn (1834–1923) devised systems for representing the meanings of categorical propositions in diagrams.[4] Here we consider just Venn's method. Consider first a universal affirmative proposition:

(14) All cats like tuna.

---

[4]Euler's version, dating from 1768, treats *some* as meaning *some and not all*. In English, this is plausible when *some* is stressed—as in *Some people claim to have been abducted by aliens*—but not when it is unstressed, as in *I met some friends of his at the party Saturday night*. Even when *some* is stressed, the *not all* portion can be cancelled and so cannot be part of the meaning. (Consider *Some people who work here are crazy. In fact, now that I think of it, all are; you have to be crazy to work here!* This is not a contradiction.) Venn presented his method in his *Symbolic Logic* (London, 1881).

This appears to say that everything that is a cat is also a thing that likes tuna; that is, that the extension of *cat* is included in the extension of *likes tuna*. Using circles to represent the extensions of F and G, therefore, we can represent the meaning of *All F are G* in the following diagram. Shading a portion of a diagram indicates that it is empty, that is, that nothing occupies that portion.

All F are G

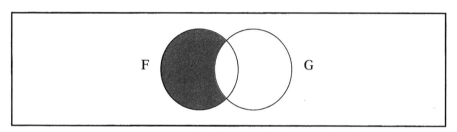

The diagram shows that the Fs are a subset of the Gs by shading a part of the F circle. Shading the portion of the F circle outside the G circle indicates that nothing is an F without being a G or, in other words, that all F are G.

> Shading a region indicates that it is empty—that is, that there is nothing in it.

The box in the background, which may be omitted in all but a few cases in this and the next chapter, represents the **universe of discourse,** the domain of objects under discussion in the discourse.

UNIVERSAL NEGATIVE PROPOSITIONS

The universal negative proposition

(15) No Republicans are Communists.

says that the set of Republicans and the set of Communists have no members in common. It says that the extensions of the terms Republican and Communist are disjointed. Accordingly, we can represent the meaning of *No F are G* by shading out the intersection of the F and G circles.

No F are G

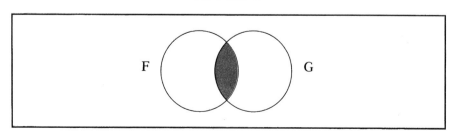

PARTICULAR AFFIRMATIVE PROPOSITIONS

A particular affirmative proposition such as

(16) Some professors enjoy giving exams.

asserts that the set of professors has at least one member in common with the set of those who enjoy giving exams: The extension of *professor* overlaps the extension of *enjoys giving exams*. We may represent a particular affirmative *Some F are G* as follows, where

An *X* indicates that the region it occupies is nonempty—that is, that there is at least one thing in that region.

If a region is shaded, we know that it is empty. If it contains an *X*, we know that it is nonempty. If it is unshaded but contains no *X*, we have no information; there may or may not be anything in the region.

Some F are G

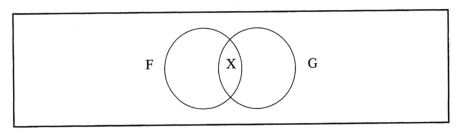

The *X* in the intersection of the circles indicates that something is both an F and a G.

PARTICULAR NEGATIVE PROPOSITIONS

Finally, a particular negative proposition such as

(17) Some steel factories are not used any longer.

asserts that there are things that are steel factories but that are not used any longer. To represent a particular negative proposition *Some F are not G*, we can draw diagrams of this sort:

Some F are not G

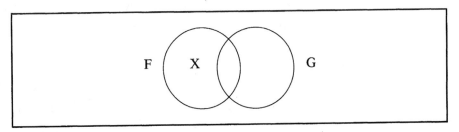

Here the *X* appears in the portion of the F circle outside the G circle. This indicates that something is an F but not a G.

P R O B L E M S

Express as categorical propositions and diagram, labeling circles clearly.

1. No victor believes in chance. (Friedrich Nietzsche)

2. He who feels punctured must once have been a bubble. (Lao-tzu)

3. Every science has been an outcast. (Ralph W. Ingersoll)

4. No man is a hero to his valet. (Madame de Cornuel)

5. All prejudices may be traced to the intestines. (Friedrich Nietzsche)

6. There is no location in Britain more than 65 miles from the sea. (John McKinney)

7. There are men who are happy without knowing it. (Vauvenargues)

8. All intelligent thoughts have already been thought. (Goethe)

9. Who has the fame to be an early riser may sleep till noon. (James Howell)

10. All that time is lost which might be better employed. (Jean-Jacques Rousseau)

11. Nothing endures but personal qualities. (Walt Whitman)

12. Those who know do not tell, those who tell do not know. (Lao-tzu)

13. Only an optimist can win in playing the game of business. (J. P. Morgan)

14. . . . nobody has money who ought to have it. (Benjamin Disraeli)

15. There is nothing good or evil save in the will. (Epictetus)

16. Whatever men aspire to, they deem best. (Petronius)

17. The trees that are slow to grow bear the best fruit. (Molière)

18. It is only shallow people who do not judge by appearances. (Oscar Wilde)

19. No man was ever wise by chance. (Seneca)

20. He who acts, spoils; he who grasps, lets slip. (Lao-tzu)

4.4 IMMEDIATE INFERENCE

There are logical relations among the four types of categorical propositions and among common variants of them. Our diagrams indicate, for example, that particular affirmative and universal negative propositions are closely related. When we diagram *Some F are G,* we place an *X* in the region we shade in diagramming *No F are G. Some F are G* says that there is something in the intersection of the extensions of F and G; *No F are G* says that the intersection is empty.

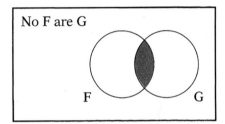

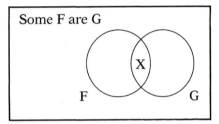

These propositions therefore directly contradict each other. They are *contradictories.*

> Two statements or propositions are **contradictories** if and only if they always disagree in truth value.

Some pigs chew coal and *No pigs chew coal,* for instance, always have opposite truth values. Any circumstance making one true makes the other false.

The same holds of *All F are G* and *Some F are not G.* These, too, always disagree in truth value. If *All F are G* is true, then *Some F are not G* must be false, and vice versa. *All ducks swim* and *Some ducks don't swim,* for example, are contradictories. If one is true, the other must be false. When we diagram a particular negative, we place an *X* in the region we shade in diagramming the corresponding universal affirmative.

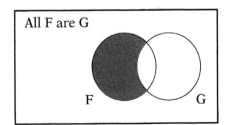

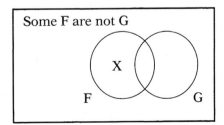

Universal affirmative and particular negative propositions, then, are also contradictories.

CONTRADICTORIES

CATEGORICAL PROPOSITION	CONTRADICTORY
All F are G	Some F are not G
No F are G	Some F are G
Some F are G	No F are G
Some F are not G	All F are G

CONVERSION

Other important logical relations concern variants of categorical propositions derived by switching or substituting terms. The first operation on propositions, which Aristotle used extensively in his method of evaluating syllogisms, is *conversion.*

> The **converse** of a categorical proposition is the proposition that results from switching its predicates.

CATEGORICAL PROPOSITION	CONVERSE
All F are G	All G are F
No F are G	No G are F
Some F are G	Some G are F
Some F are not G	Some G are not F

When is the converse equivalent to the original? When the original is an E or I proposition—when it has the form *No F are G* or *Some F are G.* The statements in the following pairs, for example, are equivalent:

(18) a. No foods that are high in fat are low in calories.
     b. No foods that are low in calories are high in fat.

(19) a. Some of her ancestors came to this country on the Mayflower.

   b. Some who came to this country on the Mayflower were
       her ancestors.

But universal affirmative and particular negative propositions are not in general equivalent to their converses. The statements in these pairs are not equivalent:

(20) a. Everyone in this room is studying logic.
     b. Everyone who is studying logic is in this room.

(21) a. Some people are not my relatives.
     b. Some of my relatives are not people.

We can demonstrate these relationships in Venn diagrams:

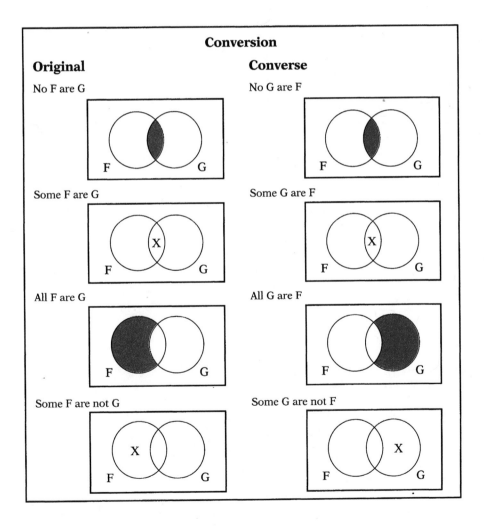

> **Conversion.** Universal negative and particular affirmative propositions are equivalent to their converses.
>
> No F are G ⟺ No G are F
>
> Some F are G ⟺ Some G are F

Thus, given a universal negative proposition, for example (*No pigs fly,* say), we can immediately infer its converse *(No flying things are pigs).*

COMPLEMENTS

Medieval logicians introduced two other relations among categorical propositions. They go beyond the framework developed so far, using the notion of the *complement* of a term.

> The **complement** *nonF* of a predicate *F* applies to exactly those things (in the universe of discourse) to which F does not apply.

If we think of F as having the set of Fs as its extension, then nonF has as its extension the set of things that are not F. (Its extension, in other words, is the complement of the extension of F.)

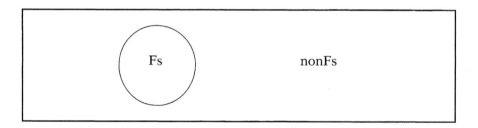

The universe of discourse is important, both because set theory defines complements in terms of a background set and because, in ordinary language, we almost always speak of complements relative to the objects under discussion. We might speak of children or behaviors or eating patterns or tissue growths as normal or abnormal, for example, but we rarely if ever have occasion to draw such a distinction among all entities, period.

Many complements are formed from other terms by adding *not* or a prefix, such as *a-, in-, im-, un-, ab-, non-, il-,* and so on.

TERM (F)	COMPLEMENT (NONF)	UNIVERSE OF DISCOURSE
mortal	immortal	living things
legal	illegal	actions
legitimate	illegitimate	children, actions
complete	incomplete	tasks, collections, passes
adequate	inadequate	performances
fair	unfair	actions

Of course, there are terms whose meanings are similarly opposed without a prefix that signals the opposition. But it's important not to confuse the complement of F with terms that are true of some, but not all, the things of which F is false. *Happy* and *unhappy*, for example, are not complements, for it's possible to be neither happy nor unhappy. Similarly, *positive* and *negative* are not complements; it is possible to be neither. The same is true of such pairs as *winner* and *loser, loser* and *finder, hero* and *coward, noble* and *base, fat* and *thin, smart* and *stupid, beautiful* and *ugly,* and so on.

Note that the complement of the complement of a term applies just to objects in the extension of the original term. For this reason, we can cancel (or add) pairs of nons: *nonnonF* and *F* are interchangeable. So, the complement of *unfair* is *fair.*

The complements of more complex terms are tricky to form, since we must be careful to negate the entire content of the term. The complement of *fair and complete hearing* is not *unfair and incomplete hearing* or *unfair and incomplete nonhearing* but something like *not a fair and complete hearing* or *something other than a fair and complete hearing.*

CONTRAPOSITION

We can now define two other operations on categorical propositions. The first is *contraposition.*

The **contrapositive** of a categorical proposition results from

(1) switching the order of the terms and
(2) replacing each term with its complement.

Proposition	Contrapositive
All F are G	All nonG are nonF
No F are G	No nonG are nonF
Some F are G	Some nonG are nonF
Some F are not G	Some nonG are not nonF

Of these, only universal affirmatives and particular negatives are equivalent to their contrapositives. The following pairs are equivalent:

(22) a. All horses are animals.
     b. Anything that isn't an animal isn't a horse.
(23) a. Some things that are legal are not fair.
     b. Some unfair things are not illegal.

But these plainly are not:

(24) a. Nothing perfect is unfinished.
     b. Nothing finished is imperfect.
(25) a. Some people are nondrivers.
     b. Some drivers are nonpeople.

We can see the equivalences easily by constructing Venn diagrams.

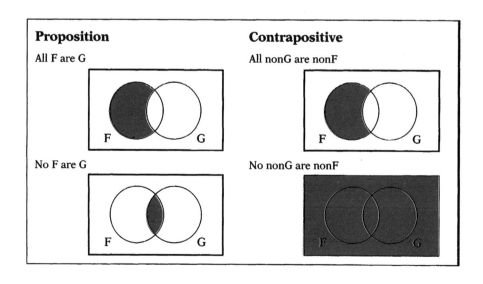

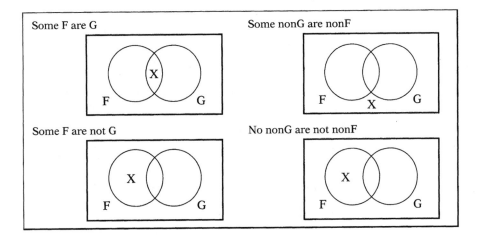

So, we can state a principle of contraposition:

Contraposition. Universal affirmative and particular negative propositions are equivalent to their contrapositives.

All F are G ⇔ All nonG are nonF

Some F are not G ⇔ Some nonG are not nonF

OBVERSION

The other relation involving complements is *obversion*.

The **obverse** of a categorical proposition results from

(1) changing its quality (from affirmative to negative or vice versa) and

(2) replacing its predicate term with its complement.

CATEGORICAL PROPOSITION	OBVERSE
All F are G	No F are nonG
No F are G	All F are nonG
Some F are G	Some F are not nonG
Some F are not G	Some F are nonG

Obversion always yields an equivalent proposition. Consider, for example, the following pairs of propositions, which illustrate obversion for each type of proposition.

(26) a.   All humans are mortal.
     b.   No humans are immortal.

(27) a.   No administrators are helpful.
     b.   All administrators are unhelpful.

(28) a.   Some kinds of killing are legal.
     b.   Some kinds of killing are not illegal.

(29) a.   Some laws passed by Congress are not constitutional.
     b.   Some laws passed by Congress are unconstitutional.

Again, we can see this clearly from Venn diagrams:

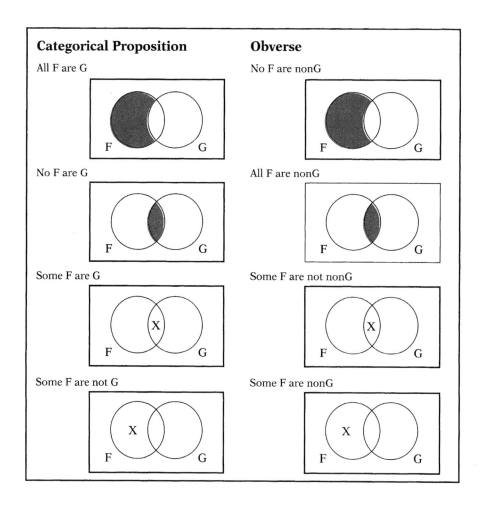

This allows us to state a principle of obversion:

> **Obversion.** Any categorical proposition is equivalent to its obverse.
> All F are G ⇔ No F are nonG
> No F are G ⇔ All F are nonG
> Some F are G ⇔ Some F are not nonG
> Some F are not G ⇔ Some F are nonG

P R O B L E M S

Express each of the following as a categorical proposition. Then, give *(a)* its converse, *(b)* its obverse, *(c)* its contrapositive, and *(d)* its contradictory. Which are equivalent to the original proposition?

1. A bird in the hand is worth two in the bush.

2. Blessed are the peacemakers.

3. Some books are worth reading twice.

4. A stitch in time saves nine.

5. He jests at scars who never felt a wound.

6. Not one of the Greeks at Thermopylae escaped.

7. I do not believe some things she said.

8. Only the good die young.

9. Among those who voted against the motion were some Republicans.

10. None but the brave deserve the fair.

Use Venn diagrams to determine whether these immediate inferences are valid or invalid.

11. All that I know, I learned after I was thirty. (Georges Clemenceau)

    ∴ I know nothing that I learned before I was thirty.

12. No man was ever wise by chance. (Seneca)

    ∴ It is not true that some have been wise by chance.

13. There are men who are happy without knowing it. (Vauvenargues)

    ∴ Some who are happy and do not know it are men.

14. He who feels punctured must once have been a bubble. (Lao-tzu)

    ∴ He who was once a bubble feels punctured.

15. All who remember, doubt. (Theodore Roethke)

    ∴ Anyone who is free of doubt is also free of memory.

16. There is no detail that is too small. (George Allen)

    ∴ Nothing that is too small is a detail.

17. Only the shallow know themselves. (Oscar Wilde)

    ∴ It is not true that some who know themselves are profound.

18. . . . and now nothing will be restrained from them, which they have imagined to do. (Genesis 11:6)

    ∴ Everything they have imagined to do will be possible for them.

19. Art never expresses anything but itself. (Oscar Wilde)

    ∴ Only art is expressed by art.

20. None deserves praise for being good who has not spirit enough to be bad. (La Rochefoucauld)

    ∴ None has spirit enough to be bad who does not deserve praise for being good.

▪ 4.5   THE ARISTOTELIAN SQUARE OF OPPOSITION

Aristotle and his medieval followers summarized logical relations among categorical propositions in **the square of opposition.**[5] The form the square takes differs, depending on whether an additional form of immediate inference should count as valid. Do universal propositions have **existential import?** That is, do they imply the existence of objects satisfying their terms? Do they, in short, imply the corresponding particular propositions: Does *All F are G* entail *Some F are G*? Does *No F are G* entail *Some F are not G*? Aristotelians say yes. Modern logicians usually say no.

The square summarizes the relations between A, E, I, and O propositions with the same subject and predicate terms in the Aristotelian view:

---

[5]The original square appears in Aristotle's *De Interpretatione,* Chapter 10. Diagrams such as those in the text first appear in medieval works, such as the *Tractatus,* written around 1230 by Peter of Spain (1205?–1277), who became Pope John XXI. That work became the standard logic text for the Middle Ages and Renaissance; it enjoyed 166 editions between the thirteenth and seventeenth centuries.

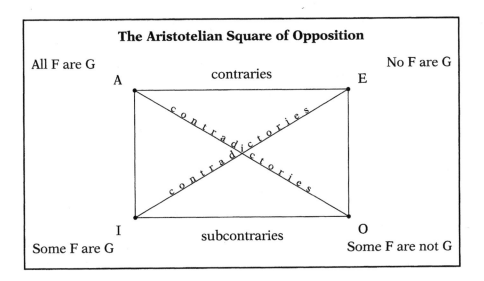

Contradictories, as we have seen, always disagree in truth value. **Contraries** are never both true but can both be false, and **subcontraries** are never both false, though they can both be true.

> Two propositions are **contraries** if and only if they can never both be true but can both be false.
>
> Two propositions are **subcontraries** if and only if they can never both be false but can both be true.

So, in Aristotle's view, *All men are mortal* and *No men are mortal* are contraries. They can never both be true but can both be false. *Some animals are pets* and *Some animals are not pets* are subcontraries. They can never both be false but could both be true.

Medieval logicians called particular propositions **subalterns** of universal forms of the same quality. The relation they had in mind, in our terminology, is implication. They held that *All F are G* implies *Some F are G* and that *No F are G* implies *Some F are not G*.

Using the square of opposition, we can use the truth value of one categorical proposition to determine the truth values of others with the same subject and predicate terms. Suppose that *All F are G* is true. What can we conclude about the corresponding E, I, and O forms? A and O forms are contradictories; so, if *All F are G* is true, *Some F are not G* must be false. A and E forms are contraries; they cannot both be true. So, *No F are G* must also be false. Finally, A forms imply I forms; so, if *All F are G* is true, so is *Some F are G*.

To take an English example: Suppose that *All professors are human* is true. Then *Some professors are human* must also be true, but *No professors are human* and *Some professors are not human* must be false.

The square does not allow us to infer truth values for all the other propositions in every case. Suppose that *No F are G* is false. Then its contradictory, *Some F are G*, must be true. But we can conclude nothing about the corresponding A and O forms; they could be either true or false. If it is false that no pigs eat milo, then it is true that some pigs do. But we cannot infer whether or not all do.

Aristotle's square concisely expresses a variety of logical relations. If one knows the truth value of a categorical proposition, one can deduce the truth value of other propositions with the same subject and predicate terms:

ARISTOTELIAN SQUARE: TRUTH VALUES

	A	E	I	O
If A is true:	true	false	true	false
If A is false:	false	?	?	true
If E is true:	false	true	false	true
If E is false:	?	false	true	?
If I is true:	?	false	true	?
If I is false:	false	true	false	true
If O is true:	false	?	?	true
If O is false:	true	false	true	false

But do the relations the square posits really hold?

ARISTOTLE'S ASSUMPTION

Certainly, A and O propositions are contradictories. Under no circumstances can *All F are G* and *Some F are not G* have the same truth value. Similarly, E and I propositions are contradictories: *No F are G* and *Some F are G* always disagree in truth value. This much is uncontroversial.

Every other relation in the square of opposition, however, is disputed by modern logicians.[6] The key question is whether the terms in a true

---

[6]Actually, the first to see a problem in the square was Peter Abelard (1079–1142), who recognized that if *All F are G* and *Some F are not G* both assume the existence of Fs, they could not strictly be contradictory. (If there are no Fs, both are false.) He continued to hold that universal propositions imply the corresponding particular propositions but

categorical proposition may have empty extensions. To see this, consider Venn diagram representations of alleged contraries. We can try to diagram *All F are G* and *No F are G* together, for example, to see whether they can both be true:

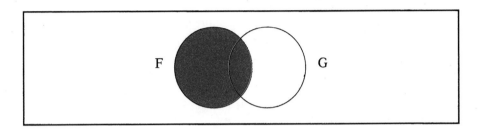

Diagramming A and E propositions together leads to no contradiction. Never are we forced to shade and mark with an X the same region. So, according to our diagram technique, *All F are G* and *No F are G* can be true at the same time: Both propositions are true *if there are no Fs*. The same diagram also shows *(a)* that *Some F are G* and *Some F are not G* can both be false—indeed, both are false if the extension of F is empty—and, so, are not subcontraries; *(b)* that *All F are G* does not imply *Some F are G*, for the former can be true while the latter is false; and, for the same reason, *(c)* that *No F are G* does not imply *Some F are not G*. All remaining relations on the square, then, depend on whether terms such as F may have empty extensions.

Our Venn diagram technique, and the modern interpretation of categorical propositions underlying it, allows empty extensions. Aristotle's logic does not. I shall call the hypothesis that no terms have empty extensions *Aristotle's assumption*. (Because assuming the truth of both *All F are G* and *No F are G* requires that the extension of F be empty, one might think that, to maintain the Aristotelian square, it is enough to assume that no subject terms have empty extensions. This is not quite right, however, if we wish to preserve conversion for E propositions. *No F are G* and *No G are F* are equivalent. And equivalent statements should have the same implications. So, *No F are G* must, in an Aristotelian reading, imply not only *Some F are not G* but also *Some G are not F*.)

There is some evidence from ordinary language in Aristotle's favor. If someone tells us that all John's children are asleep, we assume that John has children. If we hear that nobody assigned to the Alpha project had access to all the files, we assume that somebody was assigned to the

distinguished *Not all F are G* (the strict contradictory of *All F are G*), which, he held, does not imply the existence of Fs, from *Some F are not G*, which does. Later logicians did not follow him in making this distinction.

Alpha project. If Smith declares that all current military interventions in foreign countries are successful and Jones insists that none are, we ordinarily take them to be contradicting each other. All this accords with the Aristotelian square, but not with the modern understanding of categorical propositions reflected in Venn diagrams.

We thus face a choice. We must give up the contrary, subcontrary, and subaltern relations represented in the square of opposition or revise the Venn diagram technique to capture Aristotle's assumption. The next section of this chapter explores the former option. The rest of this section explores the latter.

To reflect Aristotle's approach to categorical propositions, we must capture his assumption that no terms have empty extensions. The general principle for diagrams, therefore, must be that all circles are nonempty. That does not mean that all *regions* are nonempty. Universal propositions will still require us to shade regions, declaring them empty. But circles as a whole cannot be empty.

This principle requires us to change the diagrams for universal propositions to give them existential import. The standard Venn diagram method represents universal propositions by shading areas to indicate that they are empty. To change the method so that universal propositions have existential import, we need, in addition, to place an X in other areas to indicate that they are nonempty. (We will circle these Xs to indicate that they are special to the Aristotelian interpretation; this allows us to see when Aristotle's assumption makes a difference.)

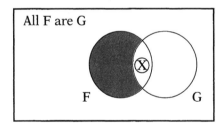

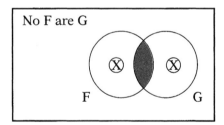

Diagrams for particular propositions do not need to change.

Diagramming universal propositions in this way shows that, so interpreted, they imply the corresponding particulars. If all F are G, the diagram indicates, there are no F that are not G, and some F are G. Similarly, if no F are G, nothing is both an F and a G, but there are F that are not G and G that are not F. The diagrams, then, give universal propositions existential import.

They also make A and E propositions contraries and I and O propositions subcontraries. We cannot consistently assume both *All F are G* and *No F are G*, for doing so would force us to shade and mark with an X both portions of the F circle.

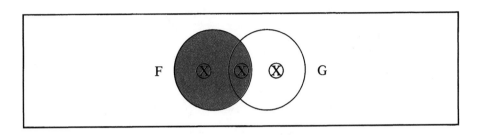

This revised Venn diagram technique, then, reflects Aristotle's assumption and the relations in the Aristotelian square of opposition.

P R O B L E M S

The terms *contradictory, contrary,* and *subcontrary* apply to all propositions, not just those of categorical form. Wherever possible, for each of the following, give *(a)* a contradictory, *(b)* a contrary, and *(c)* a subcontrary of it. (Make Aristotle's assumption.)

1. It is raining.

2. Jesse is sick today.

3. Verne doesn't like Clarence.

4. Somebody saw us.

5. Everyone left hours ago.

6. Albania and Bulgaria are both Balkan countries.

7. Rhoda or her assistant sent us the forms.

8. If you argue with me, I won't listen to you.

9. All that glitters is not gold.

10. Some, but not all of us, got away.

Answer the following according to the Aristotelian interpretation of categorical propositions.

11. If *All roads lead to Rome* is true, what may we deduce about the truth values of the following?
    (a) Some roads lead to Rome.
    (b) No roads lead to Rome.
    (c) Some roads don't lead to Rome.
    (d) Anything that leads to Rome is a road.

(e) Nothing that leads to Rome is not a road.
(f) No road fails to lead to Rome.
(g) Anything that doesn't lead to Rome is no road.
(h) Some things that don't lead to Rome aren't roads.
(i) Only roads lead to Rome.

12. What may we deduce about these propositions if *All roads lead to Rome* is false?

13. If *Nobody knows the trouble I've seen* is true, what can we deduce about the truth values of the following?
(a) Everyone knows the trouble I've seen.
(b) Somebody knows the trouble I've seen.
(c) Somebody doesn't know the trouble I've seen.
(d) Everybody is ignorant of the trouble I've seen.

14. What may we deduce about these propositions if *Nobody knows the trouble I've seen* is false?

15. If *Some actors aren't vain* is true, what can we deduce about the truth values of
(a) All actors are vain.
(b) Some actors are vain.
(c) No actors are vain.
(d) Some who aren't vain are actors.
(e) Some who are vain aren't actors.
(f) Only actors are vain.
(g) Only vain people are actors.
(h) Nobody who isn't an actor is vain.
(i) Nobody who isn't vain is an actor.

16. What can we deduce about these propositions if *Some actors aren't vain* is false?

17. If *Some beasts eat their young* is true, what can we deduce about the truth values of these propositions:
(a) No beasts eat their young.
(b) All beasts eat their young.
(c) Some beasts don't eat their young.
(d) Some that eat their young are beasts.
(e) None that eat their young are beasts.
(f) Some that don't eat their young aren't beasts.
(g) Some that aren't beasts don't eat their young.
(h) All beasts avoid eating their young.
(i) All that eat their young are beasts.
(j) Only those that eat their young are beasts.

18. What can we deduce about these propositions if *Some beasts eat their young* is false?

■ 4.6 THE MODERN SQUARE OF OPPOSITION

The key difference between Aristotelian and modern approaches to categorical propositions concerns terms with empty extensions. Aristotelians assume that the terms involved in a true proposition apply to something; modern logicians do not. As a result, modern logicians deny a number of logical relations that Aristotelians defend: that universal propositions have existential import, implying the corresponding particular propositions; that *All F are G* and *No F are G* cannot both be true; and that *Some F are G* and *Some F are not G* cannot both be false.

It may seem bizarre to maintain, as modern logicians do, that *All F are G* does not imply *Some F are G* and that *All F are G* and *No F are G* might both be true. But consider some examples of English propositions with subjects that could have empty extensions.

All shoplifters will be prosecuted.	No shoplifters will be prosecuted.
T	T
Some shoplifters will be prosecuted.	Some shoplifters will not be prosecuted.
F	F

If a store posts a sign saying that all shoplifters will be prosecuted and the sign works to deter shoplifters, then nobody will be prosecuted for shoplifting. This does not contradict what the sign says. In such a case, the propositions seem to have the truth values shown in the above square. But that shows that *All shoplifters will be prosecuted* does not imply *Some shoplifters will be prosecuted*. Similarly for the universal and particular negative propositions. It also shows that the universal affirmative and negative propositions can be true together and that the particular affirmative and negative propositions can be false together.

To take another example: Suppose that a teacher announces that all students who fail to do the homework will fail the course. This frightens the students; everyone does the homework, and everyone passes. What the teacher said remains true; but, since nobody failed the course, nobody who failed to do the homework failed the course. So, the A and E propositions are both true, while the I and O propositions are false.

A final example: The theory of relativity says that accelerating a body to the speed of light would require an infinite amount of energy. So, *All accelerations of a body to the speed of light require an infinite amount of energy* is true. But this does not imply that there are, or even could be, such accelerations. In fact, it implies that accelerating a body to the speed of light is impossible.

In Aristotle's logic, no terms may have empty extensions. Usually, this is not a problem. As we have just seen, however, sometimes it is important to reason with terms (such as *shoplifter, student who did not do the homework,* and *accelerations of a body to the speed of light*) that apply to nothing. The modern interpretation of categorical propositions, by allowing empty as well as nonempty extensions, covers a wider range of propositions and arguments than the Aristotelian interpretation. Consequently, logicians since the nineteenth century have usually rejected Aristotle's thesis that universal propositions have existential import, implying the corresponding particular propositions. They have therefore rejected the relations posited in the square of opposition except for contradictoriness.

To summarize: Modern logic accepts the principles of contradictories, conversion, contraposition, and obversion but rejects the idea that A and E propositions cannot both be true, that I and O forms cannot both be false, that A forms imply I forms, and that E forms imply O forms.

This means that we can infer, from the truth value of any given categorical proposition, the truth values of only a few other propositions with the same subject and predicate terms.

MODERN SQUARE: TRUTH VALUES

	A	E	I	O
If A is true:	true	?	?	false
If A is false:	false	?	?	true
If E is true:	?	true	false	?
If E is false:	?	false	true	?
If I is true:	?	false	true	?
If I is false:	?	true	false	?
If O is true:	false	?	?	true
If O is false:	true	?	?	false

P R O B L E M S

Answer the questions at the end of section 4.5, rejecting Aristotle's assumption and assuming the modern interpretation of categorical propositions.

CHAPTER 5

SYLLOGISMS

Aristotle used the term *syllogism* roughly as we use *argument*. His theory, however, focuses on arguments of a restricted form. In modern usage,

A **syllogism** is an argument containing three statements: two premises and one conclusion. A **categorical syllogism** is a syllogism made up of categorical propositions and containing three terms, each appearing in two propositions.

Each of the following is a categorical syllogism:

(1) All humans are mortal.

   No angels are mortal.

   ∴ No angels are human.

(2) Nobody who has a college degree is stupid.

   Some stupid people make a lot of money.

   ∴ Some people who make a lot of money don't have college degrees.

(3) All New York teams have high revenues.

   Some New York teams are consistent winners.

   ∴ Some teams with high revenues are consistent winners.

(4) Some of your friends aren't very nice.

   Nobody who would insult a total stranger is very nice.

   ∴ Some of your friends would insult a total stranger.

(5) Some logic books avoid syllogisms.

Some books that avoid syllogisms are worthless.

∴ Some logic books are worthless.

Some are valid, but others are not. In this chapter we develop ways of telling which is which.

The theory of categorical syllogisms is fairly simple. It has been subsumed by modern predicate logic. Nevertheless, the theory is worth studying, for it illustrates various logical techniques in particularly clear ways.

■ 5.1 STANDARD FORM

A categorical syllogism contains exactly three general terms.

The **major term** of a syllogism is the conclusion's predicate. The **minor term** is the conclusion's subject. The **middle term** appears in both premises.

(Because the rest of this chapter deals only with categorical syllogisms, we drop the adjective and speak simply of syllogisms.) Each premise contains the middle term and one other. The **major premise** contains the major term; the **minor premise** contains the minor term. A syllogism is in **standard form** if it is stated in this form:

Major premise

Minor premise

∴ Conclusion

Consider, for example, the syllogism

(6) All cities with rent control have high   All R are H
occupancy rates.

Some cities with high occupancy rates   Some H are P
have many homeless people.

∴ Some cities with rent control have   ∴ Some R are P
many homeless people.

The major term is P *(have many homeless people);* the minor term is R *(cities with rent control).* H *(cities having high occupancy rates)* appears in both premises and is therefore the middle term. The first premise contains the minor term and is thus the minor premise. This syllogism, consequently, is not in standard form. To put it in standard form, we may switch the order of the premises:

(7) Some H are P   I
      All R are H   A
      ∴ Some R are P   I

A syllogism's **mood** is a list of three letters signifying the form of the major premise, minor premise, and conclusion. (Recall that A stands for universal affirmative; I, particular affirmative; E, universal negative; and O, particular negative.) This syllogism has mood IAI.

The *figure* of a syllogism depends on the placement of its middle term. In **first figure** syllogisms, the middle term is the subject of the major premise and the predicate of the minor premise. In **second figure** syllogisms, it is the predicate of both premises. In **third figure,** the middle term is the subject of both premises. In **fourth figure,** finally, the middle term is the predicate of the major premise and the subject of the minor premise. Syllogism (7) is first figure: The middle term, H, is the subject of the major premise and the predicate of the minor premise.

This chart summarizes the figures according to the position of the middle term, M, in standard form (where P is the major and S the minor term):

	FIRST FIGURE		SECOND FIGURE	
MAJOR PREMISE	(M)	P	P	(M)
MINOR PREMISE	S	(M)	S	(M)
CONCLUSION	S	P	S	P
	THIRD FIGURE		FOURTH FIGURE	
MAJOR PREMISE	(M)	P	P	(M)
MINOR PREMISE	(M)	S	(M)	S
CONCLUSION	S	P	S	P

Applying this to the examples at the beginning of the chapter, (1) and (2) are in standard form, for the major premise of each is stated first.

(1) All humans are mortal.   A
     No angels are mortal.   E
     ∴ No angels are human.   E

Human is the major term, so the first premise is the major premise. The middle term, *mortal,* is the predicate of both premises, so (1) is second figure, mood AEE.

(2) Nobody who has a college degree is stupid.   E
     Some stupid people make a lot of money.   I
     ∴ Some people who make a lot of money don't   O
        have college degrees.

People who have a college degree is the major term. The middle, *stupid people,* is predicate of the major premise but subject of the minor premise. Syllogism (2) is thus fourth figure, mood EIO.
     Syllogisms (3), (4), and (5) are not in standard form.

(3) All New York teams have high revenues.   A
     Some New York teams are consistent winners.   I
     ∴ Some teams with high revenues are consistent winners.   I

The major term, *consistent winners,* appears in the second premise, making it the major. The middle term, *New York teams,* is subject of both premises. The syllogism is therefore third figure, mood IAI.

(4) Some of your friends aren't very nice.   O
     Nobody who would insult a total stranger is very nice.   E
     ∴ Some of your friends would insult a total stranger.   I

The major term, *would insult a total stranger,* is again in the second premise. The middle term, *very nice,* is predicate of both premises. Syllogism (4) is thus second figure, mood EOI.

(5) Some logic books avoid syllogisms.   I
     Some books that avoid syllogisms are worthless.   I
     ∴ Some logic books are worthless.   I

The second premise is the major, for it contains the major term, *worthless.* The middle term, *books that avoid syllogisms,* is subject of the major and predicate of the minor, making (5) first figure, mood III.
     One of the oldest but least illuminating ways of telling whether a syllogism is valid is to memorize the valid forms. In standard form, a syllogism may have any of four conclusions (A, I, E, or O). It may be of any of four figures, and each premise may be of any of the four forms. The total number of standard form syllogisms is thus 4^4, or 256. In a modern interpretation, only 15 are valid. In an Aristotelian interpretation,

24 are. Medieval logic students memorized the valid forms with the help of a verse:[1]

> Barbara Celarent Darii Ferio *Baralipton*　　1st figure
> Celantes Dabitis *Fapesmo* Frisesomorum;　　4th figure
> Cesare Cambestres Festino Barocho; *Darapti*　　2nd figure
> 　　　　　　　　　　　　　　　　　　　　(except Darapti)
>
> *Felapto* Disamis Datisi Bocardo Ferison.　　3rd figure

The first three vowels give the valid moods. (The consonants give directions for reducing syllogisms to first figure; see section 5.5.) The verse thus tells us what the valid moods are. (Those valid only under the Aristotelian interpretation are in italics.) Or, at least, it gives 19 of the 24 valid moods; 5 valid subaltern moods (later called *Celaront, Celantos, Barbari, Cesaro,* and *Cambestros,* after the forms from which their conclusions follow by subalternation) were not recognized until later.

VALID MOODS

FIRST FIGURE	AAA, EAE, AII, EIO, *AAI, (EAO)*
FOURTH FIGURE	AEE, IAI, *EAO,* EIO, *(AEO, AAI)*
SECOND FIGURE	EAE, AEE, EIO, AOO, *(EAO, AEO)*
THIRD FIGURE	*AAI, EAO,* IAI, AII, OAO, EIO

The medievals did not distinguish the first from the fourth figure. That explains why the fourth figure comes out of order in the verse and why the order of the premises must be reversed for fourth figure moods. The valid moods they omitted are in parentheses in the chart.

This does allow us to determine validity. The following table records the figure and mood of the syllogisms we have seen so far. Checking the chart allows us to judge validity.

---

[1]This verse is sometimes attributed to William of Sherwood; it appears in his *Introduction to Logic* and in Peter of Spain's *Tractatus,* both of which were written in the early thirteenth century. L. M. de Rijk has traced the verse back to about 1200.

Medieval logicians followed Aristotle in maintaining that there were only three figures—mixing the first and the fourth together. This prevented them from distinguishing major and minor terms in the modern way. (Aristotle made no such distinction, calling both *extremes*.) Peter of Spain, for example, simply defines the major premise to be whichever premise is stated first. John Philoponus, a sixth-century Byzantine scholar, first defined the major term as the predicate of the conclusion in his *Commentary on Aristotle's Prior Analytics,* but his definition became generally accepted only in the seventeenth century—for example, by Antoine Arnauld (1612–1694) in *Logic, or The Art of Thinking* (the *Port Royal Logic,* 1662) and by Gottfried Wilhelm Leibniz (1646–1716) in *De Arte Combinatoria* (1666).

SYLLOGISM	FIGURE	MOOD	VALID?	MEDIEVAL NAME
(1)	2nd	AEE	Yes	Cambestres
(2)	4th	EIO	Yes	Frisesomorum
(3)	3rd	IAI	Yes	Disamis
(4)	2nd	EOI	No	
(5)	1st	III	No	
(6)	1st	IAI	No	

But this method seems unilluminating, for it does not tell us *why* a syllogism is valid or invalid. It gives no insight into the success or failure of a syllogism. For that, we must turn to other methods.

P R O B L E M S

For each of the syllogisms in the problem set for section 5.2 (p. 195), (1) state its major, minor, and middle terms; (2) state its figure and mood; and (3) say whether it is valid, using the chart of valid moods.

■ 5.2 VENN DIAGRAMS

As explained in Chapter 4, Venn devised a way to represent the meanings of categorical propositions in diagrams. It extends to a procedure for evaluating syllogisms. Every syllogism contains three terms. Consequently, we begin evaluating a syllogism by constructing a diagram showing a circle for each term:

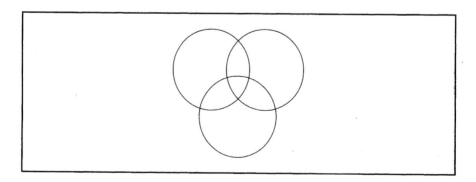

It is important that these circles overlap in this way. Otherwise, the diagram would not reflect all possible relationships between the extensions of the syllogism's terms.

Second, we label each circle with a term. Third, we diagram the syllogism's premises, beginning with universal premises. To diagram a universal proposition, ignore the circle for the term uninvolved in the proposition.

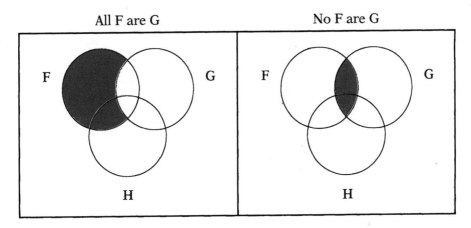

The situation for particular propositions, however, is more complicated. Consider a particular affirmative, *Some F are G*. With two circles, we put an *X* in the intersection of the circles labeled F and G.

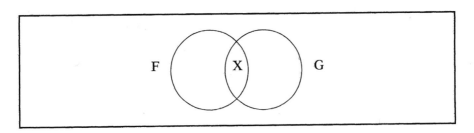

With three circles, however, this region is divided by the circle for the uninvolved term (say H). If we place the *X* outside the uninvolved H circle, we indicate that something is an F, a G, and not an H.

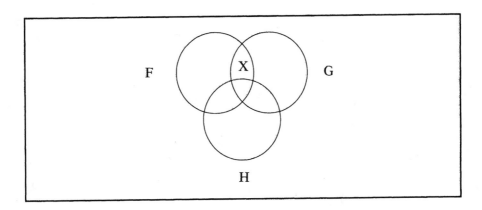

If the relevant portion of the diagram were already shaded—if, for example, we had already diagrammed the categorical proposition *No F are H*—this would be fine. But *Some F are G* asserts nothing about H; so, except in such situations, this is wrong.

If we put the *X* inside the H circle, we indicate that something is an F, a G, and an H.

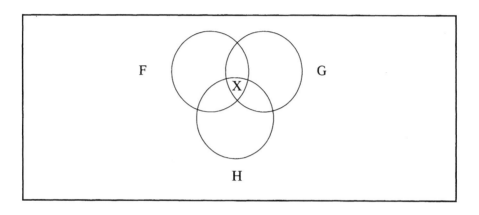

Again, this would be fine if we already knew, for example, that all G are H. In general, however, this is too much; the proposition says nothing about H. To remain uncommitted about whether the thing that is both F and G is also H, we can "straddle the fence": We can place an *X* on the line that divides the intersection of the F and G circles. This indicates that something is on one side or the other of the line.

We can represent *Some F are G*, on a diagram with three circles, in this fashion:

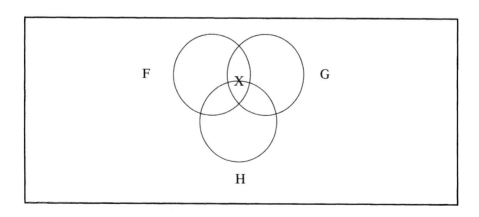

Similarly, we can represent *Some F are not G* as follows:

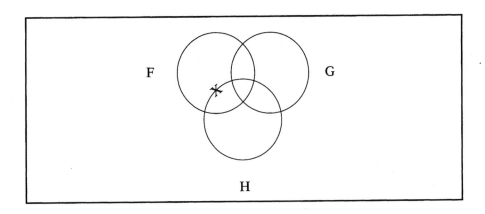

The *X* here means that we do not know whether any Fs are H. We know only that some Fs are not G. In both cases, the *X* crosses the circle for the term not present in the proposition being represented. It indicates that something is in the larger region made up of both areas into which it extends.

After diagramming both premises, we check to see whether the diagram guarantees the truth of the conclusion. If it does, the syllogism is valid. If it does not, the argument form is not valid. An *X* straddling a line between unshaded regions is a sure sign of invalidity.

Why does this method work? An argument is valid if the truth of its premises guarantees the truth of its conclusion. A diagram reflects the meanings of the premises (in the modern interpretation of categorical propositions, at any rate). It shows how the extensions of the terms must relate if the premises are true. If the syllogism is valid, the diagram shows that the conclusion must be true as well.

Consider the examples of syllogisms from the beginning of this chapter.

(1) All humans are mortal.   All H are M

   No angels are mortal.   No A are M

   ∴ No angels are human.   ∴ No A are H

To test this syllogism for validity, we construct a diagram with three intersecting circles, labeling them H, M, and A, and begin to diagram the premises. The first premise, a universal affirmative proposition, results in this diagram:

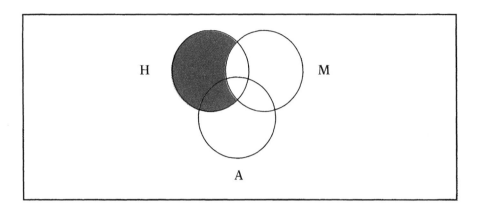

After diagramming the second, we obtain:

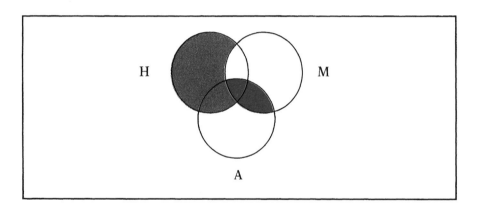

Does this diagram guarantee the truth of the conclusion, *No A are H?* Yes: The entire intersection of the A and H circles is shaded. If the premises are true, then nothing is both an A and an H. So, if the premises are true, the conclusion must be true as well. The syllogism is valid.

Our second example was

(2) Nobody who has a college degree is stupid.   No C are S

    Some stupid people make a lot of money.   Some S are M

   ∴ Some people who make a lot of money   ∴ Some M are
      don't have college degrees.   not C

We begin, once again, by constructing a three-circle diagram. After representing the first premise, we have

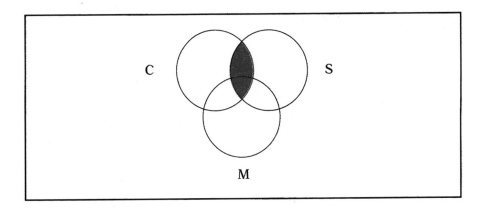

After diagramming the second, we obtain

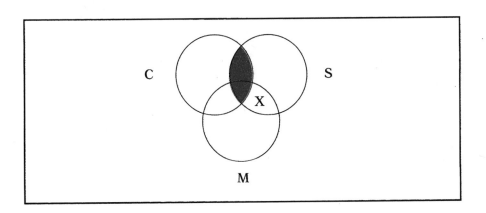

Does this guarantee the truth of the conclusion, *Some M are not C?* Clearly, the answer is yes: The diagram indicates that there is an M that is not a C.

To evaluate our third example:

(3) All New York teams have high revenues.	All N are R
Some New York teams are consistent winners.	Some N are C
∴ Some teams with high revenues are consistent winners.	∴ Some R are C

Diagramming the first premise results in

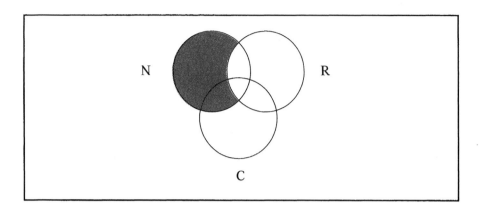

Diagramming the second produces this diagram:

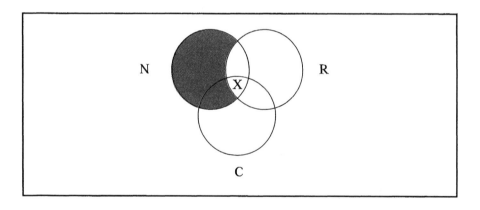

The result guarantees the conclusion's truth, so the syllogism is valid. The diagram indicates that something is an R, a C, and an N.

Evaluating our fourth example, we find it invalid:

(4) Some of your friends aren't very nice.   Some F are not N

    Nobody who would insult a total   No I are N
    stranger is very nice.

    ∴ Some of your friends would insult a   ∴ Some F are I
        total stranger.

Representing the meaning of the second, universal negative premise yields

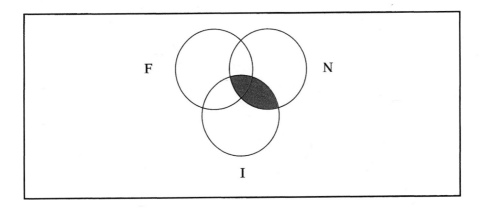

Representing the meaning of the first results in

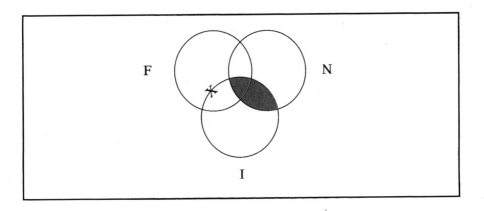

This does not guarantee that some F are I. The diagram indicates that something is an F without being an N, but it may or may not be an I. We have no guarantee that something occupies the intersection of the F and I circles. The syllogism is therefore invalid. As noted, an *X* straddling a line between unshaded regions is a mark of invalidity.

Finally, our fifth example is also invalid:

(5) Some logic books avoid syllogisms.   Some L are A

Some books that avoid syllogisms are   Some A are W
worthless.

∴ Some logic books are worthless.   ∴ Some L are W

Our method directs us to begin by diagramming universal premises. In this case, however, there is no universal premise. Our diagram is sure to contain *X*s straddling lines between unshaded regions. We can already conclude, then, that the syllogism is invalid. In general,

No syllogism with two particular premises is valid.[2]

Diagramming the first premise yields

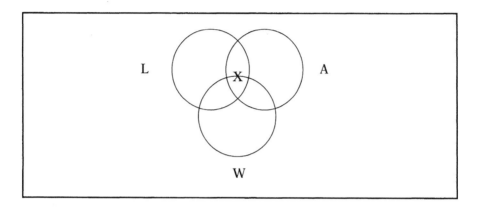

Diagramming the second yields

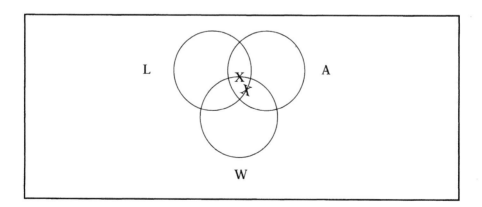

Nothing guarantees that anything occupies the intersection of the L and W circles. So, the syllogism is invalid.

---

[2]Aristotle first recognized this (*Prior Analytics* 41b7). Medieval logicians typically listed it among their rules for syllogisms.

To summarize the Venn diagram method:

1. Construct a diagram consisting of three intersecting circles.
2. Label each circle with a term.
3. Diagram the premises, starting with universal propositions, if there are any. (If both premises are particular, the syllogism is invalid.)
4. Check to see whether the diagram guarantees the truth of the conclusion. If it does, the syllogism is valid. If not, it is not valid.

PROBLEMS

Evaluate the following syllogisms for validity.

1. All G are H
   All F are G
   ∴ All F are H

2. All G are H
   Some F are G
   ∴Some F are H

3. No G are H
   No F are G
   ∴ Some F are not H

4. Some G are H
   All F are G
   ∴ Some F are H

5. Some G are not H
   All F are G
   ∴ Some F are not H

6. Some G are not H
   Some F are not G
   ∴ Some F are not H

7. All H are G
   All F are G
   ∴ All F are H

8. All H are G
   Some F are not G
   ∴ Some F are not H

9. No H are G
   All F are G
   ∴ No F are H

10. No H are G
    Some F are not G
    ∴ Some F are H

11. Some H are G
    Some F are G
    ∴ Some F are H

12. Some H are not G
    All F are G
    ∴ Some F are not H

13. Some H are not G
    No F are G
    ∴ Some F are not H

14. All G are H
    All G are F
    ∴ All F are H

15. All G are H
    No G are F
    ∴ No F are H

16. No G are H
    All G are F
    ∴ No F are H

17. No G are H
    Some G are F
    ∴ Some F are not H

18. No G are H
    Some G are not F
    ∴ Some F are H

19. Some G are not H
    All G are F
    ∴ Some F are not H

20. Some G are not H
    No G are F
    ∴ Some F are not H

21. All H are G
    All G are F
    ∴ All F are H

22. All H are G
    No G are F
    ∴ No F are H

23. All H are G
    Some G are F
    ∴ Some F are H

24. No H are G
    No G are F
    ∴ All F are H

25. Some H are G
    All G are F
    ∴ Some F are H

The following are arguments of syllogistic form from Arnauld and Nicole's *Port Royal Logic,* many of which make ethical and religious points. Evaluate them for validity.

26. No liar is believable. Every upright man is believable. Therefore, no upright man is a liar.

27. The infinite divisibility of matter is incomprehensible. The infinite divisibility of matter is most certain. Therefore, there are some most certain things which are incomprehensible.

28. There are some wicked men with great wealth. All wicked men are miserable. Therefore, there are some miserable men with great wealth.

29. No folly is eloquent. Some folly is expressed syllogistically. Therefore, some syllogisms are not eloquent.

30. All the miracles of nature are ordinary. All that is ordinary fails to catch our attention. Therefore, there are things that fail to capture our attention but that are miracles of nature.

31. All the evils of this life are transitory evils. No transitory evil is to be feared. Therefore, no evil that is to be feared is an evil of this life.

32. Some fools speak the truth. Whoever speaks the truth deserves to be imitated. Therefore, there are those who deserve to be imitated even though they are fools.

33. No unfortunate person is happy. Some happy persons are poor. Therefore, there are poor people who are not unfortunate.

34. Nothing that is followed by a just regret is ever to be wished for. There are some pleasures which are followed by a just regret. Therefore, there are some pleasures which are not to be wished for.

35. Whoever lets those he should support die of hunger is a murderer. All the rich who do not give alms in the time of public need let those they should support die of hunger. Therefore, all the rich who do not give alms in the time of public need are murderers.

 ## 5.3   DISTRIBUTION

Before Euler, Venn, and others developed diagram techniques, there were three ways of evaluating syllogisms. The first, Aristotle's method, involved reducing syllogisms to a few special forms. A syllogism is valid, according to this method, if it can be transformed into a paradigmatically valid form. (See section 5.5.) The second we have already seen: Memorize the valid forms. The third method involved applying rules for validity. A syllogism is valid, in this method, if it obeys all the rules. This section develops the method of rules and presents a modern set of rules for validity.

The key notion behind the rules for validity is **distribution.** Medieval logicians, starting around A.D. 1200, developed the doctrine of distribution as a theory about the semantics of general terms. In particular, they held that the distribution of a term indicates whether it refers to all or only part of a set. A proposition *distributes* a term occurring in it if and only if its truth depends on the entire extension of the term.[3] Thus, a

---

[3]The doctrine of distribution apparently originated in the late twelfth century. The term appears, for example, in the anonymous late-twelfth-century *Syncategoremata Monacensia.* Peter of Spain (1205?–1277), who became Pope John XXI, devoted the last chapter of his *Tractatus*, the standard logic textbook throughout the later Middle Ages and Renaissance, to distribution. The general definition of distribution, however, appears only later, in the writings of the fourteenth-century Pseudo-Scot (so called because his writings were initially attributed to John Duns Scotus). Peter of Spain says merely that distribution is the multiplication of a common term effected by a universal sign.

term is distributed in a proposition if the truth of the proposition depends on the entire extension of the term. It is undistributed there if the truth of the proposition depends on only part of its extension. In *All people are mortal,* for example, *people* is distributed; the proposition refers significantly to all people, for everyone is relevant to the truth value of the proposition. In *Some people are crazy,* however, *people* is undistributed; the truth of the proposition depends on only some people. The people who are not crazy are irrelevant to the proposition's truth value. In general, a categorical proposition distributes a term if it asserts of every member of the term's extension that it is included in or excluded from the other term's extension.

A proposition's quantity determines the distribution of its subject term. Universal propositions have distributed subjects; particular propositions have undistributed subjects. A proposition's quality determines the distribution of its predicate term. Negative propositions have distributed predicate terms, while affirmative propositions have undistributed predicate terms. To summarize:

Universal Affirmative:	All	F^D	are	G^U
Particular Affirmative:	Some	F^U	are	G^U
Universal Negative:	No	F^D	are	G^D
Particular Negative:	Some	F^U	are not	G^D

(D: Distributed; U: Undistributed)

The concept of distribution supplies simple rules for judging the validity of syllogisms. These rules are mechanical and effective and offered the most popular way of evaluating syllogisms for hundreds of years. The method works because facts about distribution reflect important features of determiners, such as *every, all, no,* and *some.* A proposition distributes a term if its truth depends on the entire extension of the term. And anything said of the entire extension is said of every subset of it.[4] So, replacing a distributed term with a term whose extension is a subset of that of the original preserves the truth of the proposition. If all roads lead to Rome, then all bumpy roads lead to Rome.

---

[4]This is a form of the principle the medievals called *dici de omni* (to be said from all) or *dictum de omni et nullo* (said from all and none). It is at least as old as the doctrine of distribution, appearing, for example, in the anonymous twelfth-century logic text *Abbrevatio Montana:* "If anything is predicated of a universal whole universally, it is also predicated of each part of it." By the thirteenth century, logicians such as Roger Bacon saw the theory of syllogisms as depending on, and thus subsidiary to, the principle. This attitude gradually led logicians to deemphasize the theory of syllogisms in favor of investigating other logical relationships.

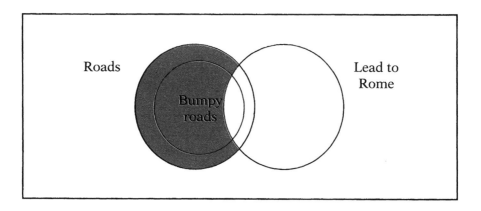

Similarly, only part of the extension of an undistributed term is relevant to the proposition's truth. Anything said of part of the extension is also said of part of any larger extension. So, replacing an undistributed term with a term whose extension includes that of the original preserves truth. If some bumpy roads ought to be repaired, then some roads ought to be repaired.

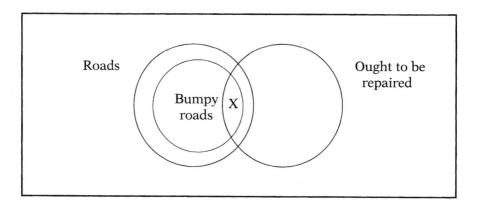

It is no accident, then, that the concept of distribution explains logical relations among categorical propositions.

P R O B L E M S

Mark the terms in these categorical propositions as distributed or undistributed.

1. All F are G.

2. No F are G.

3. Some F are G.

4. Some F are not G.

5. Every player on the team is thinking about becoming a free agent.

6. Someone I know is looking for you.

7. Barbara hasn't told somebody she ought to have told.

8. Nobody saw where Joe went.

9. Anybody could play this game better than Harry.

10. The cats have hidden my socks somewhere I haven't looked.

The following propositions are not categorical. But it still makes sense to ask whether they distribute their subject and predicate terms. Mark each term as distributed or undistributed.

11. I met a few friends for dinner.

12. Only Republicans campaigned on the issue.

13. Many of the students failed the midterm.

14. Most of her friends were there.

15. Few congressmen are eager to pursue an investigation.

16. At most two countries will vote against us.

17. Kim talks to hardly anybody in her family.

18. Joan usually calls before 10.

■ 5.4 RULES FOR VALIDITY

There are five rules for validity that any valid syllogism must satisfy:[5]

1. The middle term must be distributed at least once.
2. No term may be distributed in the conclusion if not distributed in the premises.

---

[5]Rule 3 occurs in Aristotle (*Prior Analytics* 41b6). Peter of Spain and other thirteenth-century logicians stated rules for syllogisms—including 3, 4, and the Aristotelian version of 5, stated later—but never used distribution to devise a complete set, mostly because they lacked a precise definition of distribution.

Rules 1 and 2 were first formulated by Pseudo-Scot, who devised the definition of distribution given in the text, probably in the early fourteenth century. Antoine Arnauld and Pierre Nicole adopted them in their influential *Logic, or The Art of Thinking* in 1662.

The fifth rule distinguishes the modern theory from the traditional. In its absence, the traditional theory needs to require that if the conclusion is negative, a premise must be negative. This is true in the modern theory but does not need to be stated separately.

3. At least one premise must be affirmative.
4. If a premise is negative, the conclusion must be negative.
5. If both premises are universal, the conclusion must be universal.

We may rewrite these rules as a procedure for evaluating syllogisms.

1. Is the middle term distributed at least once?
    If not, the syllogism is not valid. It commits the fallacy of the **undistributed middle.**
2. Is any term distributed in the conclusion but undistributed in a premise?
    If so, the syllogism is not valid. It commits the fallacy of **illicit major** or **minor** (depending on which term is guilty).
3. Are both premises negative?
    If so, the syllogism is not valid. It commits the fallacy of **exclusive premises.**
4. Is there a negative premise with an affirmative conclusion?
    If so, the syllogism is not valid. It commits the fallacy of **drawing an affirmative conclusion from negative premises.**
5. Are both premises universal but the conclusion particular?
    If so, the syllogism is not valid. It commits the **existential fallacy.**

If a syllogism conforms to all five rules, it is valid. To see how to apply this procedure, consider the syllogisms from the beginning of this chapter.

Here is the first, with the superscripts D and U included to indicate the distribution of each occurrence of a term:

(1) All humans are mortal.   All H^D are M^U

No angels are mortal.   No A^D are M^D

∴ No angels are human.   ∴ No A^D are H^D

The middle term, M, is distributed once, so rule 1 is satisfied. The major term, H, is distributed in both occurrences; so is the minor term, A. So, the syllogism satisfies rule 2. The first premise is affirmative, so rule 3 is satisfied. There is a negative premise and a negative conclusion, so rule 4 is satisfied. Finally, all three propositions are universal, satisfying rule 5. The syllogism, therefore, is valid.

Consider the next example, another valid syllogism.

(2) Nobody who has a college degree is stupid.   No C^D are S^D

Some stupid people make a lot of money.   Some S^U are M^U

∴ Some people who make a lot of money   ∴ Some M^U are
    don't have college degrees.   not C^D

The middle term, S, is distributed in the first premise, so rule 1 is satisfied. The major term, C, is distributed in the conclusion but also in the major premise, so rule 2 is satisfied. There is an affirmative premise, a negative premise, and a negative conclusion, so rules 3 and 4 are satisfied. Finally, there is a particular premise, so rule 5 is satisfied as well.

To evaluate our third example:

(3) All New York teams have high revenues.   All N^D are R^U

     Some New York teams are consistent winners.   Some N^U are C^U

     ∴ Some teams with high revenues are consistent winners.   ∴ Some R^U are C^U

Rule 1 requires that the middle term, N, be satisfied at least once; it is. The major and minor are undistributed everywhere, so rule 2 is satisfied. The premises and conclusion are affirmative and so obey rules 3 and 4. Finally, the conclusion and one premise are particular, satisfying rule 5. The syllogism is therefore valid.

The fourth example fails to be valid:

(4) Some of your friends aren't very nice.   Some F^U are not N^D

     Nobody who would insult a total stranger is very nice.   No I^D are N^D

     ∴ Some of your friends would insult a total stranger.   ∴ Some F^U are I^U

The middle term, N, is distributed in at least one premise—in fact, in both—satisfying rule 1. Neither term in the conclusion is distributed, satisfying rule 2. But both premises are negative, violating rule 3. The syllogism is therefore invalid, committing the fallacy of exclusive premises. It also violates rule 4, since it has a negative premise but an affirmative conclusion. Once one rule is violated, it makes no difference how many others are satisfied. Violating one rule is enough to make a syllogism invalid.

Finally, the last example:

(5) Some logic books avoid syllogisms.   Some L^U are A^U

     Some books that avoid syllogisms are worthless.   Some A^U are W^U

     ∴ Some logic books are worthless.   ∴ Some L^U are W^U

This violates rule 1, committing the fallacy of undistributed middle. It obeys the other rules, but, again, that makes no difference.

A valid syllogism satisfies all five rules. Violating even one rule condemns a syllogism to invalidity.

P R O B L E M S

Use the rules of this section to evaluate the following syllogisms for validity.

1. All G are H
   All F are G
   ∴ All F are H

2. All G are H
   No F are G
   ∴ Some F are not H

3. All G are H
   Some F are not G
   ∴ Some F are not H

4. No G are H
   All F are G
   ∴ No F are H

5. No G are H
   Some F are G
   ∴ Some F are not H

6. Some G are H
   Some F are not G
   ∴ Some F are not H

7. All H are G
   No F are G
   ∴ No F are H

8. All H are G
   No F are G
   ∴ Some F are not H

9. All H are G
   Some F are not G
   ∴ Some F are not H

10. No H are G
    All F are G
    ∴ Some F are not H

11. No H are G
    Some F are not G
    ∴ Some F are not H

12. Some H are not G
    No F are G
    ∴ Some F are H

13. Some H are not G
    Some F are G
    ∴ Some F are not H

14. All G are H
    Some G are F
    ∴ Some F are H

15. All G are H
    Some G are not F
    ∴ Some F are not H

16. No G are H
    No G are F
    ∴ No F are H

17. Some G are H
    All G are F
    ∴ Some F are H

18. Some G are H
    No G are F
    ∴ Some F are not H

19. Some G are not H
    All G are F
    ∴ Some F are not H

20. Some G are not H
    No G are F
    ∴ Some F are H

21. Some G are not H
    Some G are F
    ∴ Some F are not H

22. All H are G

    Some G are not F

    ∴ Some F are not H

23. No H are G

    All G are F

    ∴ Some F are not H

24. No H are G

    Some G are not F

    ∴ Some F are not H

25. Some H are G

    Some G are not F

    ∴ Some F are not H

Use the rules of this section to evaluate the following syllogistic arguments, taken from Lewis Carroll's *Symbolic Logic,* for validity.

26. I admire these pictures. When I admire anything I wish to examine it thoroughly. Therefore, I wish to examine some of these pictures thoroughly.

27. All soldiers can march; some babies are not soldiers. So, some babies cannot march.

28. His songs never last an hour; a song that lasts an hour is tedious. So, his songs are never tedious.

29. Some candles give very little light. Candles are meant to give light. Therefore, some things that are meant to give light give very little.

30. All who are anxious to learn work hard. Some of these boys work hard. So, some of these boys are anxious to learn.

31. Ill-managed business is unprofitable. Railways are never ill-managed. Therefore, all railways are profitable.

32. All wasps are unfriendly; no puppies are unfriendly. So, puppies are not wasps.

33. No monkeys are soldiers; all monkeys are mischievous. So, some mischievous creatures are not soldiers.

34. No frogs are poetical; some ducks are unpoetical. So, some ducks are not frogs.

35. Every eagle can fly. Some pigs cannot fly. So, some pigs are not eagles.

Use the rules for validity and the definitions of *syllogism, figure,* and *mood* to explain why these facts hold.

36. Every syllogism with mood EIO is valid.

37. The conclusion of every valid second figure syllogism is negative.

38. Every valid syllogism with a particular affirmative conclusion has exactly one particular affirmative premise.

39. The conclusion of every valid third figure syllogism is particular.

40. The only valid syllogism with a universal affirmative conclusion is in first figure, with mood AAA.

41. Every valid syllogism whose minor premise is particular negative is in second figure.

42. Every valid syllogism has at least one universal premise.

43. The only valid syllogism with an affirmative conclusion in the fourth figure has mood IAI.

44. Every valid syllogism with a particular conclusion has one particular premise.

45. No two valid syllogisms with contradictory minor premises have the same major premise.

■ 5.5 REDUCTION

Aristotle developed the method of *reduction* to show syllogisms to be valid. The method constitutes the first *natural deduction system:* a technique for establishing validity by constructing proofs.

A *natural deduction system* is a set of *rules of inference,* which allow us to deduce propositions from other propositions. Central to the concept of a natural deduction system is the idea of *proof.* Proofs are extended arguments. All proofs in this chapter are *hypothetical;* they begin with premises (also called *assumptions* or *hypotheses*). The conclusion depends on these premises. Such proofs show that the conclusion is true, not unconditionally, but only if the assumptions are true. These proofs thus show the validity of an argument.

A *proof* in a natural deduction system is a series of *lines.* On each line appears a categorical proposition. Each proposition in a proof *(a)* is an assumption or *(b)* derives from propositions on previously established lines by a rule of inference. In the system of this chapter, the last line of a proof is its *conclusion;* the proof *proves* that proposition *from* the assumptions.

Each line of a proof, then, looks like this:

PROOF LINES

Number	Proposition	Justification

The rules of inference in this chapter allow us to derive propositions in a proof if other propositions of certain kinds occupy already-established lines. The most important distinction between rules, however, is between those that work only in one direction and those that work in both directions. Rules working in one direction are *rules of implication*. Those that work in both directions are *rules of replacement* (or *equivalence*). They are also sometimes called *invertible*, for they rely on the equivalence of propositions.

Aristotle showed that all valid syllogisms reduce to first-figure syllogisms. That is, he showed that all valid syllogisms may be proved valid by assuming the validity of four first-figure syllogisms. Our system in this chapter simplifies Aristotle's method by adding a rule. That allows us to take just two forms of the syllogism as fundamental:

BARBARA	DARII
All M are P	All M are P
All S are M	Some S are M
∴ All S are P	∴ Some S are P

We take these as basic rules of implication and abbreviate them with the letters *B* and *D* in giving justifications:

	BARBARA			DARII	
n.	All M are P		n.	All M are P	
m.	All S are M		m.	Some S are M	
p.	All S are P	B, n, m	p.	Some S are P	D, n, m

Lines derived by these rules have justifications consisting of *B* or *D* and a line number or letter.

We adopt two rules of equivalence, which we have already seen in Chapter 4:

CONVERSION

Some F are G ⇔ Some G are F	C, n
No F are G ⇔ No G are F	C, n

OBVERSION

No F are G ⇔ All F are nonG	O, n
Some F are not G ⇔ Some F are nonG	O, n

Lines derived by these rules have justifications consisting of *C* (or *O*) and a line number or letter. Aristotle did not use obversion; he used two additional basic forms (Celarent and Ferio) and a method of indirect proof, often called reduction *per impossibile*.

Finally, in the Aristotelian interpretation only, another rule of implication allows the derivation of four syllogistic forms rejected as invalid in the modern interpretation. Recall that the Aristotelian interpretation assumes that terms have nonempty extensions and counts universal propositions as implying corresponding particular propositions. In particular, it takes *All F are G* to imply *Some F are G*.

SUBALTERNATION

n.	All F are G	
m.	Some F are G	S, n

This rule holds only on the Aristotelian interpretation.

Here is a simple proof of the validity of one of our examples:

(3) All New York teams have high revenues.

Some New York teams are consistent winners.

∴ Some teams with high revenues are consistent winners.

To begin the proof, we list the premises, without justifications, and place the conclusion on the right on the line of the last premise to indicate what we must show. When we reach the conclusion by applying rules to the premises, the proof is complete.

1. All N are R
2. Some N are C   / ∴ Some R are C

Now, we apply rules. This syllogism is very close to the form Darii. In fact, to match the form, we need only convert the second premise:

1. All N are R
2. Some N are C   / ∴ Some R are C
3. Some C are N   C, 2

Now, we can apply Darii to the propositions on lines 1 and 3:

1. All N are R
2. Some N are C   / ∴ Some R are C
3. Some C are N   C, 2
4. Some C are R   D, 1, 3

We have almost reached the conclusion. All we need to do is apply conversion again:

1. All N are R
2. Some N are C   / ∴ Some R are C
3. Some C are N   C, 2
4. Some C are R   D, 1, 3
5. Some R are C   C, 4

The proof is now complete; we have applied rules to the premises to derive the conclusion. The proof shows that (3) is valid.

Obversion is useful for deriving negative conclusions. Consider example (1):

(1) All humans are mortal.

   No angels are mortal.

   ∴ No angels are human.

We begin the proof by writing the premises and then the conclusion on the side:

1. All H are M
2. No A are M   /∴ No A are H

To do anything with the negative premise, we must use obversion to make it affirmative. Barbara and Darii are our basic rules of implication and both are affirmative. All negative propositions must therefore be obverted into equivalent affirmative propositions. If we apply obversion immediately in this case, however, we get *All A are nonM*, which does not form a syllogism when combined with *All H are M*. We do better to apply conversion first and then obversion:

1. All H are M
2. No A are M   /∴ No A are H
3. No M are A   C, 2
4. All M are nonA   O, 3

Obversion always switches a term to its complement, or vice versa. This example illustrates a useful strategy: Apply obversion to get the complement of the term that does not appear in the other premise. Here, H and M appear in the first premise. So, use obversion to reach a proposition involving nonA. Any other use will yield propositions that do not form a syllogism.

Now we can apply Barbara:

1. All H are M	
2. No A are M	/∴ No A are H
3. No M are A	C, 2
4. All M are nonA	O, 3
5. All H are nonA	B, 4, 1

Finally, we use obversion to return to a negative conclusion with the original terms and convert to get the conclusion.

1. All H are M	
2. No A are M	/∴ No A are H
3. No M are A	C, 2
4. All M are nonA	O, 3
5. All H are nonA	B, 4, 1
6. No H are A	O, 5
7. No A are H	C, 6

Example (2) is of the form called *Frisesomorum* in medieval times:

(2) Nobody who has a college degree is stupid.

Some stupid people make a lot of money.

∴ Some people who make a lot of money don't have college degrees.

Again, we begin by stating the premises and listing the conclusion on the right:

1. No C are S	
2. Some S are M	/∴ Some M are not C

As with (1), we must use obversion to convert the first premise into an affirmative proposition. Doing so immediately, however, leaves us something that is not a syllogism. Since S and M appear in the other premise, we want them to remain as terms. We want to apply obversion to obtain a proposition containing nonC. We may do this by converting first and then obverting:

1. No C are S	
2. Some S are M	/∴ Some M are not C
3. No S are C	C, 1
4. All S are nonC	O, 3

We still cannot apply Darii, but we could if we were to convert the second premise.

1. No C are S
2. Some S are M   /∴ Some M are not C
3. No S are C   C, 1
4. All S are nonC   O, 3
5. Some M are S   C, 2
6. Some M are nonC   D, 4, 5

Now, we can obvert to obtain the conclusion and finish the proof.

1. No C are S
2. Some S are M   /∴ Some M are not C
3. No S are C   C, 1
4. All S are nonC   O, 3
5. Some M are S   C, 2
6. Some M are nonC   D, 4, 5
7. Some M are not C   O, 6

The proofs we have seen work in both modern and Aristotelian interpretations of categorical propositions. To close, then, let us derive a conclusion that follows only on the latter interpretation using the subalternation rule.

(8) All terrorists use violence.

All terrorists have political motivations.

∴ Some who use violence have political motivations.

We begin, as usual, by listing the premises and the conclusion on the right:

1. All T are V
2. All T are P   / ∴ Some V are P

Now, to get a particular conclusion, we must either apply Barbara to our universal premises and then use subalternation or use subalternation first and then apply Darii. We cannot do the former here, for the premises do not fit the Barbara pattern, and universal affirmatives do not convert. So, we must use subalternation first. Darii requires that the major premise be universal and the minor particular, so we must apply subalternation to the minor—here, the first premise, which contains the minor term V.

1. All T are V
2. All T are P   / ∴ Some V are P
3. Some T are V   S, 1

Before we can apply Darii, we must switch the terms of the particular proposition using conversion.

1. All T are V
2. All T are P   / ∴ Some V are P
3. Some T are V   S, 1
4. Some V are T   C, 3

Finally, we can apply Darii to complete the proof.

1. All T are V
2. All T are P   / ∴ Some V are P
3. Some T are V   S, 1
4. Some V are T   C, 3
5. Some V are P   D, 2, 4

P R O B L E M S

Using reduction, show that the following are valid. If any are valid only in the Aristotelian interpretation, indicate so.

1. All M are P; Some M are S; ∴ Some S are P   Datisi

2. No M are P; All S are M; ∴ No S are P   Celarent

3. All M are S; Some P are M; ∴ Some S are P   Dabitis

4. No M are P; Some S are M; ∴ Some S are not P   Ferio

5. Some M are P; All M are S; ∴ Some S are P   Disamis

6. No M are S; All P are M; ∴ No S are P   Celantes

7. No P are M; Some S are M; ∴ Some S are not P   Festino

8. No P are M; All S are M; ∴ No S are P   Cesare

9. No P are M; All S are M; ∴ Some S are not P   Cesaro

10. No M are P; All M are S; ∴ Some S are not P   Felapto

11. Some M are not P; All M are S; ∴ Some S are not P Bocardo

12. All P are M; No S are M; ∴ Some S are not P   Cambestros

13. All who do not remember the past are condemned to repeat it. Some who fail to study history do not remember the past. So, some who fail to study history are condemned to repeat it.

14. All acts of reflection have the potential for increasing self-knowledge. Anything that can increase self-knowledge is worth doing. So, some things worth doing are acts of reflection.

15. Nothing in nature belongs to the space of reasons. All brain activity occurs in nature. Therefore nothing that belongs to the space of reasons is brain activity.

16. No moral actions are done from hate. All acts that maximize utility are moral. Hence, some acts that maximize utility are not done from hate.

17. No thoughts are independent of the movements of matter, but some thoughts are products of human freedom. It follows that some products of human freedom are not independent of the movements of matter.

18. Everything that God wills reflects His nature. Some evil occurrences do not reflect God's nature. Thus, some evil occurrences are not willed by God.

19. No products of human imagination and toil are worth only the value assigned to them by the market. Everything worth only the value the market assigns to it is worth exactly what someone else will pay for it. Hence some things that are worth exactly what someone else will pay for them are not products of human imagination and toil.

20. Everything that can be willed as universal law is a moral duty. Keeping a promise can always be willed as a universal law. Therefore keeping a promise is sometimes a moral duty.

Because it includes obversion, the natural deduction system of this section can demonstrate the validity of some arguments that are not syllogisms. Prove that the following are valid, marking any that succeed only in the Aristotelian interpretation. (Let nonnonF = F. Arguments 37 through 44 are from Lewis Carroll.)

21. All H are G; All F are nonG; ∴ All F are nonH

22. No G are nonH; No F are H; ∴ No F are G

23. No H are nonG; No G are F; ∴ All F are nonH

24. No H are nonG; Some F are not G; ∴ Some F are nonH

25. No F are H; No F are nonG; ∴ Some nonH are G

26. All G are nonH; No G are F; ∴ Some nonF are not H

27. Some nonG are H; All nonG are F; ∴ Some F are H

28. All H are G; Some nonG are F; ∴ Some F are nonH

29. No H are G; No nonG are nonF; ∴ No H are nonF

30. All nonH are nonG; All F are G; ∴ All F are H

31. Don likes only sports shown on television. Some sports that are not shown on television could never appeal to viewers. Hence, Don dislikes some sports that could never appeal to viewers.

32. Some students in this class read this section carefully, for anybody who did not read this section carefully would misunderstand this problem, and some students in this class understand it.

33. Every storm that does not drop rain on the innocent does not drop rain on the guilty. Some welcome storms drop rain on the innocent, therefore, because no storms that drop rain on the guilty are unwelcome.

34. Some who do not remember the past are not condemned to repeat it, and everyone who is not ignorant remembers the past. Some ignorant people, therefore, are not condemned to repeat the past.

35. No filial and fraternal people like to offend their superiors, and nobody who does not like to offend superiors likes to stir up rebellion. (Confucius) So, no one who likes to stir up rebellion is filial and fraternal.

36. No incompetent students are incapable of failing. No lazy students are competent. So, all lazy students are capable of failing.

37. No son of mine is dishonest. People always treat an honest man with respect. Therefore, no son of mine ever fails to be treated with respect.

38. No misers are unselfish. None but misers save eggshells. So, no unselfish people save eggshells.

39. A prudent man shuns hyenas. No banker is imprudent. Thus, no banker fails to shun hyenas.

40. Improbable stories are not easily believed. None of his stories are probable. So, none of his stories are easily believed.

41. No wheelbarrows are comfortable. No uncomfortable vehicles are popular. So, no wheelbarrows are popular.

42. None of my boys are conceited; none of my girls are greedy. Thus, no conceited child of mine is greedy. (Take one sex as the complement of the other.)

43. None of my boys are clever; none but a clever boy could solve this problem. So, none of my boys could solve this problem.

44. None of my boys are learned; some of my boys are not choristers. So, some unlearned boys are not choristers.

PART III

SYMBOLIC LOGIC

CHAPTER 6

PROPOSITIONAL LOGIC

Propositional logic takes statements or propositions as basic units and examines relationships between them that pertain to reasoning. This chapter focuses on connections between statements—specifically, between their truth values. The truth values of compound statements formed with certain words and phrases—called **truth-functional connectives**—depend entirely on the truth values of the smaller statements they connect.

6.1 CONNECTIVES

Logicians use *statement* in a special sense, in which one statement may be a part of another. Within the single statement

(1) When the going gets tough, the tough get going.

are two other statements,

(2) a. The going gets tough.
   b. The tough get going.

Example (1) results from combining these two shorter statements. Example (1) is, therefore, a *compound* statement; (2)a and (2)b are its *components*.

> One statement is a **component** of another if *(a)* it is a proper part of that statement, and *(b)* replacing it with another statement yields something meaningful.

A statement is **compound** (or **molecular**) if it has components and **simple** (or **atomic**) if it does not.

Words like *when*, which link statements together to form compounds, are *connectives*.

A **connective** is a word or phrase that forms a single compound statement from one or more component statements.

Examples of connectives forming a compound from two or more component statements follow:

although
and
because
before
but
however
if
or
though
unless
when

Some connectives form a compound from a single statement:

can
could
may
maybe
might
must
necessarily
not
of course
possibly
should

The **truth value** of a statement is *truth,* if it is true, and *falsehood,* if it is false. In propositional logic we assume that every statement has one and only one truth value. We assume, in other words, that every statement is either true or false, but not both.

Some connectives form compound statements whose truth values depend solely on the truth values of their components. These connectives are *truth-functional.*

> A connective is **truth-functional** if and only if the truth values of the component statements the connective joins always completely determine the truth value of the compound statement formed by the connective.

If a connective is truth-functional, then two compounds formed from it have the same truth value whenever their corresponding components match in truth value. To determine the truth value of the compound, we need to know only whether the component statements are true or false.

Consider the statement

(3) Brazil isn't a Spanish-speaking country.

which contains the component

(4) Brazil is a Spanish-speaking country.

If the component is true, the compound is false: If it's true that Brazil is a Spanish-speaking country, it's false that it isn't a Spanish-speaking country. If the component is false, the compound is true: If it's false that Brazil is a Spanish-speaking country, then it's true that it isn't. The truth value of (4), in other words, completely determines the truth value of (3). This does not depend on any special feature of the statement *Brazil is a Spanish-speaking country.* Any statement would behave similarly. *Not* is thus a truth-functional connective.

In contrast, consider

(5) John thinks that Brazil is a Spanish-speaking country.

The truth value of the component statement doesn't determine the truth or falsehood of the statement as a whole. That Brazil is a Spanish-speaking country does not guarantee that John thinks so. Similarly, that Brazil isn't a Spanish-speaking country implies nothing about whether John thinks it is.

We can summarize these examples in a small table. We consider the possible truth values of the statements and see whether, in every case, they determine the truth value of the compound. (*T* abbreviates *true,* and *F* abbreviates *false.*)

BRAZIL IS A SPANISH-SPEAKING COUNTRY	BRAZIL ISN'T A SPANISH-SPEAKING COUNTRY	JOHN THINKS THAT BRAZIL IS A SPANISH-SPEAKING COUNTRY
T	F	T or F
F	T	T or F

Because *John thinks that Brazil is a Spanish-speaking country* could be either true or false, whether Brazil is a Spanish-speaking country or not, *John thinks that* is not truth-functional.

We can draw the same contrast with binary connectives. Consider the two connectives *and* and *after*. The statements

(6) George arrived and Joan left.

and

(7) George arrived after Joan left.

for example, both contain the components *George arrived* and *Joan left*. If both components are true, then it is true that George arrived *and* Joan left. But the truth of both components does not similarly fix a truth value for (7). George may have arrived before or after Joan left. So, *after* is not truth-functional.

That the truth values of the components sometimes suffice to determine the compound's truth value is not enough to guarantee truth-functionality. They must suffice in *every* case if the connective is truth-functional. The truth value of the components sometimes fixes the value of the compound formed from them by *after:* If George did not arrive at all, he certainly did not arrive after Joan left. *After*, nevertheless, is not truth-functional.

GEORGE ARRIVED	JOAN LEFT	GEORGE ARRIVED AND JOAN LEFT	GEORGE ARRIVED AFTER JOAN LEFT
T	T	T	T or F
T	F	F	F
F	T	F	F
F	F	F	F

A connective is truth-functional if and only if it is possible to fill in every entry under the compound statement with a determinate truth value. If, as with *after*, there is any row that cannot be filled in with just a single truth value, then the connective is not truth-functional.

PROBLEMS

Are these connectives truth-functional? Why, or why not?

1. or

2. it is true that

3. but

4. because

5. it is obvious that

6. John realizes that

7. can

8. should

9. allegedly

10. however

11. when

12. it is probable that

13. it is necessarily true that

14. in spite of the fact that

15. nevertheless

 6.2   TRUTH FUNCTIONS

Truth-functional connectives, in English or in the symbolic language we are about to develop, yield compounds whose truth values depend solely on the truth values of their components. To understand the meanings of compounds, we need to specify the meanings of those connectives. Each truth-functional connective represents a corresponding function from truth values into truth values, called a **truth function.**[1] Such a function takes as inputs the truth values of the component statements and yields the truth value of the compound statement.

This chapter presents a few commonly used truth functions. Any truth function at all can be defined in terms of them alone. In fact, a single

---

[1] Where n is any number, an n-ary function relates n inputs to a unique output. The inputs are called **arguments** of the function; the outputs are its **values.** Truth functions are simply functions whose inputs and outputs are truth values. For any n, there are 2^{2^n} n-ary truth functions.

truth function suffices to define this chapter's truth functions and, therefore, any truth function at all.[2]

The first and simplest function we define yields the truth value of a compound formed from one component sentence. Reflecting the meaning of the truth-functional connective *not*, it is called **negation,** and we use the tilde, ~, to represent it. (In this definition, *p* is any formula—that is, any symbolic representation of a statement.)

p	~p
T	F
F	T

Negation transforms the truth value of the component statement into its opposite.

As we have seen, an English connective that has this effect is the logical particle *not*. Other English expressions having much the same impact are *it is not the case that, it is false that,* and, often, the prefixes *un-, dis-, a-, im-,* and so on.

The second function yields truth values of compounds formed from two component sentences. Called **conjunction,** we represent it with the dot.

p	q	(p · q)
T	T	T
T	F	F
F	T	F
F	F	F

A conjunction is true if and only if both its components—called **conjuncts**—are true.

Functioning in this way are most grammatical conjunctions: *and, both . . . and, but, though,* and *although*. These do not all have the same meaning, but they affect truth values in the same way. English treats conjunction as a **multigrade** connective, one that can take two, three, or more statements and combine them into a single compound statement.

---

[2]Actually, there are two: Sheffer's stroke, meaning "not both," and nondisjunction, meaning "neither . . . nor." Both were discovered by the American philosopher Charles Sanders Peirce (1839–1914), though Emil Post, in 1920, was the first to prove their functional completeness. See the exercises at the end of this section for precise definitions of these connectives.

The logical · always links two statement letters, but in English we can say *John brought the mustard, Sally brought the pickles, and I brought the hot dogs.*[3]

The third function also yields the truth value of a compound formed from two component sentences. Reflecting the meaning of *or,* and represented by ∨, it is called **disjunction.**

p	q	(p ∨ q)
T	T	T
T	F	T
F	T	T
F	F	F

A disjunction is true just in case one or both of its components—called *disjuncts*—is true.

Corresponding to this function in English are *or, either . . . or,* and, as we shall see, *unless.*[4] *Or,* like *and,* is multigrade in English.

The fourth function is represented by ⊃ and is called the **conditional:**

p	q	(p ⊃ q)
T	T	T
T	F	F
F	T	T
F	F	T

Here, for the first time, the order of the components makes a difference. The first component of a conditional is its **antecedent;** the second is its **consequent.** A conditional is true if and only if it does not have a true antecedent and false consequent.

---

[3]Some writers see *and* as ambiguous between a purely logical sense, captured by the table for conjunction, and another sense that includes relations in time. They point out that *Moe fell down and got up* describes a different sequence of events from that described by *Moe got up and fell down.* This is true, but it has nothing to do with the meaning of *and.* It stems instead from the way that verb tenses lead us to represent sequences of events. (We expect successive sentences in the simple past tense, for example, to report a sequence of events in order.) To see this, note that the difference remains without *and: Moe fell down. He got up.* and *Moe got up. He fell down.* also represent different sequences of events.

[4]Some writers find the English *or* ambiguous between an inclusive reading ("p or q or both") and an exclusive reading ("p or q but not both"). The disjunction truth function corresponds to the inclusive sense. I do not discuss this issue in the text, because I do

English expressions we symbolize as p ⊃ q include:

q if p

if p then q

p only if q

q so long as p

q provided that p

q assuming that p

q on the condition that p

Because the direction of the conditional matters, symbolizing expressions of these forms requires care. In English, it is easy to confuse the antecedent with the consequent.

At first it may seem surprising that *p only if q* and *If p, then q* are both correlates of the conditional p ⊃ q. But these statements are equivalent:

(8) a.   A number is even only if it divides by two.
    b.   If a number is even, it divides by two.

We can symbolize both in the same way because they are true in exactly the same circumstances. More generally, *p only if q* is equivalent to *If not q, then not p*, as these pairs of sentences demonstrate:

(9) a.   The patient will survive only if we operate.
    b.   If we do not operate, the patient will die.

(10) a.   I'll go only if you come with me.
     b.   If you don't come with me, I won't go.

(11) a.   You will pass the course only if you do your homework.
     b.   If you don't do your homework, you won't pass the course.

And *If not q, then not p* is equivalent to *If p, then q* according to our definition of the conditional. (This principle is called *transposition:* both p ⊃ q and ~q ⊃ ~p come out false in only one circumstance, where p is true and q is false.)

---

not believe the ambiguity exists. In my view, *or* is always inclusive. Purported examples of exclusive *or* fall into two categories: *(a)* Sentences of the form p *or* q, where p and q are incompatible (for example, *Jane is in Pittsburgh or Philadelphia right now*). We can conclude that Jane is not in both cities but because Pittsburgh and Philadelphia do not overlap and people can't be in two places at once, not because *or* has a special meaning in the sentence. This is not to deny that an argument might rest on these additional factors or that we might want to represent them explicitly in our logical symbolism. But, in doing so, we would be representing additional information, not a different sense of *or.* *(b)* Sentences that combine disjunction with permission (e.g., *You may have a cookie or a lollipop*). Analysis of these "free choice permission" sentences is notoriously difficult, but, however they should be analyzed, they are not exclusive disjunctions: *You may have a cookie or a lollipop* implies *You may have a cookie!*

Admittedly, *If not q, then not p* and *If p, then q* often do not sound equivalent in English. The conditional truth function matches the English correlates of the conditional in many respects. Nevertheless, it omits some features of English conditionals. Consider these statements, all of which, according to our definitions of negation and the conditional, we would symbolize as equivalent formulas:

(12) a. The patient will survive only if we operate.
      b. If we do not operate, the patient will die.
      c. If the patient survives, we operate.
      d. We will not operate only if the patient dies.

The first two are plainly equivalent. The last two, however, sound bizarre. Example (12)c makes it sound as if we operate only when the patient is out of danger, and (12)d makes it sound as if we operate as long as we have a live body to operate on. The conditional does not capture these differences. Whatever follows the English word *if* states a condition on the other component. Often, this condition is causal: What follows *if* is a cause or causal factor, and the other component is the effect. (For further discussion of causation, see Chapter 11.) The truth-functional rendering of the conditional cannot capture the English conditional's causal meaning.

The English connective *unless* raises similar issues. Consider the statement

(13) The patient will die unless we operate.

This is equivalent to (9)a and b. In general, we can interpret *p unless q* as *p, if not q*. This is equivalent to *p or q* according to our definitions of the connectives. So, we can symbolize *unless* as a disjunction.

But, as with *if* and *only if,* we cannot express causal aspects of the meaning of *unless* in our symbolic language. Example (9) seems to assert that there is a causal connection between the operation and the patient's chances for survival. Our logical symbolism does not capture this. The formula p ∨ q is true if either p or q is true. But we would not normally count (9) true, just because the operation was performed or just because the patient died.[5]

Finally, the **biconditional** is a truth function symbolized by ≡.

---

[5]I would like to thank Robert L. Causey for suggesting the examples in the text. These problems with the truth-functional rendering of the conditional have been recognized since the third century B.C., when the Greek logician Philo of Megara—a classmate of Zeno (336–264 B.C.), the founder of Stoicism—first proposed the analysis reflected in our table. Diodorus, Philo's teacher, held that conditionals involve necessity, and Chrysippus (279–206 B.C.), the third head of the Stoic school and widely acclaimed as the greatest logician of his time, held that the sort of necessity involved is specifically logical, as opposed, for instance, to causal necessity. The controversy among their followers became so intense that Callimachus wrote that "even the crows on the roofs caw about the nature of conditionals."

p	q	(p ≡ q)
T	T	T
T	F	F
F	T	F
F	F	T

Biconditionals are true just in case their components agree in truth value. Such English expressions as *if and only if, when and only when,* and (in one of its uses) *just in case* correspond to the biconditional.

The symbols ~, ·, ∨, ⊃, and ≡ are in common use. But no logical notation is standard. This table shows other symbols that have been used as logical connectives. The last in each case is so-called Polish notation, which requires no parentheses.

TRUTH FUNCTION	OUR SYMBOL	OTHER SYMBOLS
Negation	~p	-p, ¬p, p′, p̄, Np
Conjunction	p · q	p ∧ q, pq, p & q, Kpq
Disjunction	p ∨ q	p ⊻ q, Apq
Conditional	p ⊃ q	p → q, Cpq
Biconditional	p ≡ q	p ↔ q, p ~ q, Epq

Every truth function can be defined in terms of these five truth functions alone. Indeed, not even all five are necessary. Negation and conjunction alone suffice; so do negation and disjunction or negation and the conditional. Any set of truth functions that, like {~, ·}, {~, ∨}, and {~, ⊃}, allows us to define every other truth function is **functionally** (or **expressively**) **complete.**

P R O B L E M S

Careful inspection of the tables defining the meanings of the connectives reveals that some may be defined in terms of others. For each schema on the left, find an equivalent schema using just the connectives on the right.

1. ~(p ∨ q)   ~, ·

2. ~(p · q)   ~, ∨

3. (p ⊃ q)   ~, ∨

4. ~(p ⊃ q)   ~, ·

5. (p ⊃ q)   ~, ·

6. (p ∨ q)   ~, ·

7. (p ∨ q)   ~, ⊃

8. (p · q)   ~, ∨

9. (p ≡ q)   ·, ⊃

10. (p ≡ q)   ~, ·, ∨

Two binary connectives are functionally complete by themselves: /, "not both," and ↓, "neither . . . nor," defined by these tables:

p	q	p / q	p ↓ q
T	T	F	F
T	F	T	F
F	T	T	F
F	F	T	T

For each schema on the left, find an equivalent using just the connective on the right.

11. ~p   /

12. ~p   ↓

13. ~(p · q)   /

14. ~(p ∨ q)   ↓

15. (p · q)   /

16. (p ∨ q)   ↓

17. ~p · ~q   /

18. p ∨ q   /

19. p ⊃ q   ↓

20. p ≡ q   ↓

▮ 6.3   SYMBOLIZATION

We construct a symbolic language to evaluate arguments. We can determine whether English arguments are valid by translating them into symbolic notation and then judging the validity of the translations.

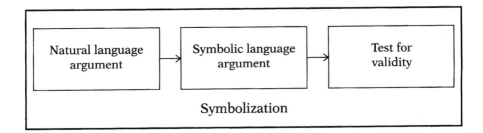

To use a symbolism, therefore, we must develop a way to translate arguments into it.

To symbolize a statement in our symbolic language,

1. Identify statement connectives.
2. Identify components containing no connectives, and replace each distinct component with a distinct **statement letter,** a capital letter of the alphabet. (A record of which statement letter symbolizes which statement component is called a **dictionary.**)
3. Replace statement connectives with symbolic connectives.
4. Use the structure of the statement to determine grouping.

The goal of symbolization is to devise a formula that is true exactly when the corresponding statement is true and false exactly when that statement is false. In the process, we want to represent, in our symbolism, as much of the logical structure of the corresponding statement as possible.

Several factors complicate these steps. The first step—identify statement connectives—relies on the correlation between truth functions and certain English statement connectives. Since English connectives do not always match their closest symbolic counterparts exactly, this step can produce distortions in meaning, particularly with *if*.

Consider the paradoxical extended argument given by the ancient skeptic Sextus Empiricus, discussed in Chapter 1:

(14) If Socrates died, he died either when he was living or when he was dead.

But he did not die while living; for assuredly he was living, and as living he had not died.

Nor when he died; for then he would be twice dead.

∴ Socrates did not die.

The second and third premises contain arguments that support them. Omitting those, we obtain the argument

(15) If Socrates died, he died either when he was living or when he was dead.

But he did not die while living.

Nor when he died.

∴ Socrates did not die.

It is easy to identify truth-functional statement connectives:

(16)   *If* Socrates died, he died *either* when he was living *or* when he was dead.

But he did *not* die while living.

Nor when he died.

∴ Socrates did *not* die.

Step two—identify and replace components—is complicated because English arguments try not to be repetitive or dull. Rarely will an author use a component statement twice in the same passage. Even if the meaning is the same, the wording may vary. To allow for this, we judge components to be the same if they clearly have the same meaning.

For example, in the argument from Sextus, we find the statements

(17)   a.   Socrates died.
     b.   he died when he was living.
     c.   (he died) when he was dead.
     d.   he did die while living.
     e.   (he did die) when he died.
     f.   Socrates did die.

How many statements should we distinguish? Statements (17)a and f differ in wording but plainly have the same meaning. The same is true of (17)b and d and of (17)c and e. So, it seems reasonable to count three atomic statements. Replacing them according to the dictionary

S: Socrates died.

L: Socrates died while he was living.

D: Socrates died while he was dead.

we obtain

(18)   *If* S, *either* L *or* D.

But not L.

Nor D.

∴ *not* S.

At this point it is easy to replace English connectives with symbolic connectives. The *But* that begins the second premise corresponds to a conjunction. But it does not link components of the same statement; it links the first two premises. So, there is no need to symbolize it. (We can always think of the premises as conjoined; our method for evaluating arguments takes that for granted.) Ignoring that conjunction, then, we obtain

(19) S, ⊃ L ∨ D

   ~ L

   ~ D

   ∴ ~ S

This is not yet an adequate symbolic representation of the argument, since the first premise is not a formula. It contains a comma, and its grouping is unclear.

The fourth step—determining the grouping—introduces further issues. Our symbolic language avoids ambiguity by using parentheses as grouping indicators. English does not, and so natural language statements are sometimes ambiguous. But English has some devices for making grouping clear.

One such device is the use of commas. The English statement

(20) John will come and Fred will stay only if you sign.

has no very clear grouping. Using the dictionary

J: John will come

F: Fred will stay

S: You sign

we can symbolize this statement as either $(J \cdot F) \supset S$ or $J \cdot (F \supset S)$. But a comma can make it clear that *and* is the main connective:

(21) John will come, and Fred will stay only if you sign.

This should be symbolized as $J \cdot (F \supset S)$. A comma emphasizes a break in the statement, marking the combination of two different main clauses. Commas, therefore, tend to suggest that the nearest connective has some priority.

English also has cooordinate phrases, such as *either . . . or*, *both . . . and*, and *if . . . then*. The coordinate phrase in

(22) Either Bill brought Mary and Susan brought Sam or Susan brought Bob.

makes the intended grouping clear. Using the dictionary

M: Bill brought Mary

S: Susan brought Sam

B: Susan brought Bob

we should symbolize it as $(M \cdot S) \vee B$. The coordinated connective takes priority over any connective appearing within the coordinated sections.

Finally, English allows a device that logicians call "telescoping" and linguists call "reduction." The statement

(23) Susan brought Sam or Susan brought Bob.

for example, "reduces" to the shorter statement

(24) Susan brought Sam or Bob.

Similarly,

(25) Fred likes Wanda and Kim likes Wanda.

reduces to

(26) Fred and Kim like Wanda.

These techniques are combined in the example from Sextus Empiricus. The first premise of the argument is

(27) If Socrates died, he died either when he was living or when he was dead.

The comma after *died* emphasizes the break between *If Socrates died* and the rest of the statement. The coordinate phrase *either . . . or* acts to group *when he was living or when he was dead* as a unit, as does the telescoping of *he died when he was living or he died when he was dead* to *he died when he was living or when he was dead*. All these cues suggest that the proper grouping is $S \supset (L \vee D)$.

The three steps thus lead us to symbolize Sextus's skeptical argument as

(28) $S \supset (L \vee D)$

  $\sim L$

  $\sim D$

  $\therefore \sim S$

To see a more complex example, consider this passage from the *Magna Carta*, which limits inheritance taxes:

(29) If any of our earls, or barons, or others who hold of us in chief by military service, shall die, and at the time of his death his heir shall be of full age, and owe a relief, he shall have his inheritance by the ancient relief. . . .

First, we identify English connectives:

(29) *If* any of our earls, *or* barons, *or* others who hold of us in chief by military service, shall die, *and* at the time of his death his heir shall be of full age, *and* owe a relief, he shall have his inheritance by the ancient relief. . . .

Second, we identify atomic statement components and replace distinct components with distinct statement letters.

Dictionary: E: An earl dies

B: A baron dies

M: Another who holds of us in chief by military service dies

F: At the time of death the heir is of full age

R: At the time of death the heir owes a relief

H: The heir has his inheritance by the ancient relief

(30) *If* E, *or* B, *or* M, *and* F, *and* R, H

Third, we replace English with symbolic connectives to obtain

(31) E ∨ B ∨ M, · F, · R, ⊃ H

The only problem is determining where the ⊃ should go. Here it is in the only position where the English word *then* would make sense.

Finally, we determine grouping. The telescoping of disjuncts in *If any of our earls, or barons, or others who hold of us in chief by military service, shall die* tells us that they should be grouped together:

(32) (E ∨ B ∨ M) · F, · R, ⊃ H

Now (E ∨ B ∨ M) is not a formula; although *or* is multigrade, ∨ is not. We must group the disjunction as ((E ∨ B) ∨ M) or (E ∨ (B ∨ M)). It makes no difference which we choose, for they are equivalent.

Our placement of the arrow tells us that it must be the main connective:

(33) (((E ∨ B) ∨ M) · F, · R) ⊃ H

Finally, the comma after *of full age* suggests that we should group F with the disjunction

(34) ((((E ∨ B) ∨ M) · F) · R) ⊃ H

although the symbolization

(35) (((E ∨ B) ∨ M) · (F · R)) ⊃ H

is equivalent.

P R O B L E M S

Symbolize each of these statements in our symbolic language, using the dictionary provided.

1. Life is either daring adventure or nothing. (Helen Keller)
   A: Life is daring adventure; N: Life is nothing.

2. God is, and all is well. (John Greenleaf Whittier)
   A: God is; W: All is well.

3. Wealth serves a wise man, but commands a fool. (Thomas Fuller)
   A: Wealth serves a wise man; C: Wealth commands a fool.

4. . . . if there is no love, nothing is possible. (Erich Fromm)
   A: There is love; P: Something is possible.

5. Life is not a spectacle or a feast; it is a predicament. (George Santayana)
   A: Life is a spectacle; F: Life is a feast; P: Life is a predicament.

6. Life is a misery if you do not get more than you deserve. (Harry Oppenheimer)
   A: Life is a misery; G: You get more than you deserve.

7. If you ain't got no boots, it is tough to lift yourself up by your bootstraps. (Jack Kemp)
   A: You have boots; T: It is tough to lift yourself up by your bootstraps.

8. You can't expect to win unless you know why you lose. (Benjamin Lipson)
   A: You expect to win; K: You know why you lose.

9. This is a world of action, and not of moping and droning in. (Charles Dickens)
   A: This is a world of action; M: This is a world of moping; D: This is a world of droning in.

10. Opposition enflames the enthusiast, never converts him. (Johann Schiller)
    A: Opposition enflames the enthusiast; C: Opposition converts the enthusiast.

11. Though he treads upon the tiger's tail, it does not bite him. *(I Ching)*
    A: He treads upon the tiger's tail; B: The tiger bites him.

12. If you do not know how to spend money, you are not rich. (Mohammed Mannei)
    A: You know how to spend money; R: You are rich.

13. One either meets or one works. One cannot do both at the same time. (Peter Drucker)
    A: One meets; W: One works.

14. I never give them hell. I just tell the truth and they think it is hell. (Harry Truman)
    A: I give them hell; T: I tell the truth; K: They think it is hell.

15. God cannot alter the past, but historians can. (Samuel Butler)
    A: God can alter the past; H: Historians can alter the past.

Symbolize each of these statements in our symbolic language. Devise and display your own dictionary, exposing as much truth-functional structure as possible.

16. It does not depend on size, or a cow would catch a rabbit. (Pennsylvania German proverb)

17. I am opposed to millionaires, but it would be dangerous to offer me the position. (Mark Twain)

18. The trouble ain't that there is too many fools, but that the lightning ain't distributed right. (Mark Twain)

19. Democracy means government by discussion, but it is only effective if you can stop people talking. (Clement Richard Attlee)

20. Thrift is care and scruple in the spending of one's means. It is not a virtue, and it requires neither skill nor talent. (Immanuel Kant)

21. The idea that all wealth is acquired through stealing is popular in prisons and at Harvard. (George Gilder)

22. One will not go too far wrong if one attributes extreme actions to vanity, average ones to habit, and petty ones to fear. (Friedrich Nietzsche)

23. Doing is overrated, and success undesirable, but the bitterness of failure even more so. (Cyril Connolly)

24. Ambition has its disappointments to sour us, but never the good fortune to satisfy us. (Benjamin Franklin)

25. Though men pride themselves on their great actions, often they are not the result of any great design but of chance. (La Rochefoucauld)

26. The race is not always to the swift, nor the battle to the strong— but that's the way to bet. (Damon Runyan)

27. If other people are going to talk, conversation becomes impossible. (James McNeill Whistler)

28. The Christian ideal has not been tried and found wanting. It has been found difficult and left untried. (G. K. Chesterton)

29. If all the year was playing holidays, to sport would be as tedious as to work. (Shakespeare)

30. I am not sure just what the unpardonable sin is, but I believe it is a disposition to evade the payment of small bills. (Kin Hubbard)

■ 6.4   A SYMBOLIC LANGUAGE

A central strategy of logic is constructing symbolic or artificial languages.[6] Armed with such a language, we can evaluate arguments as valid or invalid by symbolizing them—by translating them into the symbolic language—and then testing the resulting **symbolic argument** for validity. In the preceding section, we began using a symbolic language that allows us to evaluate arguments depending on truth-functional connectives. In this section, we develop it more rigorously.

To use any language, natural or symbolic, we need to understand its *syntax* and its *semantics*. The **syntax** of a language consists of its vocabulary and grammar. The vocabulary consists of the meaningful signs of the language. In a natural language, it is a collection of words and idioms. The grammar of a language specifies how to put the meaningful signs of the language together to form statements. (In symbolic languages, we refer to these as **formulas** rather than statements.) The **semantics** of a language is its theory of meaning. It indicates what the signs in the vocabulary mean and also how the meanings of statements depend on the meanings of those signs. So far, we have been giving a semantics for, or for *interpreting*, items in our symbolic language in two ways. We have been specifying dictionaries that correlate symbolic letters with English statements. And we have been linking statements to truth values and connectives to truth functions.

The syntax of our symbolic language is simple. Vocabulary items fall into three basic categories: *statement letters, connectives,* and *parentheses.*

VOCABULARY

Statement letters:	uppercase letters, with or without numerical subscripts
Connectives:	~ · ∨ ⊃ ≡
Parentheses:	( )

The symbol ~ is the *tilde;* ·, the *dot;* ∨, the *wedge;* ⊃, the *horseshoe;* and ≡, the *triple bar.*

The grammar allows us to combine these vocabulary elements to form formulas. In presenting the rules, we must distinguish carefully between the symbolic language that we define and the language, namely English,

---

[6]Aristotle was also the first logician to use symbols. (He used letters for terms.) The Stoics made the practice common, using numbers to stand for sentences. Nevertheless, no one developed a fully symbolic logical language until the nineteenth century, when Englishman George Boole (1815–1864) saw that logic could profit from mathematical analysis and used symbols for logical connectives.

that we use to define it. We call the artificial language the **object language;** the language we use to discuss it is the **metalanguage.** In the metalanguage, we use lowercase variables, such as p, q, r, and so on, to range over items in the object language. They are in the metalanguage but stand for items in the object language. In particular, we assume that p and q stand for formulas.

Metalanguage:	p, q, r, . . .
Object language:	A, B, C, ~A, C ∨ D, . . .

A string of symbols containing a lowercase variable is not a formula of the symbolic language but part of the metalanguage. We call letters such as p and q **statement variables** (or **schematic letters**) and strings such as (p ⊃ q) **statement forms** (or **schemata**).[7] Arguments symbolized in the object language (e.g., A ∨ B; ~A; ∴ B) are **symbolic arguments;** in the metalanguage, patterns such as p ∨ q; ~p; ∴ q are **argument forms.**

Before defining the object language precisely, we want to stress two points about statement and argument forms. First, they are *forms;* they are more abstract than statements. A statement variable stands for any statement, simple or complex. So, ~p is a form, not only of ~A, for example, but also of ~(A · ~B), ~((A ⊃ B) ⊃ C), and ~((A ∨ B) ≡ (A · B)). And all the following arguments have the form of *modus ponens:* p ⊃ q, p, ∴ q.

A ⊃ B	~A ⊃ (B ∨ C)	(A · ~B) ⊃ ~(C ≡ ~(D ⊃ E))
A	~A	A · ~B
∴ B	∴ B ∨ C	∴ ~(C ≡ ~(D ⊃ E))

Second, argument forms are valid if and only if all arguments of that form are valid.

FORMATION RULES

1. Any statement letter is a formula.

2. If p is a formula, then ~p is a formula.

3. If p and q are formulas, then (p · q), (p ∨ q), (p ⊃ q), and (p ≡ q) are formulas.

4. Every formula can be constructed by a finite number of applications of these rules.

---

[7]The singular form of *schemata* is *schema.* We assume that sentence connectives and other vocabulary items of our symbolic language are names of themselves. The string (p ⊃ q) thus refers to the formula resulting from concatenating a left parenthesis, the formula p, an occurrence of the conditional connective, the formula q, and a right parenthesis. We cannot refer to this formula by using ordinary quotation: '(p ⊃ q)' signifies the string of symbols quoted, that is, a left parenthesis, followed by the letter 'p', followed by an occurrence of the conditional, followed by the letter 'q', followed by a right parenthesis.

A formula is **atomic** if and only if it contains no connectives. Atomic formulas are statement letters. We adopt the convention of dropping the outside parentheses of a formula, since they do no further work.

The definition of a formula allows us to tell how a formula can be built from statement letters. Any formula constructed in building a formula is a **subformula** of it.

There is an important theoretical difference between the syntax of our symbolic language and the syntax of a natural language. As we have seen in Chapter 3, some English statements, such as *They watched the fireworks explode on the porch,* can be constructed in more than one way. As a result, they are ambiguous. No such ambiguities arise in our symbolic language. Every formula can be constructed in one and only one way.

Two additional concepts, *scope* and *main connective,* prove useful.

> The **scope** of a connective occurrence in a formula is the connective occurrence itself, together with the subformulas (and any parentheses) it links.

Specifically, the scope of ~ in a formula or subformula of the form ~p is ~p; the scope of · in a formula or subformula of the form (p · q) is (p · q); and similarly for the other connectives.

> The **main connective** of a formula is the connective occurrence in the formula with the largest scope.

The main connective is always the connective occurrence having the entire formula as its scope.

CONNECTIVE OCCURRENCE	FORMULA	SCOPE	MAIN CONNECTIVE
~	(A ⊃ ~B)	~B	⊃
~	~(A ⊃ B)	~(A ⊃ B)	~
∨	(A ∨ (B · C))	(A ∨ (B · C))	∨
·	(A ∨ (B · C))	(B · C)	∨

P R O B L E M S

Classify each of these as *(a)* a formula, according to the official formation rules, *(b)* a formula, given our convention about dropping outside parentheses, *(c)* a statement form in the metalanguage, or *(d)* none of these. If you answer *(a)* or *(b)*, identify the main connective.

1. ~A

2. A · B

3. (A ∨ B)

4. ~a ∨ ~b

5. ~(p · ~q)

6. ~(G)

7. (~A)

8. (~F ∨ ~H)

9. (·J ∨ ·R)

10. (K ∨ M) ⊃ C

11. (S ∨ (B ⊃ C))

12. T ∨ B ⊃ U

13. (V ∨ B ⊃ W)

14. (A ∨ Y) ⊃ C

15. B ∨ (A ⊃ (C ⊃ (Q ∨ ~B)))

16. The rules for forming formulas of propositional logic admit no ambiguity. Give an example of a set of rules that would allow ambiguity, together with an ambiguous formula constructed according to those rules.

■ 6.5 LOGICAL PROPERTIES OF STATEMENTS

We use logic primarily to investigate logical connections among statements. Nevertheless, we can also use it to classify individual statements. The following statements, depending on what the facts are, could be either true or false. It is possible to conceive of cases in which they would be true and other cases in which they would be false.

(36) The next president will be a Democrat.

(37) Nicholas is flying to Santa Fe this weekend.

Such statements are *contingent:*

> A statement is **contingent** if and only if it is possible for it to be true and possible for it to be false.

Contingent statements could be true, given the right set of circumstances. Of course, they could also be false, depending on the facts.

Some statements, in contrast, cannot help being true. It is simply impossible for them to be false. Such statements are called *tautologies.*

> A statement is a **tautology** (or **tautologous**, or **logically true**) if and only if it is impossible for it to be false.

If you doubt that there are any statements that cannot be false, no matter what the facts may be, then try to imagine circumstances in which these statements are false.

(38) A rose is a rose.

(39) When you're hot, you're hot.

These statements are true in every possible circumstance. They also seem to say very little. But not all tautologies are so straightforward. Statement (40), for example, is logically true:

(40) If everyone loves a lover and Sam does not love Jeanne, then Jeanne does not love Greg.

Tautologies can be useful. They often set up the structure of an argument.

Tautologies and contingent statements are **satisfiable:** They could be true. Some statements could never be true. They are false, regardless of the facts. These statements are *contradictory* (or *contradictions*).

> A statement is **contradictory** if and only if it is impossible for it to be true. (Otherwise, the statement is *satisfiable.*)

Here are some examples of contradictions:

(41) Seventeen is both odd and even.

(42) Nobody's seen the trouble I've seen.

In no conceivable circumstance could these statements be literally true. Try, for example, to imagine a situation in which seventeen is both odd and even. And, since *I've* seen the trouble I've seen, somebody (namely, me) has indeed seen the trouble I've seen.

Contradictions tend to signal that we should interpret some terms generously, since we assume that our colleagues in communication are trying to say something that could be true. Hearing (42), then, we tend to read *nobody* as *nobody else*. Contradictions nevertheless often fulfill important functions in arguments.

Since every statement is either tautologous, contingent, or contradictory, the terms introduced in this section divide statements into three groups, as shown in the following diagram.

Statements

Tautologous (true in every circumstance)	**Contingent** (true in some circumstances, false in others)	**Contradictory** (false in every circumstance)

Satisfiable
(true in some circumstances)

P R O B L E M S

Classify these statements as tautologous, contradictory, or contingent.

1. Austin is in Texas.

2. Some cats are calicos.

3. All calicos are calicos.

4. Some calicos are not calicos.

5. I know everything I know.

6. Some cats eat fish and some do not.

7. Some cats eat fish and do not eat fish.

8. Most cats eat fish and most do not.

9. There are many trees in Cibola National Forest.

10. Some bachelors are unhappy.

11. All bachelors are unmarried.

12. A clever lie will not work unless it's clever.

13. . . . what we are, we are. . . . (Alfred, Lord Tennyson)

14. A lie is a lie, no matter how ancient; a truth is a truth, though it was born yesterday. (American proverb)

15. Nay, Sir, argument is argument. (Samuel Johnson)

16. . . . if it does not work, it does not work. (Prince Charles)

17. Bigness is bigness in spite of a hundred mistakes. (Jawaharlal Nehru)

18. All babies are young. (Benjamin Spock)

19. When people are out of work, unemployment results. (Calvin Coolidge)

20. Our past has gone into history. (William McKinley)

21. The nobles are to be considered in two different manners; that is, they are either to be ruled so as to make them entirely dependent on your fortunes, or else not. (Niccolo Machiavelli)

22. Sudden death, though fortunately it is rare, is frequent. *(British Medical Journal)*

23. Nobody goes there anymore; it's too crowded. (Yogi Berra)

24. "Everybody that hears me sing it—either it brings tears to their eyes, or else—"
    "Or else what?" said Alice, for the Knight had made a sudden pause.
    "Or else it doesn't, you know." (Lewis Carroll)

Sometimes classifying a statement as tautologous, contingent, or contradictory is not easy. Classify these statements and explain your classification. Would another classification be defensible?

25. All trees in Cibola National Forest are trees.

26. I'll see you when I see you.

27. Some people know nobody.

28. The death of Francis Macomber was a turning point in his life. (student blooper, compiled by Jim Mattison)

29. The difference between a king and a president is that a king is the son of his father, but a president is not. (student blooper, compiled by Jim Mattison)

30. He is audibly tan. (Fran Lebowitz)

31. Just because everything is different does not mean anything has changed. (Irene Peter)

32. It is a foolish man that hears all he hears. (Austin O'Malley)

33. If there's one major cause for the spread of mass illiteracy, it is the fact that everybody can read and write. (Peter de Vries)

34. There are two kinds of people in the world: those who divide the world into two kinds of people, and those who do not. (H. L. Mencken)

35. All seven of the patients who died never completely recovered. *(Medicine)*

36. Say that statement A implies another statement B. What can we conclude about B, if A is *(a)* a tautology? *(b)* contingent? *(c)* satisfiable? *(d)* contradictory?

37. Say that statement A implies another statement B. What can we conclude about A, if B is *(a)* a tautology? *(b)* contingent? *(c)* satisfiable? *(d)* contradictory?

38. Say that statement A is equivalent to statement B. What can we conclude about A, if B is *(a)* a tautology? *(b)* contingent? *(c)* satisfiable? *(d)* contradictory?

The Englishman William of Ockham (1285–1349), perhaps the most influential philosopher and logician of the fourteenth century, recorded eleven rules of logic in a chapter of his *Summa Totius Logicae*. Four use concepts from this section. Say whether each is true and explain why.

39. The contingent does not follow from the tautologous.

40. The contradictory does not follow from the satisfiable.

41. Anything whatsoever follows from the contradictory.

42. The tautologous follows from anything whatsoever.

6.6   TRUTH TABLES FOR STATEMENTS

The definitions of the logical connectives specify the truth values of complex formulas on the basis of the values of their components. This allows us to compute the truth value of a formula of any length, given the values of the atomic formulas—that is, statement letters—appearing in the formula. That, in turn, lets us classify formulas as tautologous, contingent, or contradictory.

To take an example, suppose that P and Q are true but R is false. What is the truth value of the formula $((P \supset Q) \supset R)$? Its main connective is the second horseshoe. The antecedent of the conditional is another conditional, both of whose components are true; the definition of the conditional indicates that this smaller conditional, P ⊃ Q, is true. The larger conditional, then, has a true antecedent, but its consequent, R, is false. The definition of the conditional tells us that the larger conditional is false.

Approaching this problem more systematically, we might list the values of the statement letters first and then proceed to generate the values of subformulas until we reach the formula as a whole. This would lead us to the table on the right and, perhaps, to the more compressed table below it:

P Q R	(P ⊃ Q)	((P ⊃ Q) ⊃ R)
T T F	T T T	T T T F F
P Q R	((P ⊃ Q) ⊃ R)	
T T F	T T T **F** F	

A boldface letter in the tables represents the truth value of the entire formula above it. In constructing these tables, we begin with the statement letters. We then assign values to subformulas, working our way, gradually, to the formula itself. This has the effect that we work from inside parentheses out. Whenever two subformulas are equally "deep" inside parentheses, it makes no difference which we attack first. Our definitions thus allow us to compute the value of a formula, given truth values for its statement letters. This is the central idea behind the method of *truth tables.*

A **truth table** is a computation of the truth value of a formula or set of formulas under each combination of truth values for its statement letters.[8] Recall that tautologies are true in all circumstances; contingent statements, true in some and false in others; and contradictory statements, false in all circumstances. In our symbolic language, there is a direct and useful way to represent circumstances. Our symbolic language is a language of truth functions. All its connectives are truth-functional. This means that it uses nothing about a statement beyond its truth value. So, in our symbolic language, we can represent circumstances by combinations of truth values. To survey all possible circumstances within truth-functional propositional logic, we just need to survey all possible combinations of truth and falsehood. In analyzing formulas and symbolic arguments for validity, therefore, we will speak about possible combinations of truth values.

---

[8]Building on the work of German mathematician Gottlob Frege (1848–1925), the American logician Emil Post and the Austrian philosopher Ludwig Wittgenstein (1889–1951) developed the truth table method independently in 1921.

Truth tables classify formulas and their corresponding statements as tautologous, contingent, or contradictory. If a formula is a tautology, then the main column of a truth table for it will be a string of Ts. If it is contradictory, the main column will be a string of Fs. If it is contingent, finally, the main column will be a string containing both Ts and Fs. Truth tables amount simply to several simple tables, of the kind we just saw, done at once. But they allow us to evaluate any formula of propositional logic as valid, contradictory, or contingent.

A truth table for a formula consists of four elements:

1. a list of the formula's statement letters,
2. the formula itself,
3. a list of the possible combinations of truth values of the statement letters, and
4. a computation of the formula's truth value on each combination.

Truth tables have this form:

STATEMENT LETTERS	FORMULA
Combinations of truth values	Computation

The top of a table lists the formula to be evaluated, preceded by the statement letters in the formula.

To see how this works, consider an example.

(43) Either Paul is behind the movement or he is not.

Is the formula $P \vee \sim P$ valid, contradictory, or contingent? We can set up column heads of the table by listing the statement letters in the formula, followed by the formula.

P	$P \vee \sim P$

Under the statement letters, the table lists the possible combinations of their truth values. If there is one statement letter, there are only two possibilities—truth and falsehood. Two letters have four possible combinations, three letters have eight possible combinations, and so on. (In general, n statement letters have 2^n combinations.)

Returning to our example, we have only one statement letter. Listing the two possible combinations :

P	$P \vee \sim P$
T	
F	

Now, we copy the truth values for each letter under its occurrences in the formula.

P	P ∨ ~P
T	T T
F	F F

Then, we begin searching for subformulas as far inside parentheses as possible. Negations of single statement letters are good places to start, since the value of the negated letter depends on nothing but the value of the letter itself. As a rule, compute values for negations of single statement letters first. Then, compute values for subformulas, working from inside parentheses out.

In the example, we begin with the negation

P	P ∨ ~P
T	T F T
F	F T F

The last computation should be for the main connective of the entire formula. So, we can conclude our truth table by computing the value for the disjunction. Since the ∨ is the formula's main connective, we are now computing the truth value of the entire formula on each combination of truth values.

P	P ∨ ~P
T	T **T** F T
F	F **T** T F

This tells us that the formula P ∨ ~P is true, whether P is true or false. That tells us that P ∨ ~P is a logical truth. In fact, it is an instance of a well-known logical principle called *the law of excluded middle*, p ∨ ~p.

In summary, to compute the truth value of a formula under all circumstances,

1. Copy the truth value of each statement letter under its occurrences in the formula.
2. Compute the values of negations of single statement letters.
3. Compute values of subformulas, working from inside parentheses out.

Under each statement letter and connective of the formula, a completed table will have a column of Ts and Fs. These represent the truth value the formula or subformula has under each combination of its components' truth values. The column for the entire formula itself—the last to be filled in—is the table's *main column*. It specifies the truth value of the formula on each combination. In the previous table, and throughout this chapter, the main column is in boldface type.

> The **main column** of a truth table is the column under the main connective of the formula at the top of the table.

A tautology is true on all combinations of its components' truth values; the main column of a table for it, therefore, should contain all Ts. A contradictory formula is false on all such combinations; the main column of a table for it, therefore, should contain all Fs. The main columns of tables for contingent formulas, finally, contain both Ts and Fs. To summarize:

MAIN COLUMN	FORMULA
All Ts	Tautologous
All Fs	Contradictory
Ts and Fs	Contingent

P R O B L E M S

Calculate the truth values of these formulas on the interpretations listed.

1. $(A \lor B) \cdot (A \lor \sim C)$ (A and B false; C true)

2. $(A \supset (B \equiv C)) \supset (B \equiv \sim C)$ (A and C true; B false)

3. $((C \supset B) \cdot \sim A) \equiv (\sim B \cdot A)$ (A true; B and C false)

4. $\sim((A \cdot \sim B) \lor (B \supset \sim C))$ (A false; B and C true)

5. $(A \equiv \sim C) \equiv (C \equiv B)$ (A, B, and C true)

6. $\sim(A \lor B) \supset \sim(A \cdot C)$ (A, B, and C false)

7. $(A \equiv (B \cdot C)) \lor (A \supset B)$ (A and B true; C false)

8. $(\sim A \equiv \sim B) \equiv \sim C$ (A and C true; B false)

9. $\sim(\sim(\sim A \supset \sim B) \lor \sim C)$ (A true; B and C false)

10. $\sim((\sim B \lor \sim A) \supset \sim C)$ (A false; B and C true)

11. $(A \equiv \sim(B \lor C)) \equiv \sim A$ (A and B false; C true)

12. $(A \cdot \sim B) \equiv (A \supset \sim C)$ (A and C false; B true)

13. $(\sim(A \vee C) \equiv \sim B) \supset \sim B$ (A, B, and C false)

14. $\sim((A \equiv \sim B) \equiv \sim C)$ (A, B, and C true)

15. $A \vee \sim(\sim(A \supset \sim A) \cdot A)$ (A true)

Display truth tables for these formulas and state whether they are tautologous, contradictory, or contingent.

1. $P \supset P$

2. $P \supset \sim P$

3. $P \equiv \sim P$

4. $\sim(P \cdot \sim P)$

5. $\sim(\sim P \supset Q)$

6. $P \supset (Q \supset P)$

7. $(Q \supset P) \supset Q$

8. $(\sim P \supset P) \equiv P$

9. $\sim P \supset (Q \supset P)$

10. $P \supset (Q \cdot P)$

11. $(P \supset Q) \vee (Q \supset P)$

12. $(P \supset Q) \cdot \sim(Q \supset P)$

13. $(P \supset Q) \cdot (P \cdot \sim Q)$

14. $(P \cdot Q) \supset (P \vee Q)$

15. $(P \vee Q) \equiv \sim(\sim P \cdot \sim Q)$

16. $(Q \vee R) \supset ((P \supset Q) \vee (P \supset R))$

17. $(Q \cdot R) \equiv ((P \equiv Q) \vee (P \equiv R))$

18. $(P \vee (Q \supset R)) \equiv ((P \vee Q) \supset (P \vee R))$

19. $(P \cdot (Q \vee R)) \equiv ((P \cdot Q) \vee (P \cdot R))$

20. $(P \supset (Q \vee R)) \equiv ((P \supset Q) \vee (P \supset R))$

Which of these principles, advanced by the Stoics, are tautologies?

21. $((P \supset \sim Q) \cdot P) \supset \sim(P \supset Q)$

22. $((\sim P \supset Q) \cdot \sim Q) \supset \sim(P \supset Q)$

Boethius (480–524), philosopher and Roman consul, translated Aristotle's logical works into Latin and advanced logical principles of his own, including these concerning the conditional. Are they tautologies in modern logic?

23. $(P \supset {\sim}Q) \equiv {\sim}(P \supset Q)$

24. $((P \supset (Q \supset R)) \cdot (Q \supset {\sim}R)) \supset {\sim}P$

25. Suppose that p is a tautology. What can we conclude about the logical properties of
(a) ~p; (b) p ∨ q; (c) p · q; (d) p ⊃ q; (e) q ⊃ p; (f) p ≡ q?

26. Suppose that p is contradictory. What can we conclude about the logical properties of
(a) ~p; (b) p ∨ q; (c) p · q; (d) p ⊃ q; (e) q ⊃ p; (f) p ≡ q?

27. Suppose that p is contingent. What can we conclude about the logical properties of
(a) ~p; (b) p ∨ q; (c) p · q; (d) p ⊃ q; (e) q ⊃ p; (f) p ≡ q?

28. Suppose that p and q are both tautologies. What can we conclude about the logical properties of
(a) ~p; (b) p ∨ q; (c) p · q; (d) p ⊃ q; (e) q ⊃ p; (f) p ≡ q?

29. Suppose that p and q are both contradictory. What can we conclude about the logical properties of
(a) ~p; (b) p ∨ q; (c) p · q; (d) p ⊃ q; (e) q ⊃ p; (f) p ≡ q?

30. Suppose that p and q are both contingent. What can we conclude about the logical properties of
(a) ~p; (b) p ∨ q; (c) p · q; (d) p ⊃ q; (e) q ⊃ p; (f) p ≡ q?

■ 6.7   TRUTH TABLES FOR SYMBOLIC ARGUMENTS

Our symbolic language allows us to evaluate arguments as valid or invalid by symbolizing them and then testing the resulting *symbolic argument* for validity. Recall the diagram presented earlier in this chapter:

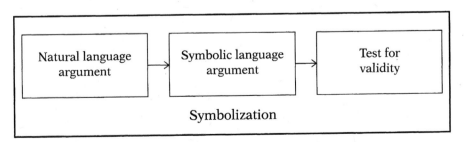

First, we symbolize statements or arguments in our symbolic language. Second, we evaluate the results for validity or other logical properties.

Sequences of statements, when symbolized in our symbolic language, become sequences of formulas, or *symbolic arguments.*

> A **symbolic argument** consists of a finite sequence of formulas, called its **premise formulas,** together with another formula, its **conclusion formula.**

The argument from Sextus Empiricus, for instance, we symbolized as

(44) $S \supset (L \vee D)$

    ~L

    ~D

    ∴ ~S

This *corresponds to* Sextus's argument.

Under what circumstances is a symbolic argument valid? An argument is valid if its conclusion is true in every circumstance in which its premises are true. Similarly,

> A symbolic argument is valid if its conclusion formula is true in every circumstance in which its premise formulas are true.

Truth tables let us calculate the truth values of formulas under every circumstance. So, a truth table can give us the information we need to evaluate a symbolic argument.

We can set up a single table that computes values for each of the premise formulas, and the conclusion formula, separately. We can then use the table to evaluate the symbolic argument. We

1. list the statement letters appearing in the symbolic argument;
2. beneath them, list all possible combinations of their truth values;
3. list each premise formula, and then the conclusion formula, as column heads; and
4. compute the value of each formula.

The computation produces a main column for each formula.

> A symbolic argument is valid just in case no row of the table makes the premise formulas all true but the conclusion formula false.

Here are some sample symbolic arguments corresponding to simple English arguments.

(45) People will buy the product if the advertising reaches them. The advertising will reach them; so, people will buy the product.

(46) A ⊃ B

A

∴ B

(47) The cost of insurance will go down only if damage awards decrease. Those awards will decrease if and only if legislation to limit them passes. Hence, the cost of insurance will go down if legislation to limit damage awards passes and those awards decrease.

(48) C ⊃ D

D ≡ L

∴ (L · D) ⊃ C

It is easy to evaluate them for validity.

A B	A	(A ⊃ B)	B
T T	T	T T T	T
T F	T	T F F	F
F T	F	F T T	T
F F	F	F T F	F

Argument (46) is valid; on no row are the premises all true while the conclusion is false. It is an instance of the best-known logical rule, *modus ponens*: that from p and p ⊃ q we may infer q.

This table evaluates (48).

C D L	(C ⊃ D)	(D ≡ L)	((L · D) ⊃ C)
T T T	T T T	T T T	T T T T T
T T F	T T T	T F F	F F T T T
T F T	T F F	F F T	T F F T T
T F F	T F F	F T F	F F F T T
F T T	F T T	T T T	T T T F F
F T F	F T T	T F F	F F T T F
F F T	F T F	F F T	T F F T F
F F F	F T F	F T F	F F F T F

Argument (48) is invalid; the combination

C	D	L
F	T	T

makes the conclusion formula false but the premise formulas all true. Transferring this information to the English argument (47): Even given the truth of the premises, it could happen that legislation to limit damage awards passes, that those awards decrease, but that nevertheless the cost of insurance does not go down. (The companies may keep the savings as increased profit, the number of accidents may increase, or. . . .)

When a truth table shows a symbolic argument to be invalid, it specifies a combination of truth values making the premise formulas true and the conclusion formula false. The table indicates that there is such a combination and, moreover, tells us what it is. One such row is enough to show the symbolic argument to be invalid. But if there are several, it specifies them all.

IMPLICATION

In a valid symbolic argument, the premise formulas imply the conclusion formula. Consequently, truth tables can evaluate implication. To find out whether a set of formulas $\{p_1, \ldots, p_n\}$ implies a formula q, we can construct a table for the symbolic argument

p_1

.

.

.

p_n

∴ q

To determine whether

(49) Either the filmmakers had a very low opinion of their audience, or they were just incompetent.

implies

(50) If the filmmakers had a very low opinion of their audience, they were not incompetent.

for example, we can ask whether L ∨ I implies L ⊃ ~I. We list the letters L and I, write the four possible combinations of truth values below them, and write the two formulas as column heads.

L	I	L ∨ I	L ⊃ ~I
T	T	T **T** T	T **F** FT
T	F	T **T** F	T **T** TF
F	T	F **T** T	F **T** FT
F	F	F **F** F	F **T** TF

In the first row, L ∨ I comes out true, but L ⊃ ~I is false. Thus, L ∨ I does not imply L ⊃ ~I. The combination assigning truth to both L and I makes L ∨ I true but L ⊃ ~I false. The implication between the English statements does not hold, then, because the filmmakers might have had a very low opinion of their audience *and* been incompetent.

Equivalence is implication in both directions. Two formulas are equivalent if they have the same truth value in each combination of their components' truth values.

To show that p is equivalent to ~~p, for example, we can construct the table

p	p	~~p
T	**T**	T**F**T
F	**F**	F**T**F

The table shows that p and ~~p agree in truth value under any circumstance; they are equivalent.

STRENGTHS AND LIMITATIONS

Truth tables are simple but remarkably powerful tools for solving problems in propositional logic. They can evaluate arguments for validity, assess the logical properties of formulas, and determine implication and equivalence. They achieve all this, furthermore, in a clear and easily understandable way. Truth tables constitute a *decision procedure* for validity in our symbolic language.

> A **decision procedure** for a property is a mechanical method for determining, in a finite number of steps, whether any given thing has or does not have that property.

Decision procedures are completely mechanical; using them involves no more than following rules. They require no ingenuity or creativity. A decision procedure, applied to an object, always gives a *yes* or *no* answer after a finite time. Truth tables, clearly, are decision procedures for validity, equivalence, and other logical properties. We can construct tables by following rules; the tables always have a finite size. If there is a decision procedure for a property, the property is **decidable.** In our symbolic language, formula and symbolic argument validity, formula contradictoriness and contingency, implication, and equivalence are all decidable.

The truth table technique's clarity comes at a price. When a formula or symbolic argument is very complex, the number of possible combinations may be very large. Truth tables do not always reach a decision within a reasonable amount of time. The size of a truth table grows exponentially with the number of statement letters involved; someone who began a truth table for a problem with 139 statement letters at the moment of the big bang would still not be finished.[9] In the next chapter, therefore, we develop a method that answers questions about validity, implication, and so on, without examining every possible way of assigning truth values to statement letters.

Truth tables are decision procedures for formula and symbolic argument validity in our symbolic language. How does the validity of its symbolization in our symbolic language, however, bear on the validity of an argument in natural language? Recall that our strategy is twofold:

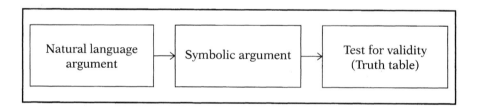

Truth tables can tell us whether any given symbolic argument is valid or invalid in our symbolic language. What can we infer from this verdict about the validity of the original English argument?

The connectives of our symbolic language do not match English connectives in every respect. The symbolic connectives approximate their natural language counterparts closely, but the fit is not exact. So, especially when an argument contains conditionals, we need to check whether its premises and conclusion would be true under the same conditions as the formulas symbolizing them.

Suppose that there is no problem about the match between English and symbolic connectives. Then, if a symbolic argument resulting from our process of symbolization is valid in our symbolic language, the original argument it symbolizes is valid. But an invalid symbolic argument

---

[9]Christopher Cherniak, *Minimal Rationality* (Cambridge: Bradford Books), p. 89.

does not show that the original argument was invalid. The symbolic languages logicians use are only partial theories of logical relations. Logic aims at a comprehensive theory that would take account of all such relations, but no such theory yet exists. So, in particular, our symbolic language accounts for only some cases of validity. Many English arguments depend on logical relationships that propositional logic does not capture. Consider, for example, the argument

(51) All captains are officers.

Some of the troops caught behind enemy lines are captains.

∴ Some of the troops caught behind enemy lines are officers.

This argument contains no statement connectives: We can symbolize it in our symbolic language, at best, as

(52) C

T

∴ O

This is a paradigm of an invalid symbolic argument. We can conclude, not that the English argument is invalid, but that it is *invalid relative to our symbolic language.* It is valid relative to the Aristotelian language of syllogisms discussed in Chapters 4 and 5.

Therefore, in drawing conclusions about natural language arguments, "valid" truth table verdicts merit more trust than "invalid" verdicts. The same holds of other concepts. "Contradictory" verdicts merit more trust than "Satisfiable" verdicts, and "Implies" and "Are Equivalent" verdicts merit more trust than their opposites. It is always possible that to establish the validity of an argument, the equivalence of two statements, and so on, we would have to use more machinery than the logical theory we're working in provides.

P R O B L E M S

First, symbolize each argument in our symbolic language, displaying your dictionary. Then display a truth table and say whether the corresponding symbolic argument is valid.

1. Larry is either a knave or a fool. Larry is a knave, so he's no fool.

2. If I'm right, then I'm a fool. But if I'm a fool, I'm not right. Therefore, I'm no fool.

3. Unless I'm mistaken, I'm a fool. But if I am a fool, I must be mistaken. So I'm mistaken.

4. Hank will believe you only if he's incredibly gullible. Hank will not believe you. Therefore, he is not incredibly gullible.

5. If Kim does well on her GREs, she'll get into her first-choice school. She will get into her top choice. It follows that she'll do well on her GREs.

6. The members of this department agree when and only when the issue at hand is of no significance. This issue is significant, but the members nevertheless agree. Therefore, the end of the earth is at hand.

7. Sarah knows that the company is unlikely to promote her. She will look for another job if and only if she's unhappy. She's bound to be unhappy if she knows that she's unlikely to be promoted. So, Sarah will look for another job.

8. If nobody catches on to Rudy's scam, he'll make a lot of money, but other people will lose a lot of cash. Rudy will not make a lot of money. It follows that somebody will figure out Rudy's scam.

9. Nothing can be conceived as greater than God. If God existed in our imaginations, but not in reality, then something would be conceivable as greater than God (namely, the same thing, except conceived as existing in reality). Therefore, if God exists in our imaginations, He exists in reality. (Anselm)

10. Either we ought to philosophize or we ought not. If we ought, then we ought. If we ought not, then also we ought (to justify this view). Hence in any case we ought to philosophize. (Aristotle)

Certain argument forms, such as *modus ponens,* are so common that they have well-recognized names. Some are basic rules of the deduction system presented in Chapter 8. Determine whether each of the following is valid.

11. Simplification: p · q, ∴ p

12. Conjunction: p, q, ∴ p · q

13. Addition: p, ∴ p ∨ q

14. Disjunctive syllogism: p ∨ q, ~q, ∴ p

15. *Modus tollens:* p ⊃ q, ~q, ∴ ~p

16. Denying the antecedent: p ⊃ q, ~p, ∴ ~q

17. Affirming the consequent: p ⊃ q, q, ∴ p

18. Hypothetical syllogism: p ⊃ q, q ⊃ r, ∴ p ⊃ r

During the Watergate hearings (July 26, 1974), the House Judiciary Committee held a brief debate on the meaning of *unless.* Representative McClory moved "to postpone for ten days further consideration of whether sufficient grounds exist for the House of Representatives to exercise constitutional powers of impeachment unless by 12 noon, eastern daylight time, on Saturday, July 27, 1974, the president fails to give his unequivocal assurance to produce forthwith all taped conversations subpoenaed by the committee. . . ." This led to a number of interpretations and attempted restatements of the motion. Which of the following are equivalent to the original?

19. "If he fails tomorrow, we get ten days." (Representative Latta)

20. "If he complies, we do not." (Representative Latta)

21. ". . . you get ten days providing the president gives his unequivocal assurance to produce the tapes by tomorrow noon." (Representative Latta)

22. ". . . there is no postponement for ten days if the president fails to give his assurance. . . ." (Representative McClory)

23. "If he fails to give it [assurance], there is only a twenty-four hour or there is only a twenty-three-and-a-half-hour day [i.e., there is no postponement]." (Representative McClory)

Consider the statement *If a fetus is a person, it has a right to life.* Which of the following sentences follow from this? Which imply it?

24. A fetus is a person.

25. A fetus has a right to life.

26. A fetus is a person only if it has a right to life.

27. If a fetus has a right to life, then it is a person.

28. A fetus has a right to life only if it is a person.

29. If a fetus isn't a person, it doesn't have a right to life.

30. If a fetus doesn't have a right to life, it isn't a person.

31. A fetus isn't a person only if it doesn't have a right to life.

32. A fetus doesn't have a right to life only if it isn't a person.

33. A fetus doesn't have a right to life unless it is a person.

34. A fetus isn't a person unless it has a right to life.

35. A fetus is a person unless it doesn't have a right to life.

36. A fetus has a right to life unless it isn't a person.

Chrysippus (279–206 B.C.) regarded the following symbolic arguments as basic to propositional logic. In some cases, however, he thought of connectives as having meanings different from those we have associated with them. Which of these are valid in our symbolic language?

37. P ⊃ Q; P; ∴ Q.

38. P ⊃ Q; ~Q; ∴ ~P.

39. ~(P · Q); P; ∴ ~Q.

40. P ∨ Q; P; ∴ ~Q.

41. P ∨ Q; ~Q; ∴ P.

42. Cicero (106–43 B.C.), the famous Roman orator, summarized Stoic logic by citing seven principles: the five of Chrysippus above, a repeat of the third, and ~(P · Q); ~P; ∴ Q. Is this valid in our symbolic language?

Sextus Empiricus, a Greek skeptic who wrote in the third century A.D., preserved some symbolic arguments, the validity of which the Stoics thought followed from their basic principles. Which of these are valid in our symbolic language?

43. P ⊃ (P ⊃ Q); P; ∴ Q.

44. (P · Q) ⊃ R; ~R; P; ∴ ~Q.

45. P ⊃ Q; P ⊃ ~Q; ∴ ~P.

46. P ⊃ P; ~P ⊃ P; ∴ P.

Display truth tables to determine whether the formulas in each of the following pairs are equivalent in our symbolic language. If they are not, say whether either formula implies the other.

47. ~(Q ∨ P)   and   ~Q ∨ ~P

48. ~(Q · P)   and   ~Q · ~P

49. ~(Q ≡ P)   and   ~Q ≡ P

50. ~(Q ⊃ P)   and   Q · ~P

51. Q · P   and   (Q ∨ P) · (Q ⊃ P)

52. Q ≡ P   and   (Q ≡ P) ≡ Q

The following equivalences are rules in the deduction system of Chapter 8. Display truth tables to show that the formulas in each pair are equivalent.

53. ~~P   and   P   (Double Negation)

54. P ∨ P   and   P   (Tautology)

55. P · P   and   P   (Tautology)

56. ~(Q · P)   and   ~Q ∨ ~P   (DeMorgan)

57. ~(Q ∨ P)   and   ~Q · ~P   (DeMorgan)

58. (P ⊃ Q)   and   (~P ∨ Q)   (Material Implication)

59. P ⊃ Q   and   ~Q ⊃ ~P   (Transposition)

60. P ⊃ Q   and   P ⊃ (P · Q)   (Absorption)

61. P · Q   and   Q · P   (Commutation)

62. P ∨ Q   and   Q ∨ P   (Commutation)

63. Q ≡ P   and   (Q ⊃ P) · (P ⊃ Q)   (Material Equivalence)

64. Q ≡ P   and   (P · Q) ∨ (~P · ~Q) (Material Equivalence)

Classify these statements as correct or incorrect about our symbolic language, and explain.

65. Symbolic arguments having tautologous conclusion formulas are valid.

66. Some symbolic arguments with contradictory premise formulas are not valid.

67. There is a formula that implies every other formula.

68. There is a formula that is equivalent to every other formula.

69. Any formula that follows from a satisfiable formula is satisfiable.

70. Any formula that implies a contingent formula is not valid.

71. Any formula that follows from a contingent formula is contingent.

72. Any formula that follows from a tautology is also a tautology.

73. Any formula that implies a tautology is also a tautology.

74. All contradictory formulas imply one another.

75. All contingent formulas imply one another.

76. All tautologies imply one another.

77. Some symbolic arguments with contradictory formulas as conclusion formulas are valid.

78. Any formula that implies its own negation is contradictory.

79. Any formula implied by its own negation is a tautology.

CHAPTER 7

SEMANTIC TABLEAUX

The method of truth tables constitutes a decision procedure for the validity of both formulas and symbolic arguments. But truth tables can be very large. Englishman Charles Dodgson (1832–1898), better known under his pen name *Lewis Carroll*, devised "Froggy's Problem." The puzzle is an argument with 18 distinct letters, requiring a table of 262,144 rows. The table as a whole would contain over 31 million Ts and Fs. A person filling in a symbol per second and working 40-hour weeks would take about four years to complete it! Furthermore, although truth tables serve as a decision procedure for propositional logic, they do not extend easily to other, more comprehensive logical systems. Once a symbolic language becomes powerful enough to symbolize the English words *some* and *all*, the truth table method breaks down.

In this section, we develop another decision procedure for propositional logic that offers many practical advantages. It matches intuitive ways of thinking about arguments more closely than truth tables. It evaluates arguments more efficiently. And it extends readily to more comprehensive logical systems.

The method is that of **semantic tableaux:** treelike diagrams that serve as tests for validity, implication, contradictoriness, and so on.[1] Semantic tableaux are semantic because they concern truth and falsehood. They test for validity and other logical relations by searching for combinations.

---

[1] *Tableaux* is the plural of *tableau*. (Both are pronounced tab-LOH.) They differ from truth trees in using one side for truth and the other for falsehood. Gerald Massey first developed the system used in this book. The Dutch logician E. W. Beth and the Finnish-American logician Jaacko Hintikka independently developed its forerunners in 1955 by simplifying a logical system that the German logician Gerhard Gentzen (1909–1945) devised in the 1930s. See Beth, "Semantic Entailment and Formal Derivability," in Hintikka, ed., *Philosophy of Mathematics* (Oxford: Oxford University Press, 1969), pp. 9–41; Hintikka, "Form

■ 7.1 MOTIVATION

Semantic tableaux formalize, in essence, the process of working a truth table backward. Truth tables waste time by looking at every possible combination of truth and falsehood for the relevant statement letters. In evaluating a symbolic argument, for example, most of these tell us nothing useful. We want to know whether any combination assigns truth to all premise formulas but falsehood to the conclusion formula. Combinations that assign falsehood to a premise formula, or truth to the conclusion formula, do not help.

Suppose, therefore, that we begin a truth table by assigning truth to each premise formula and falsehood to the conclusion formula. We might then work backward toward a combination of truth values for statement letters that would produce such an assignment. If we find one, we can conclude that the argument is invalid. If we cannot find one, we can conclude that it is valid.

Consider, for example, an argument from the prior chapter:

(1) The cost of insurance will go down only if damage awards decrease. Those awards will decrease if and only if legislation to limit them passes. Hence, the cost of insurance will go down if legislation to limit damage awards passes and those awards decrease.

(2) $C \supset D$

$D \equiv L$

$\therefore (L \cdot D) \supset C$

We constructed the following table to show that (1) and (2) are invalid.

C D L	$(C \supset D)$	$(D \equiv L)$	$((L \cdot D) \supset C)$
T T T	T T T	T T T	T T T T T
T T F	T T T	T F F	F F T T T
T F T	T F F	F F T	T F F T T
T F F	T F F	F T F	F F F T T
F T T	F T T	T T T	T T T F F
F T F	F T T	T F F	F F T T F
F F T	F T F	F F T	T F F T F
F F F	F T F	F T F	F F F T F

and Content in Quantification Theory," *Acta Philosophica Fennica* 8 (1955): 7–55; Gentzen, "An Investigation into Logical Deduction," in M. Szabo, ed., *The Collected Papers of Gerhard Gentzen* (Amsterdam: North-Holland, 1969). The method of this chapter is very close to Gentzen's original system, although Gentzen thought of the technique as purely syntactic.
    Similar tree methods have been developed by Richard Jeffrey and Raymond Smullyan.

In row five, the premise formulas are true while the conclusion formula is false. This row reflects the possibility that the cost of insurance does not go down, even though damage awards decrease and legislation to limit damage awards is enacted. So, the fifth row shows that the argument fails. The other rows, however, show nothing.

To work backwards, begin with the assumption that the premise formulas are true and the conclusion formula false:

C D L	(C ⊃ D)	(D ≡ L)	((L · D) ⊃ C)
	T	T	F

Now, what assignments to the statement letters, if any, could produce these values? A conditional is false only when its antecedent is true and its consequent is false. So, the conclusion formula, (L · D) ⊃ C, is false only when L · D is true and C is false.

C D L	(C ⊃ D)	(D ≡ L)	((L · D) ⊃ C)
F	T	T	T F F

Moreover, L · D is true only when both its components, L and D, are true.

C D L	(C ⊃ D)	(D ≡ L)	((L · D) ⊃ C)
F T T	T	T	T T T F F

Now we check to see whether these assignments yield a coherent row.

C D L	(C ⊃ D)	(D ≡ L)	((L · D) ⊃ C)
F T T	F T T	T T T	T T T F F

They do; C ⊃ D would be true if C were false and D true, and D ≡ L would be true if both D and L were true. Since it is possible for the premise formulas to be true while the conclusion formula is false, the argument is invalid.

If we evaluate a valid argument in this way, we find no coherent row. Consider the argument

(3) If the president faces a serious challenge from within his own party, he will not win the general election.

∴ The president will win the general election only if there is no serious challenge from within the party.

(4) C ⊃ ~L

∴ L ⊃ ~C

Again, we assume that the premise formula is true and the conclusion formula false:

C	L	C ⊃ ~L	L ⊃ ~C
		T	F

The conditional L ⊃ ~C is false only if L is true and ~C is false. And, ~C is false only if C is true.

C	L	C ⊃ ~L	L ⊃ ~C
T	T	T	T F FT

Now, we place those values throughout the row. Do we obtain a coherent assignment?

C	L	C ⊃ ~L	L ⊃ ~C
T	T	T T F T	T F F T

No. If L is true, then ~L is false. Thus, if C is true, the premise formula C ⊃ ~L is false. But we began assuming that the premise formula was true. Assigning the conclusion formula falsehood requires us to assign falsehood to the premise formula as well—which tells us that the argument is valid.

Working truth tables backward is more efficient than constructing them the usual way. But it can be tricky, especially when one must investigate several options. For instance, it is possible to define the conditional in terms of conjunction and the biconditional: $p \supset q$ is equivalent to $p \equiv (p \cdot q)$. If we try to show that $p \supset q$ implies $p \equiv (p \cdot q)$ by backward truth tables, however, things get messy.

p	q	p ⊃ q	p ≡ (p · q)
		T	F

We begin by assuming that $p \supset q$ is true and $p \equiv (p \cdot q)$ false. Analyzing the premise formula does not help much, for three combinations of truth values make a conditional true. Analyzing the conclusion formula is a little better, but there are two combinations of truth values that make a biconditional false. If p ≡ (p · q) is false, we can conclude only that p and p · q disagree in truth value. So, we must investigate two possibilities:

p	q	p ⊃ q	p ≡ (p · q)
		T	T F F
		T	F F T

The second, fortunately, is incoherent, for p · q cannot be true when p is false. The first is no better, for, if p is true and p · q false, q must be false. But in that case p ⊃ q, the premise formula, must be false as well.

p	q	p ⊃ q	p ≡ (p · q)
T	F	T T F	T F T F F
F	T	F T T	F F F T T

More complicated examples, clearly, can lead to much trickier chains of reasoning.

P R O B L E M S

Use the backward truth table technique to do problems 1–30 at the end of section 6.7, on pages 255–257.

 ## 7.2 TABLEAUX

Semantic tableaux evaluate arguments using the strategy of backward truth tables, but they do so systematically and elegantly. They assume that the premises are all true and the conclusion is false and investigate possible assignments of truth values that might lead to that result. If they find only incoherence, they conclude that the argument is valid. If they find a coherent assignment making the premises true and the conclusion false, they show the argument to be invalid.

Tableaux are (upside-down) trees with labels. At the top of each is its **root;** at the bottom are its **tips.** A path going directly from the root to a tip is a **branch.** Trees with more than one branch **split** where the paths diverge.

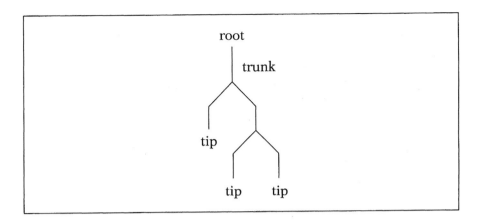

A **semantic tableau** is a tree with formulas appearing on it. A formula may appear on either the left side or the right side of a branch. The left side represents truth; the right, falsehood.

> The left side of a tableau branch is the **truth** side; we assume that formulas on the left are true. The right side is the **falsehood** side.

This, for example, is a semantic tableau set up to determine whether *modus ponens* is valid.

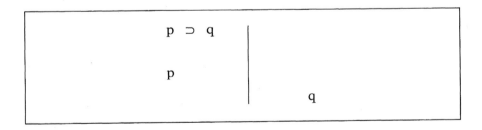

The tableau represents the information that p ⊃ q and p are true but that q is false. If that assignment of truth values is possible, the argument form is not valid. If it is impossible, the argument form is valid. Our tableau rules determine whether the initial assignment is possible or not.

The general strategy underlying the use of semantic tableaux is just that of backward truth tables. A tableau branch represents a combination or a possible circumstance. It corresponds to a row of a truth table.

A tableau begins with a set of assumptions about the truth or false-hood of certain formulas. (In this case, that p ⊃ q and p are true and q is false.) We represent these assumptions by placing formulas on the left or right side of the tableau. By way of the rules, we generate other formulas, which extend the branches. We use the tableau to build a picture of how the world could be, given the initial assumptions. We try, in other words, to describe some possible circumstances—that is, some combinations of truth values—under which the initial assumptions would hold. In this case, we try to describe a possible circumstance in which p ⊃ q and p are true but q is false.

We proceed by applying rules. As we do, we dispatch some formulas with a **check,** √, to indicate that rules have been applied to them. The check signals that we have used the information in the formula to extend the tableau. We can safely ignore dispatched formulas, because we have already taken account of the information they provide. Dispatched formulas are **dead;** undispatched formulas are **live.**

At the next step of our tableau for *modus ponens,* for example, we would apply a rule for the conditional, which, as we shall see in the next section, would yield this tableau:

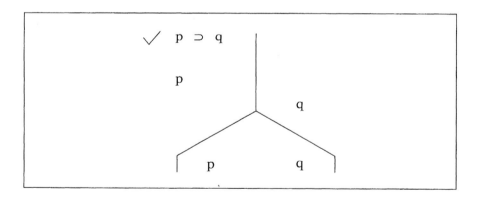

We have checked p ⊃ q to show that we have applied a rule to it. It no longer plays a role in the tableau. Additionally, we have split the tableau into two branches to indicate that we must consider two possibilities. If p ⊃ q is true, then, according to the truth table for the conditional, either p is false or q is true.

Branches that have the same formula appearing live on both sides are **closed.** Both branches of our tableau are now closed *(Cl).* The left branch has p on both sides.

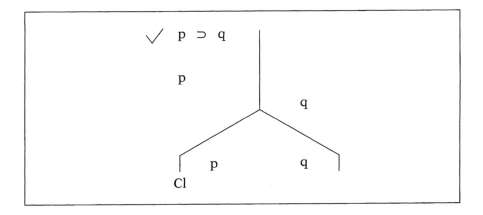

The right branch has q on both sides.

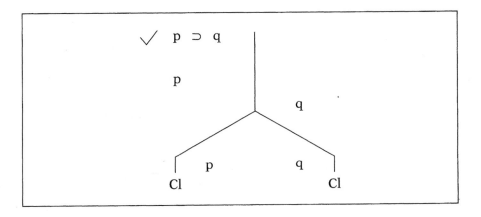

Tableaux like this, with all their branches closed, are also called *closed*.

A **tableau branch is closed** if and only if the same formula appears live on both sides of it. Otherwise, the branch is **open**.
A **tableau is closed** if and only if every branch of it is closed. Otherwise, it is **open**.

When a branch closes—when the same formula appears live on both sides—an attempt to describe a circumstance in which the formulas at the top of the tableau have the indicated truth values ends in a contradiction. The rules have forced us to assign some formula both truth and falsehood. When all branches close, every attempt ends in contradiction. When this happens, we can conclude that under no possible circumstances would the initial assumptions hold. Our tableau for *modus ponens* closes, indicating that under no circumstances can p ⊃ q and p be true while q is false. That tells us that *modus ponens* is a valid argument form.

If some branches remain open, they reveal combinations that would meet the initial assumptions. We can apply rules until only atomic formulas are live. Then we can read interpretations—in effect, rows of truth tables—from the open branches.

Closure, or dispatching all complex formulas, finishes the tableau process.

> A tableau is **finished** if and only if (1) it is closed or (2) only atomic formulas are live on it.

In constructing a tableau, we continue until it is finished—until all branches close or we run out of formulas to apply rules to, having only atomic formulas left.

All the tableaux in this text have special features. We construct them by using certain explicit rules. The tableaux will be finite. And they will be **binary** in the sense that, whenever they split, one path will divide into two. Formulas never appear as tips or as other nodes on a branch; they are always on the left or right of a branch.

We can generalize our example to see how semantic tableaux decide the validity of arguments and argument forms. A semantic tableau is a search for a combination of truth values meeting certain initial conditions. Closed branches represent contradictions; they correspond to blind alleys in the search. A branch having the same formula on both sides indicates that the formula is both true and false, an absurdity. If all branches of a tableau close, then, no matter what choices we might make in our search, we reach contradictions. In a closed tableau, all paths lead to absurdity. A closed tableau, hence, demonstrates that the initial conditions cannot be met. No combination will produce the truth value assignment that began the tableau.

The usefulness of tableaux in searching for combinations makes them ideal for evaluating symbolic arguments. A symbolic argument is valid, after all, just in case there is no combination making its premise formulas true but its conclusion formula false. This suggests an easy method: Search for such a combination. If the tableau produces one, then the symbolic argument is invalid; some truth value assignment does indeed make the premise formulas all true but the conclusion formula false. If the tableau closes, however, there is no such combination, so the symbolic argument is valid.

Tableaux force us to think backward about symbolic argument validity. We begin by assuming that a symbolic argument is invalid and see what develops from this assumption. If the tableau closes, we have reached contradictions; the symbolic argument is valid. If the tableau remains open, we have reached a combination showing that the symbolic argument is invalid.

Since the validity of symbolic arguments is a kind of implication, furthermore, the same test is a decision procedure for implication. To see whether a set of formulas implies a given formula, assume that every formula in the set is true but the given formula is false. The tableau will then search for a combination with this effect. If there is one, a branch will remain open and specify it. If not, the tableau will close.

The tableau test for validity (and implication), then, is this. Say that our symbolic argument has premise formulas $p_1, \ldots, p_n$ and conclusion formula q. We list the premise formulas on the left and place the conclusion formula on the right. If the tableau closes, the symbolic argument is valid:

Test for Argument Validity
(and Implication)

p_1

p_n

q

Closes: Valid (Argument); **Implied** (Implication)
Open: Invalid (Argument); **Not Implied** (Implication)

■ 7.3 NEGATION, CONJUNCTION, AND DISJUNCTION

Once we place some formulas on the tableau, we try to generate pictures of possible circumstances by inferring the truth values of the statement letters they contain. The strategy is the opposite of that we used in constructing truth tables. A truth table begins with a set of assignments of truth values to statement letters; the table generates values for entire formulas. Truth tables, in other words, follow a "bottom-up" strategy. The table assigns values to statement letters and gradually assigns values to more complex parts of the formula. The table ends with the formula's main connective. A tableau, in contrast, starts with an assignment of truth values to formulas and tries to find corresponding assignments to statement letters. Semantic tableaux follow a "top-down" strategy like that of backward truth tables. They begin with a value for an entire formula (or set of formulas) and, starting with the formula's main connective, assign values to parts of formulas until the tableau closes or reaches statement letters.

Tableau rules come in pairs. Since we follow a "top-down" strategy, we are interested in decomposing formulas. We always begin, therefore, by applying a rule to the main connective of a formula. The formula could be on the left or right side of a branch. It could be represented, that is, as either true or false. So, we need two rules for each connective. One will handle formulas appearing on the left having that connective as a main connective; the other will handle similar formulas on the right. In this section we develop six rules: ~L (Negation Left), ~R (Negation Right), ·L (Conjunction Left), ·R (Conjunction Right), ∨L (Disjunction Left), and ∨R (Disjunction Right).

NEGATION

The rules reflect the definitions of the truth functions. We can see this more clearly, in the case of negation, by expressing the tabular definition differently.

> ~p is true if and only if p is false.

~L (NEGATION LEFT)

This rule applies to formulas with a negation sign as the main connective, appearing on the left side of a tableau branch. The left side represents truth. So the question becomes if a formula ~p is true, what is the truth value of p? The answer, clearly, is falsehood. So this rule takes the form

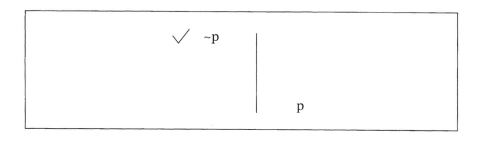

Since p must be false, we check ~p and, extending the branch, write p on the right side of it. In our statement of the rule, what is in regular type must be present for the rule to be applied; applying the rule consists in writing what is in boldface. The check beside the decomposed formula on the left signals that a rule has been applied to this formula. Formulas dispatched with a check may be ignored for the rest of the tableau.

~R (NEGATION RIGHT)

This rule applies to negated formulas appearing on the right side of a branch, which represents falsehood. If a formula ~p is false, then p must be true, so we write p on the left side of the tableau branch:

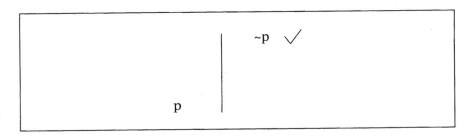

Using these rules, we can show that ~~p implies p—that is, that ~~p, ∴ p is a valid argument form. In general, p implies q if and only if p cannot be true while q is false. So, ~~p implies p if and only if ~~p cannot be true while p is false. We begin by assuming that ~~p is true and p false:

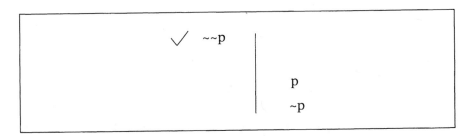

We then proceed to apply rules to see whether this is a real possibility. First, we use ~L:

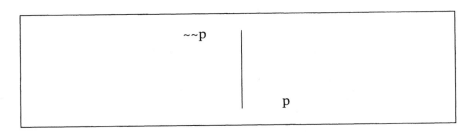

Then, we use ~R to finish the tableau:

$$
\begin{array}{c|c}
\checkmark \ \ {\sim}{\sim}p & \\
 & p \\
 & {\sim}p \\
p & \\
 & Cl
\end{array}
$$

The tableau is closed; p appears on both sides of the branch.

CONJUNCTION

Rules for conjunction also derive from that truth function's definition. Putting the prior tabular definition into a different form:

p · q is true if p is true and q is true.

The rules mirror this definition directly.

·L (CONJUNCTION LEFT)

This rule applies to formulas appearing on the left of a branch, with conjunctions as their main connectives. Under what circumstances is a formula (p · q) true? Obviously, circumstances in which p and q are both true. So we can write both p and q on the left side of the branch:

$$
\begin{array}{c|c}
\checkmark \ \ p \cdot q & \\
p & \\
q &
\end{array}
$$

·R (CONJUNCTION RIGHT)

This rule applies to formulas appearing on the right of a branch, with conjunctions as their main connectives. Under what circumstances is (p · q) false? There are two possibilities: either p or q is false. To reflect the two options, we split the branch. On one, we reflect the possibility that p is false by writing p on the right. On the other, we write q on the right to reflect the possibility that q is false. The rule thus takes the form

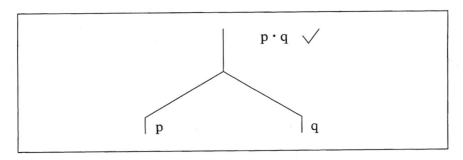

We do not need a third branch to reflect the possibility that both p and q might be false; these branches already take care of that possibility. Consider the left branch, which says that p is false. It says nothing about q; q could be either true or false. This branch alone, then, really captures two possibilities when we consider the truth value of q. It reflects the possibility that p is false and q is true as well as the possibility that p and q are both false.

To see how these rules work, we show that conjunction is **commutative,** that is, that the order of the conjuncts makes no difference to truth values. We do this by showing that p · q implies q · p. Suppose that p · q were true but q · p false:

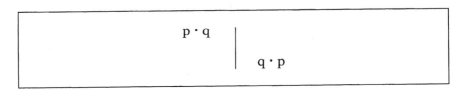

Now, apply rules. The order of applying rules makes no difference to the outcome. But, as we shall see in more detail later, tableaux are shortest when rules that require splits are applied last. So, we begin with ·L, which does not force us to split the tableau:

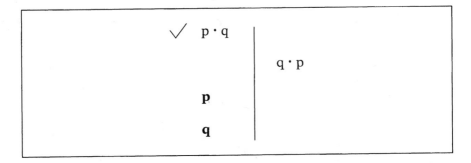

Now, we apply ·R, which does force a split. If q · p is false, there are two possibilities: Either q or p is false.

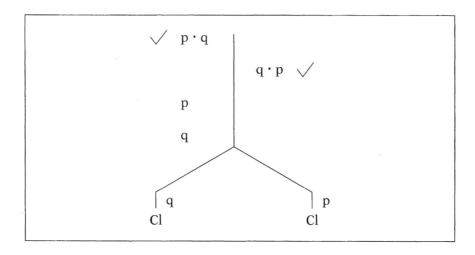

Both branches close. On the left, q appears on both sides; on the right, p does. The tableau thus closes, indicating that the initial assignment of truth values is impossible. It cannot happen that p · q is true while q · p is false. Hence, p · q implies q · p.

DISJUNCTION

The rules for disjunction reflect its definition. We can express the content of our tabular definition in the preceding chapter in somewhat different form:

> p ∨ q is true if and only if p is true or q is true.

Given this formulation, it is easy to see what the rules should be.

∨L (DISJUNCTION LEFT)

This rule, for formulas on the left with disjunctions as main connectives, tells when formulas of the form (p ∨ q) are true: whenever either p or q is true. Again we must split the branch to reflect these two possibilities:

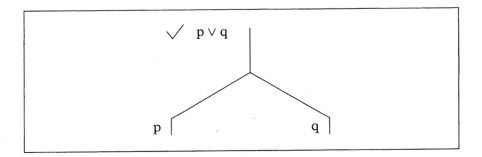

∨R (DISJUNCTION RIGHT)

This rule applies to formulas on the right with disjunctions as main connectives. When is (p ∨ q) false? When both p and q are false. Dispatching the disjunction, we write both disjuncts on the right side of the branch:

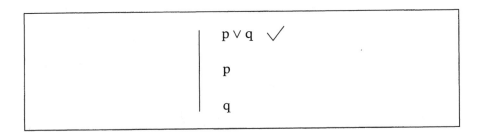

These rules allow us to show that disjunction is **associative,** that is, within disjunctions, grouping makes no difference to truth values. We show that (p ∨ q) ∨ r implies p ∨ (q ∨ r). As before, we assume that the former is true and the latter false.

$$\begin{array}{c|c} (p \vee q) \vee r & \\ & p \vee (q \vee r) \end{array}$$

We find ourselves with disjunctions on both sides of the tableau. The rule ∨R does not require us to split the tableau, so we will save time and effort by using it first. If p ∨ (q ∨ r) is false, then both p and q ∨ r must be false.

$(p \lor q) \lor r$

$p \lor (q \lor r)$ ✓

p

$q \lor r$

Now, we apply ∨R again, to q ∨ r:

$(p \lor q) \lor r$

$p \lor (q \lor r)$ ✓

p

$q \lor r$ ✓

q

r

At this point, we must apply ∨L. It forces us to split the tableau:

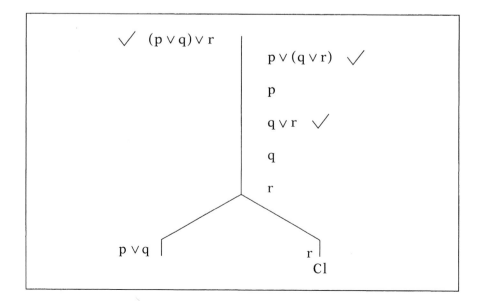

The right branch closes, for r appears on both sides. The left branch is open but still contains a live complex formula, p ∨ q, on the left. So, we finish the tableau by applying ∨L once more.

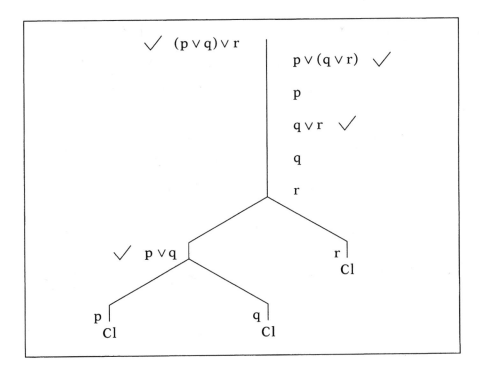

Both the resulting branches close, so the tableau itself is closed. That tells us that we cannot describe a circumstance in which (p ∨ q) ∨ r is true but p ∨ (q ∨ r) is false. It tells us, therefore, that (p ∨ q) ∨ r implies p ∨ (q ∨ r).

P R O B L E M S

Display a tableau that determines whether the formula on the left implies the one on the right.

1. P	P ∨ Q
2. ~~P	P · Q
3. ~(P · Q)	~P · ~Q
4. ~(P ∨ Q)	~P · ~Q
5. ~(P · Q)	~P ∨ ~Q
6. ~(P ∨ Q)	~P ∨ ~Q

7.	$P \lor (Q \cdot R)$	$(P \lor Q) \cdot (P \lor R)$
8.	$P \cdot (Q \lor R)$	$(P \cdot Q) \lor (P \cdot R)$
9.	$(P \cdot Q) \cdot (R \lor S)$	$(P \cdot R) \lor (Q \cdot S)$
10.	$(P \lor Q) \cdot (R \lor S)$	$((P \cdot R) \lor (P \cdot S)) \lor ((Q \cdot R) \lor (Q \cdot S))$

Symbolize these arguments in our symbolic language, and display tableaux to evaluate the resulting symbolic arguments for validity. If the symbolic argument is invalid, describe the combinations under which the premises would be true but the conclusion false.

11. I have already said that he must have gone to King's Pyland or to Mapleton. He is not at King's Pyland, therefore he is at Mapleton. (Sir Arthur Conan Doyle)

12. A man cannot serve both God and Mammon. But a man starves unless he serves Mammon; he can't serve God, moreover, unless he survives. But he won't survive unless he doesn't starve. Therefore, a man cannot serve God.

13. Peg will major in biology or mathematics. Unless she majors in biology, she won't go to medical school; she'll go into business instead. Her mother will be disappointed unless she goes to medical school. Peg will major in mathematics. Consequently, her mother will be disappointed.

14. Neither Bob nor Susan will come to the party unless you apologize. And you won't apologize unless Bob and Susan do. They won't apologize unless you are willing to change your ways. You won't change your ways unless you see profit in it. You see no profit in it. So, Susan won't come to the party.

15. Laura is in love with Mel or Ned. Ned loves Laura or Olive. Olive will be miserable unless Ned loves her. Laura will be happy unless either Ned is in love with her and she's not in love with him or she is in love with him and he's not in love with her. Ned doesn't love Olive, but Laura is still unhappy. So, Ned loves Laura, Laura loves Mel, and Olive is miserable.

■ 7.4 POLICIES

In applying tableau rules, several questions arise. First, when a rule directs us to enter a formula on a given side of a branch, where do we put it? After all, there may already be formulas under that to which we are applying the rule. We adopt the policy

1. Always write the new formula at the bottom of the branch, underneath all the formulas already appearing there.

For example, we may want to apply ·L to the second formula in the tableau on the left below. The rule tells us to enter two new formulas (the conjuncts of the conjunction) on the left. We write these below the other formulas, resulting in the tableau on the right.

(1)	(2)
P ∨ Q	P ∨ Q
Q · R	✓ Q · R
R ⊃ S	R ⊃ S
	Q
	R

Second, what happens when the tableau splits? Say that we want to apply a rule to a formula above a split in the tableau. The rule directs us to enter a new formula on the branch. But which branch? What was one branch before is now two (or four, or eight, etc.). To capture the meaning of the formula on every branch on which it appears, we must make the required entries on every such branch. We adopt the policy

2. When applying a rule to a formula, make the entries it calls for on every branch on which that formula appears.

Normally, this means that we must make the entries on each branch that extends the one on which the formula appeared. The progress on a tableau beginning with (P · Q) on the right and (Q ∨ R) on the left might take this form:

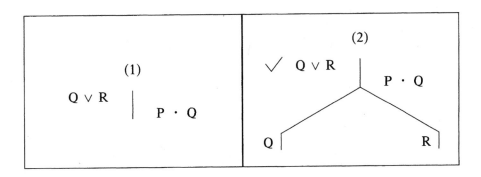

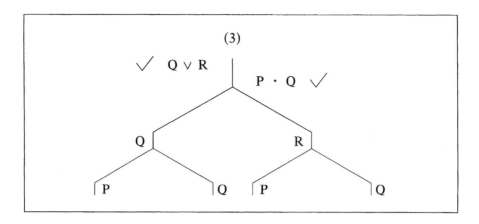

Applying the rule ∨L to (Q ∨ R) results in the second tableau. Applying ·R to (P · Q) results in the third. Notice that we had to make the entries— and perform the required split—on both branches that resulted from applying ∨L.

We follow two other policies in applying rules. The order of application of rules, within the bounds of propositional logic, makes no difference in results. But it does make a difference in efficiency. To create as few branches as possible, we adopt the policy

3. Apply rules that do not require any splitting before those that do.

So we apply ~L, ~R, ·L, and ∨R before ·R and ∨L.

In addition, closed branches merit no further interest, since they contain a contradiction. No matter what rules are applied to them, they will remain closed; they cannot depict any possible circumstance. We therefore adopt the policy

4. Abandon branches as soon as they close, marking them with the notation *Cl* to indicate why.

This, too, adds to efficiency.

 ## 7.5 THE CONDITIONAL AND BICONDITIONAL

So far we have developed rules for three of our five connectives. Rules for the conditional and biconditional are similar, though slightly less intuitive. Like the rules for negation, conjunction, and disjunction, they come in pairs. Also, like those rules, they mirror the definitions of the conditional and biconditional truth functions. We can express the definition of the conditional in this form:

p ⊃ q is true if and only if p is false or q is true.

⊃L (CONDITIONAL LEFT)

This rule applies to conditional formulas appearing on the left of a branch. Under what circumstances is a conditional formula true? The truth table definition of the conditional indicates that (p ⊃ q) is false only when p is true and q is false. So (p ⊃ q) is true whenever p is false or q is true. These two possibilities force us to split the branch.

```
        ✓  (p ⊃ q)
              /\
          p        q
```

⊃R (CONDITIONAL RIGHT)

This rule applies to conditionals on the right. These formulas, according to the tableau branch, are false. The formula (p ⊃ q) is false just in case p is true and q is false, as we have seen, so the rule asks us to enter p on the left and q on the right:

```
                    (p ⊃ q)  ✓
        p
                    q
```

These rules allow us to show that ~q ⊃ ~p implies p ⊃ q. Suppose that the former were true and the latter were false.

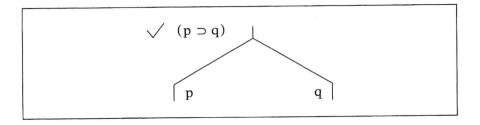

Does that reflect a real possibility? We apply rules to find out. We use ⊃R first, since it does not require a split.

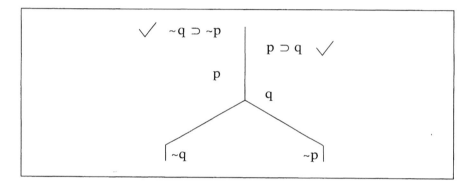

Now, we split the tableau as required by ⊃L.

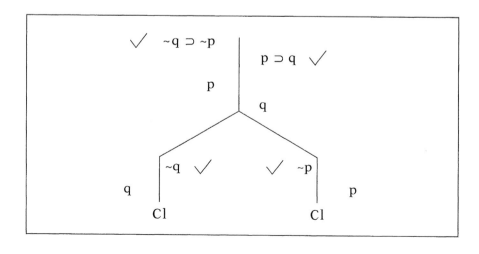

On both branches, we find a negation. We can use the rules ~R (on the left branch) and ~L (on the right branch) to finish the tableau.

This tells us that ~q ⊃ ~p implies p ⊃ q, for there could be no circumstance in which the former is true and the latter false.

The rules for the biconditional also reflect its definition:

p ≡ q is true if and only if p and q agree in truth value.

≡L (BICONDITIONAL LEFT)

This rule applies to biconditionals on the left side of a branch. If (p ≡ q) is true, p and q must have the same truth value. Both must be true, or both must be false. Since there are two possibilities, the rule splits the branch:

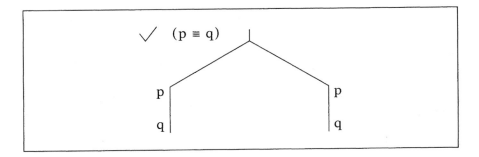

≡R (BICONDITIONAL RIGHT)

This rule applies to biconditionals on the right. If (p ≡ q) is false, then p and q must have opposite truth values. That is, either p is true and q is false or q is true and p is false. Again the rule forces splitting to reflect these two possibilities.

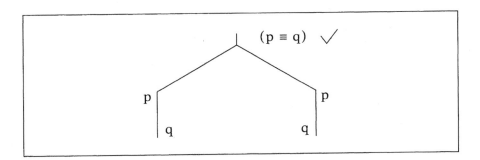

To see how these rules function, consider two examples. Suppose that

(5) Rocky will win the fight only if he keeps moving.

is true but that

(6) Rocky will win the fight if and only if he keeps moving.

is false. In the first tableau, we assume that W ⊃ M is true but W ≡ M is false. We decompose these formulas to find out what this means about the truth values of W and M alone. The order in which we apply the rules makes no difference, even to efficiency, since both conditional left and biconditional right split the tableau. We end up, therefore, with four branches; two close, and two remain open.

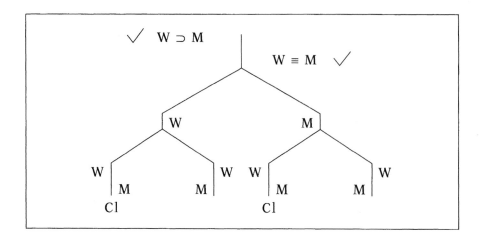

Starting with the initial condition that (W ⊃ M) is true but (W ≡ M) is false, this tableau considers a variety of possibilities for assignments to W and M. The tableau, in effect, tries to describe a circumstance in which these formulas would have this particular combination of truth values. The two branches that close do not describe possible circumstances that would make (W ⊃ M) true and (W ≡ M) false, but the open branches do. As it happens, the open branches describe the same circumstance: one in which M is true and W false. To see this, look at each open branch, and record the atomic formulas appearing on each side:

| LEFT OPEN BRANCH | True: M | False: W, W |
| RIGHT OPEN BRANCH | True: M, M | False: W |

The duplication of appearances makes no difference. These branches, therefore, describe the same combination of truth values. Both branches describe a circumstance in which M is true and W is false. (In which, that is, Rocky keeps moving but nevertheless does not win the fight.) The tableau, then, tells us the following. There is a combination of truth values for the letters W and M making (W ⊃ M) true and (W ≡ M) false. That combination assigns truth to M and falsehood to W.

A second example: Consider the inference

(7) The statement, that this sentence is true if and only if this is an excellent book, is true if and only if this sentence is true.

∴ If that sentence is true, this is an excellent book.

Is there a combination making (P ≡ Q) ≡ P true but (P ⊃ Q) false? Here, applying conditional right first helps efficiency, since it does not force us to split the tableau:

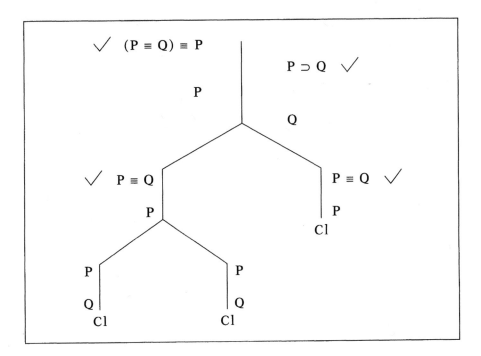

Every branch is closed, so the tableau itself is closed. Thus, there is no combination meeting our initial conditions; no combination makes (P ≡ Q) ≡ P true but (P ⊃ Q) false. No matter which way the tableau turns in trying to describe a possible circumstance meeting these conditions, it runs into a contradiction. The inference is therefore valid.

A special problem arises when a branch does not contain every statement letter involved in the tableau. Suppose that we are interested in evaluating the argument

(8) The theft will shock the art world unless the painting is returned unharmed, and quickly. If it does shock the art world,

the painting will be returned unharmed. Thus, the painting will be returned unharmed, and museums will review their security measures carefully.

Symbolizing, we obtain the symbolic argument

(9) S ∨ (R · Q)

S ⊃ R

∴ R · M

We need to know whether any combination makes S ∨ (R · Q) and S ⊃ R true but R · M false. We can place the first two formulas on the left and the last on the right and apply rules:

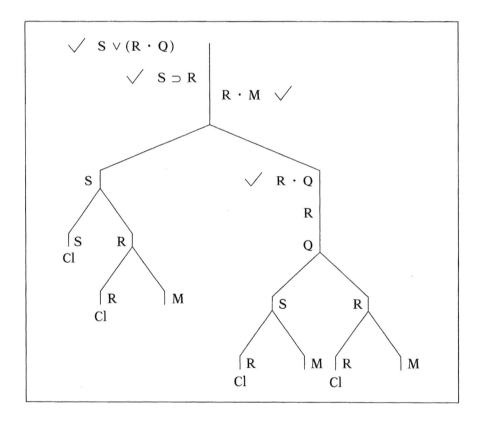

There are three open branches, signaling that there is such a combination. On the left-most open branch, S and R both appear on the left, indicating that both S and R must be true. Moreover, M appears on the right, so it must be false. But the branch says nothing about Q. Thus, any combination making S and R true and M false meets the initial conditions, no matter what it assigns Q. In terms of the English argument: The theft could shock the art world and the painting could be returned unharmed without museums carefully reviewing their security measures (whether the return is quick or not). This makes the argument invalid. The conclusion does not follow. The left-most branch thus reveals two combinations making S ∨ (R · Q) and S ⊃ R true but R · M false. One makes S true, R true, M false, and Q true. The other makes S true, R true, M false, and Q false.

P R O B L E M S

Display a tableau that determines whether the formula on the left implies the one on the right.

1. P ∨ Q	P ≡ Q
2. P ⊃ Q	~P ∨ Q
3. ~P · ~Q	~(P ⊃ Q)
4. (P ⊃ R) ∨ (Q ⊃ R)	(P ∨ Q) ⊃ R
5. P ≡ Q	(P · Q) ∨ (~P · ~Q)

Display tableaux that determine whether these symbolic argument forms are valid.

6. P; P ⊃ R; ∴ R

7. P; P ⊃ Q; ∴ P · Q

8. P ⊃ Q; ~Q; ∴ ~P

9. P ⊃ Q; P ⊃ ~Q; ∴ P

10. ~P ⊃ ~Q; P ⊃ ~Q; ∴ ~Q

11. P ⊃ Q; ~P ⊃ Q; ∴ ~~Q

12. P ⊃ Q; ~P ⊃ Q; ∴ ~Q

13. ~P ∨ Q; ~Q; ~P ≡ R; ∴ ~R ≡ Q

14. ~P ∨ Q; ~Q; ~P ⊃ R; ∴ ~R ⊃ Q

15. ~P ∨ Q; ~Q; ~(P ≡ R); ∴ ~(R ≡ Q)

16. ~P ⊃ Q; ~Q; ~P ⊃ R; ∴ ~R ⊃ P

17. (P ∨ ~Q) ⊃ R; P ≡ ~R; ∴ ~((Q · R) ∨ P)

18. (P · ~Q) ⊃ R; P ≡ ~R; ∴ (Q · R) ∨ P

19. (P ∨ ~Q) ⊃ R; P ≡ ~R; ∴ (Q · ~R) · P

20. (~Q · R) ⊃ ~P; P ⊃ R; ∴ ~(Q ≡ (~R · P))

21. (~Q ∨ R) ⊃ ~P; P ⊃ R; ∴ ~(Q ≡ (~R ∨ P))

22. (~Q · R) ≡ ~P; P ≡ R; ∴ P · (~Q ⊃ ~R)

23. $P \supset (Q \supset (R \cdot {\sim}P))$; ${\sim}P \supset {\sim}(S \vee K)$; $(M \cdot T) \equiv Q$; $\therefore S \supset (M \supset {\sim}T)$

24. $(P \equiv {\sim}Q) \equiv {\sim}R$; $R \vee (P \cdot S)$; $S \supset (P \cdot (Q \cdot T))$; $\therefore (S \cdot K) \supset (R \cdot Q)$

Symbolize the following arguments in our symbolic language, and display tableaux to evaluate the resulting symbolic arguments for validity. If the symbolic argument is invalid, describe the combinations under which the premises would be true but the conclusion false.

25. If Nathan publishes, he'll get tenure. He will not get tenure. Hence, he will not publish.

26. If Nathan publishes, he'll get tenure. He will not publish. Thus, he will not get tenure.

27. If God is all-powerful, then He can make a rock so heavy that He cannot lift it. But if God is omnipotent, then He can lift any rock that He can make. Therefore, God is not all-powerful.

28. If Geraldine signs the contract, she'll lose money. If she loses money, we lose money, too. Geraldine will sign the contract. So, we lose money.

29. If nobody objects, the council will adopt Holly's plan. Somebody will object if those affected by the plan believe it harms their interests. Thus, the council will adopt Holly's plan if those it affects do not think it will harm their interests.

30. If Johnson leaves the firm, then Jones will leave, too. Furthermore, if Jones leaves the firm, Johnson will leave. So, Robinson will leave if Johnson leaves only if Robinson will leave if Jones leaves.

31. If you want to maximize your job opportunities after graduation, you'll major in business. But you'll succeed in the top ranks of industry only if you can write effectively; you'll be able to write effectively, however, only if you major in the liberal arts. Hence, you'll succeed in the top ranks of industry only if you do not want to maximize your job opportunities after graduation.

32. There will be scorpions in the house this week if there is any building going on in the neighborhood. There will be building only if the depression in the housing market ends. If there are scorpions in the house, the cats will kill them and bring them to me as trophies. Therefore, if the housing depression ends, my cats will bring me scorpions as trophies.

33. My cat will not sing opera unless all the lights are out. If I am very insistent, then my cat will sing opera; and if I either turn all the lights out or howl at the moon, you can be sure that I am very insistent. I howl at the moon if I am not very insistent.

Therefore, my lights are out, my cat is singing opera, and I am very insistent.

34. If the president retaliates against Libya with military force, the public will be ambivalent. But if Libya directs terrorist attacks toward Americans on American soil, then the American public will not be ambivalent at all. So, Libya will not engage in terrorism against Americans on U.S. soil unless the president does not retaliate militarily.

35. If people do not conserve energy, or allow much of the environment to suffer, utilities will have no choice but to rely on nuclear power. They'll avoid relying on atomic energy, moreover, just in case it is prohibitively expensive and risky. Nuclear power will remain expensive. If it also remains risky, therefore, the environment will suffer unless people conserve energy.

36. If John attends the meeting, he'll be able to convince at least a few people to vote against the development proposal. If we can get most of the neighborhood to sign a petition, then the developers will change their proposal. Either we get most of the neighborhood to sign a petition or John will attend the meeting. So, John will be able to convince at least a few people to vote against the development proposal, unless the developers change their proposal.

37. If changes in baseball over the past decade continue, then power pitching will gradually drive out power hitting. This will happen just in case managers select players with speed over players with home run power. But they will not unless runs become very difficult to score. Finally, if power pitching does not drive out power hitting, power hitting will drive out power pitching. Consequently, either the changes in baseball over the past decade will continue or power hitting will drive out power pitching.

38. The central administration will not increase the department's budget if it does not improve its performance. But the department will not perform better if the administration does not increase its budget allocation. If the department fails to improve, then its chairman will be replaced. If the central administration refuses to increase the department's budget, it will begin to lose some good people. If the department's chairman is replaced, however, it will not lose any good people. So, the central administration will increase the department's budget if the chairman is replaced.

39. If the party maintains its current economic policy, there will be a flight of capital to other countries. But the party will not change its economic policy if it tightens its control over the

economy. If the party maintains its policy, and capital flees to other countries, then it will tighten its grip on the economy. If the party maintains its current policy, however, the nation will have to pay large amounts of foreign debt in hard currency, and this it will not do. So, the party will not tighten its grip on the economy.

Use semantic tableaux to solve these problems.

40. A brutal ax murder has been committed; you are called in to investigate. You have evidence to support each of these contentions: *(a)* If Farnsworth committed the murder, Shelby is innocent. *(b)* Shelby is guilty unless the butler's testimony has not been delivered with a clear conscience. *(c)* If Farnsworth committed the murder and Shelby is innocent, then the butler's testimony has been delivered with a clear conscience. *(d)* If the butler is not guilty of the murder, then neither is Farnsworth. You must answer these questions: (1) Is any contention implied by the other three? (2) Does your theory imply anything about the guilt or innocence of Farnsworth? (3) Does your theory imply anything about the guilt or innocence of Shelby?

41. You have been diagnosed as having a rare psychological disorder manifesting itself in intense bouts of fear and loathing whenever you are faced with a logic problem. Your psychiatrist tells you that *(a)* if you do not undergo psychoanalysis, you will not recover. But you know, given his rates, that *(b)* if you do undergo psychoanalysis, you'll be poverty-stricken. Unknown to both of you, the reasoning involved in undergoing psychoanalytic treatment will make your condition worse; if you improve at all, it will be for reasons unrelated to your treatment. So *(c)* if you undergo psychoanalysis, then you will recover only if you improve spontaneously. Do these facts imply that you will recover only if you improve spontaneously? That you will recover only if you become poverty-stricken? That if you become poverty-stricken, you'll recover?

42. Roger, a hapless accounting major, works on a take-home final exam in a course on U.S. tax law. He attempts to analyze a problem concerning the tax liability of a corporation involved in overseas shipping. Some of the fleet counts as American for tax purposes, but some does not. Roger thinks that the definition of "American vessel" goes something like this: "A ship is an American vessel if and only if (1) it is either numbered or registered in the United States or (2) if it is neither numbered nor registered in the United States and is not registered in any foreign country, then its crew members are all U.S. citizens or all employees

of corporations based in the United States." Unfortunately for Roger, this is not the right definition. Show that Roger's version implies that if a ship is registered in a foreign country, it is an American vessel.

█ 7.6 OTHER APPLICATIONS

Since semantic tableaux constitute a decision procedure for implication, they also amount to a test for equivalence. Equivalence, after all, is just implication in two directions. Equivalent formulas have the same truth value in every combination of them. This means that there should be no combination making one true but the other false. To test for this, we can do two tableaux. Say that the two formulas are p and q. One tableau starts with the assumption that p is true and q is false; the other begins by assuming that q is true and p is false. If both close, no combinations assign the formulas different truth values. If at least one tableau remains open, however, it will specify a combination making one formula true and the other false.

The test, then, is this:

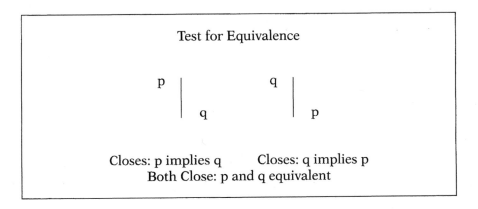

Test for Equivalence

p   q

    q   p

Closes: p implies q   Closes: q implies p
Both Close: p and q equivalent

We showed above, for example, that ~q ⊃ ~p implies p ⊃ q. To show that they are in fact equivalent, we need only to show, in addition, that p ⊃ q implies ~q ⊃ ~p. To do that, we assume that p ⊃ q is true and ~q ⊃ ~p false:

p ⊃ q

~q ⊃ ~p

We first apply ⊃R:

$$
\begin{array}{c|c}
\text{p} \supset \text{q} & \\
 & \text{~q} \supset \text{~p} \;\; \checkmark \\
\text{~q} & \\
 & \text{~p}
\end{array}
$$

Now, we can apply ~R and ~L:

$$
\begin{array}{c|c}
\text{p} \supset \text{q} & \\
 & \text{~q} \supset \text{~p} \;\; \checkmark \\
\checkmark \;\; \text{~q} & \\
 & \text{~p} \;\; \checkmark \\
\text{p} & \\
 & \text{q}
\end{array}
$$

At this point, we must split the tableau, using ⊃L:

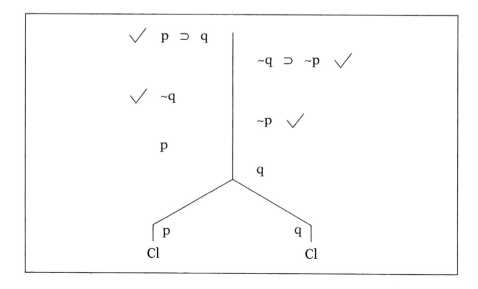

Both branches close: p appears on both sides of the left branch, and q appears on both sides of the right branch. In either case, we reach a contradiction. It is impossible, therefore, for p ⊃ q to be true while ~q ⊃ ~p is false. Since we earlier showed that ~q ⊃ ~p implies p ⊃ q, we can conclude that they are equivalent.

Tableaux also offer an elegant test for the logical truth of formulas. A tautology comes out true in every combination of it. No assignment of truth values makes a tautology false. To determine whether a formula A is a tautology, therefore, we can search for a combination making it false. If the search succeeds—if, that is, the tableau we construct remains open—then we have a combination making the formula false. If the search fails, the tableau closes, and the formula is a tautology.

TEST FOR LOGICAL TRUTH

| p

Closes: Tautologous

Open: Contingent or Contradictory

Is p ∨ ~p, for example, a tautology? We assume that it is false and see whether our assumption leads to a contradiction. We place p ∨ ~p on the right and apply ∨R and then ~R:

$$
\begin{array}{c|c}
 & p \lor {\sim}p \ \checkmark \\
 & p \\
 & {\sim}p \ \checkmark \\
p & \\
\mathrm{Cl} &
\end{array}
$$

The tableau closes; p ∨ ~p cannot be false, so it is a tautology.

Similarly, tableaux present a method for testing formulas for contradictoriness or satisfiability. A satisfiable formula can be true; a contradictory formula cannot. To test for contradictoriness, then, simply search for a combination of truth values making the formula in question true. An open branch will specify such a combination, establishing satisfiability; a closed tableau establishes contradictoriness.

Test for Contradictoriness or Satisfiability

p ⌐

Closes: Contradictory

Open: Satisfiable

Together, these tests allow us to classify any formula as tautologous, contingent, or contradictory. One test does not always determine the status of a formula: If it is contingent, then testing it for validity will result in an open tableau. This means that the formula can be false, so it must be either contradictory or contingent. To determine which, we need the other test.

Semantic tableaux, then, provide a simple and efficient way of testing symbolic arguments for validity; formulas for validity or contradictoriness; pairs of formulas for equivalence; and sets of formulas, paired with formulas, for implication. Tableaux always terminate after a finite number of steps. Their construction, furthermore, requires no ingenuity, but the mechanical following of rules. Tableaux are thus decision procedures for the logical properties and relations we have discussed. As later chapters will demonstrate, the tableau technique extends readily to more complex and comprehensive logical languages.

P R O B L E M S

Display tableaux to show whether these formulas are equivalent.

1.	~~~A	~A
2.	A · A	A ∨ A
3.	A ⊃ A	A ≡ A
4.	A ∨ B	~A ⊃ B
5.	A · ~B	~(A · B)
6.	A ⊃ ~B	~(A ⊃ B)
7.	A ≡ ~B	~(A ≡ B)
8.	(A ∨ B) ⊃ C	(A ⊃ C) · (B ⊃ C)
9.	A ≡ B	(A ⊃ B) · (~A ⊃ ~B)

These equivalences are rules in the deduction system of Chapter 8. Display tableaux to show that the formulas in each pair are equivalent.

10. ~(Q · P) and ~Q ∨ ~P (DeMorgan)

11. ~(Q ∨ P) and ~Q · ~P (DeMorgan)

12. P ⊃ Q   and   ~Q ⊃ ~P   (Transposition)

13. P ⊃ Q   and   P ⊃ (P · Q)   (Absorption)

14. Q ≡ P   and   (Q ⊃ P) · (P ⊃ Q)   (Material Equivalence)

15. Q ≡ P   and   (P · Q) ∨ (~P · ~Q)   (Material Equivalence)

16. (P · Q) ⊃ R   and   P ⊃ (Q ⊃ R)   (Exportation)

17. (P · Q) · R   and   P · (Q · R)   (Association)

18. (P ∨ Q) ∨ R   and   P ∨ (Q ∨ R)   (Association)

19. P · (Q ∨ R)   and   (P · Q) ∨ (P · R)   (Distribution)

20. P ∨ (Q · R)   and   (P ∨ Q) · (P ∨ R)   (Distribution)

Determine whether the following are tautologous, contingent, or contradictory.

21. p ⊃ p

22. p ⊃ ~p

23. p · ~p

24. p ≡ ~p

25. ~(p ∨ q)

26. p ⊃ (q ∨ ~q)

27. ((p ⊃ q) · ~((p ⊃ (q ∨ r)))

28. ((p ⊃ q) ⊃ p) ⊃ p

Symbolize these statements and determine whether they are equivalent.

29. (a) If a fetus is a human being, then abortion is wrong.
    (b) If a fetus isn't a human being, then abortion is permissible.

30. (a) Xenia will graduate only if she improves her GPA.
    (b) If Xenia doesn't improve her GPA, she won't graduate.

31. (a) If Orrin retires next year, we will promote Edna.
    (b) We won't promote Edna unless Orrin retires next year.

32. (a) If either Barry or Jane drops out of the race, Yvonne will reap a huge windfall.
    (b) If Barry drops out of the race, Yvonne will reap a huge windfall.

33. (a) If Sandra and Harold leave by 11:00, I'll be amazed.
    (b) If Harold leaves by 11:00, I'll be amazed.

34. (a) Quincy and Lou will agree only if you do.

    (b) Quincy will agree only if you do; the same holds for Lou.

35. (a) Fred will drop the suit if you settle for $50,000.

    (b) Fred will drop the suit only if you settle for $50,000.

These puzzles concern a land of knights and knaves.[2] Knights always say true things; knaves always utter falsehoods. You are a traveler in this strange land and must try to identify those you meet as knights or knaves. You encounter two people, Punch and Judy, one or both of whom speak to you. What can you deduce in each case, using semantic tableaux, about whether they are knights or knaves? (Hint: a person is a knight if and only if what he or she says is true. Suppose Judy says, "Neither of us is a knight." You cannot assume that this is true. But you can assume that Judy is a knight if and only if it is true. That is, letting J stand for "Judy is a knight" and P stand for "Punch is a knight," you can begin a tableau with $J \equiv \sim(J \vee P)$ on the left side and see what you can infer about the truth values of P and J. You should be able to conclude that Punch is a knight and Judy is a knave.)

36. Judy: Either I'm a knight or I'm not.

37. Punch: If I'm a knight, Judy's a knave.

38. Punch: We're not both knights.

39. Punch: I'm a knave unless Judy's a knight.

40. Judy: If either of us is a knight, it is Punch.

41. Punch: I'm a knight if and only if Judy's a knave.
    Judy: Punch is a knave.

42. Punch: Judy's a knight.
    Judy: At least one of us is a knave.

43. Punch: If I'm a knave, Judy's a knight.
    Judy: Punch is a knight if I am.

44. Punch: If I'm a knave, we both are.
    Judy: Either he's a knight or I'm a knave.

45. Punch: Judy's a knight if and only if her sister is.
    Judy: Unfortunately, my sister's a knave.

---

[2]These examples are based on puzzles developed by Raymond Smullyan in *What Is the Name of This Book?* (Englewood Cliffs, NJ: Prentice-Hall, 1978).

At this point, you meet three people in the land of knights and knaves. What can you deduce about their status?

46. Curly: Larry's a knave.
    Moe: Either Curly or Larry is a knave.
    Larry: If I'm a knave, they are, too.

47. Curly: Moe's a knight.
    Moe: We're all knaves.
    Larry: Curly, Moe, and their cousins are all knaves.

48. Curly: If Moe's a knight, Larry is, too.
    Moe: Larry's a knave if Curly is.
    Larry: Curly and Moe are not both knaves.

49. Curly: If any of us are knights, Larry is.
    Moe: Larry's a knave.
    Larry: I'm a knave if and only if Moe is.

50. Curly: If Moe's a knave, Larry is, too.
    Moe: If Larry's a knave, Curly is, too.
    Larry: If Moe's a knight, we all are.

CHAPTER 8

PROOF

$\mathbf{T}$ruth tables and semantic tableaux can evaluate arguments, but they say almost nothing about how to *construct* arguments. In this chapter we will develop a system designed for constructing arguments that mirrors intuitive logical thinking much more closely than our previous methods.

■ 8.1 RULES AND PROOFS

A **natural deduction system** is a set of **rules of inference,** which allow us to deduce formulas from other formulas.[1] Central to the concept of a natural deduction system is the idea of **proof.** Proofs are extended arguments. **Hypothetical** proofs begin with premises (also called **assumptions,** or **hypotheses**). The conclusion depends on them. Such proofs show that the conclusion is true, not unconditionally, but if the assumptions are true. These proofs thus show the validity of an argument form. **Categorical** proofs use no assumptions. They show that their conclusions are true outright. A categorical proof shows that a certain formula is a tautology.

A *proof* in a natural deduction system is a series of **lines.** On each line appears a formula. Each formula in a proof *(a)* is an assumption or *(b)* derives from formulas on previously established lines by a rule of inference. In the system of this chapter, the last line of a proof is its **con-**

---

[1]Gerhard Gentzen, a German logician, and Stanislaw Jaskowski, a Polish logician, independently devised the first modern natural deduction systems in 1934. The system of this chapter owes a great deal to them, to the German logician and mathematician David Hilbert (1862–1943), to the Swiss mathematician Paul Bernays, and to the American logician Irving Copi. The simple form of indirect proof introduced in the last section was inspired by W. V. O. Quine's "main method."

clusion; the proof *proves* that formula *from* the assumptions (if there are any). Conclusions proved from no assumptions at all are **theorems** of the system.

Hypothetical Proof	Categorical Proof
Premises	.
.	.
.	.
.	.
Conclusion	Conclusion (Theorem: Tautology)

The last section of this chapter discusses categorical proofs. Until then, all the proofs we consider will be hypothetical.

Here is a simple hypothetical proof:

1. p		
2. p ⊃ q	/ ∴ q	
3. q	MP, 1, 2	

All proofs are series of lines structured in certain ways. As in this example, proofs have three columns. The middle column consists of a sequence of formulas. The left column numbers these formulas; the right column provides justifications for them. If a formula derives from previously established formulas by a rule of inference, the right column says what rule of inference and what earlier lines were used. If the formula is an assumption, the right column is blank. (The / ∴ q in line 2 is not an official part of the proof; it simply reminds us of what conclusion we are trying to reach.) The "MP, 1, 2" in line 3 indicates that q follows from the formulas on lines 1 and 2 by the rule MP (*modus ponens,* one of the most important rules of our system). The example is a successful completed proof of q from the assumptions p and p ⊃ q.

Each line of a proof, then, looks like this:

Proof Lines

Number	Formula	Justification

The rules of inference in this chapter allow us to derive formulas of certain kinds in a proof if other formulas of certain kinds occupy already-established lines there. For example, our system has a rule letting us write p or q, whichever we like, if we already have p · q.

   Some rules apply only to entire formulas; some apply to parts of formulas. We shall call rules that apply only to entire formulas **rules of implication.** Those that apply to parts of formulas as well are **rules of replacement** (or **equivalence**). They are also sometimes called **invertible,** for they rely on the equivalence of formulas and so can be used in both directions.

8.2 RULES OF IMPLICATION I: CONJUNCTIONS AND CONDITIONALS

CONJUNCTIONS

The rule of **simplification** tells how we can use the information encoded in a conjunction. It reflects the reasoning manifested in the argument

   (1) Jonas dropped out of college and found work as a grave digger.

   ∴ Jonas dropped out of college.

What follows from the truth of a conjunction? The truth of both conjuncts. If p · q is true, both p and q must be true. The rule thus takes two forms:

SIMPLIFICATION (SIMP)

n.	<u>p · q</u>	
n+m.	p	Simp, n

n.	<u>p · q</u>	
n+m.	q	Simp, n

(Here, n and n+m are line numbers; n+m is simply any line after line n. The line between the premise and conclusion of this rule separates the formula deduced from what must be present earlier; it does not appear in actual proofs.) Hereafter, we abbreviate rules having two forms by using parentheses. We can write Simp as

n.	<u>p · q</u>	
n+m.	p (or q)	Simp, n

This rule tells us that, from a conjunction, we can derive each conjunct. When we apply this rule, we write a conjunct together with the explanation that the line comes by application of simplification to the formula on line n.

To apply simplification to a formula, a conjunction must be its main connective. The letters p and q in the rule are statement variables of the metalanguage; any statement or formula of the object language may substitute for them. Thus, each of the following is a legitimate application of Simp:

1. A · B		1. (A ∨ B) · ~C	
2. A	Simp, 1	2. ~C	Simp, 1

1. ~((A · ~B) ⊃ C) · (D ∨ ~E)		1. ~A · ~(B ≡ ~C)	
2. ~((A · ~B) ⊃ C)	Simp, 1	2. ~(B ≡ ~C)	Simp, 1

This, in contrast, is *not* a legitimate application, for · is not the main connective:

1. (A · B) ⊃ C	
2. B ⊃ C	Simp, 1

In general, we can move from p · (q ⊃ r) to q ⊃ r, but not from (p · q) ⊃ r to p ⊃ r. The latter inference is not valid; we would not want a rule that allowed it.

To demonstrate the use of simplification, consider the argument

(2) Fran and Gloria both attended, but Hilary did not introduce them.

∴ Gloria attended.

To begin, assume (F · G) · ~H and prove G. We begin with an assumption.

1. (F · G) · ~H	/ ∴ G

Now, we exploit the conjunction to derive a smaller conjunction, (F · G). We cannot derive G directly. The connective to which we apply the rule must be the main connective of the formula.

1. (F · G) · ~H	/ ∴ G
2. (F · G)	Simp, 1

The right column tells us that the formula on line 2 comes from line 1 by applying S. We can easily apply that same rule again to obtain G, which finishes the proof.

1. (F · G) · ~H	/ ∴ G
2. (F · G)	Simp, 1
3. G	Simp, 2

This proof is very simple. But it already illustrates four important guidelines for constructing proofs:

GUIDELINES FOR CONSTRUCTING PROOFS

1. Look first at the conclusion: What do you want to prove?

2. Find (parts of) the conclusion in the premises.

3. Simplify the premises as much as possible.

4. Try to get single letters or their negations.

The rule of **conjunction** indicates how to introduce conjunctions. It reflects the reasoning in this argument:

(3) Jack knows where the money is.

Ken knows where the money is.

∴ Jack and Ken know where the money is.

It is a very simple rule; to prove a conjunction, it says, prove each conjunct. If p is true and q is true also, then p · q is true. So, from the two formulas p and q we can derive p · q:

CONJUNCTION (CONJ)

n.	p	
m.	q	
p.	p · q	Conj, n, m

The right column indicates that p · q comes from conjoining the formulas on lines *n* and *m*. The order in which p and q occur in the proof does not matter. We could just as easily have concluded q · p.

To see how we might use conjunction in a proof, note that switching the order of conjuncts produces an equivalent formula: A · B allows us to derive B · A. Again we begin with a premise:

1. A · B	/ ∴ B · A

To show q · p, we need to separate the two conjuncts. We can derive them separately, using simplification, and put them back together in the other order, using conjunction:

1. A · B	/ ∴ B · A
2. A	Simp, 1
3. B	Simp, 1
4. B · A	Conj, 3, 2

The conjunction rule allows us to conjoin two formulas on previous lines. Like most of our rules, it operates on entire formulas. From p ⊃ q and r we can derive (p ⊃ q) · r, but not p · r or q · r.

CONDITIONALS

Some logical systems have only one rule of inference: *modus ponens*, which sanctions the inference from p and p ⊃ q to q. It stands behind such arguments as

(4) If Socrates is human, then Socrates is mortal.

Socrates is human.

∴ Socrates is mortal.

And it reflects the truth table for the conditional: p ⊃ q is true if and only if p is false or q is true. If p and p ⊃ q are both true, q must be true as well.

MODUS PONENS (MP)

n.	p ⊃ q	
m.	p___	
p.	q	MP, n, m

This rule, for example, lets us show that

(5) If Wendy understands the situation, she'll tread carefully.
   She does understand.
   ∴ Wendy will tread carefully.

is valid, because it lets us derive T from the premises U ⊃ T and U.

---

1. U ⊃ T		
2. U	/ ∴ T	
3. T	MP, 1, 2	

The next rule allows negations to help in exploiting conditionals. Suppose we have established that if the butler committed the crime, it must have occurred before dark (because he was seen in the tavern all evening, say). Suppose we also establish that the crime did not occur before dark. We can infer that the butler is innocent. The rule sanctioning this inference is called **modus tollens.**

MODUS TOLLENS (MT)

n.	p ⊃ q	
m.	~q___	
p.	~p	MT, n, m

If p is true only if q is and q is false, then p must also be false. We can use this rule to show the validity of

(6) Laura will say no only if Patrick raises objections.
   If Oliver doesn't file his report, Patrick will raise no objections.

Oliver won't file his report.

∴ Laura will not say no.

---

    1. L ⊃ P

    2. ~O ⊃ ~P

    3. ~O   / ∴ ~L

    4. ~P   MP, 2, 3

    5. ~L   MT, 1, 4

The third rule is an argument form traditionally called **hypothetical syllogism.** It allows conditionals to form chains of reasoning. It underlies such arguments as

(7) If your baby cries, my baby will wake up.

    If my baby wakes up, she'll cry.

    ∴ If your baby cries, my baby will, too.

HYPOTHETICAL SYLLOGISM (HS)

    n.   p ⊃ q

    m.   q ⊃ r

    p.   p ⊃ r   HS, n, m

This rule permits a straightforward proof of the validity of

(8) If Ernest quits, Walt will replace him.

    If Walt does, Margaret is sure to win.

    If Margaret wins, Ernest will be sorry.

    ∴ If Ernest quits, he'll be sorry.

---

    1. E ⊃ W

    2. W ⊃ M

    3. M ⊃ S   / ∴ E ⊃ S

    4. E ⊃ M   HS, 1, 2

    5. E ⊃ S   HS, 4, 3

P R O B L E M S

Show that the conclusion of each argument form is provable from the premises.

1. A · (B · C); ∴ C

2. A · D; A ⊃ (D ⊃ C); ∴ C

3. ~A ⊃ B; ~A · C; ∴ C · B

4. A ⊃ B; A ⊃ C; A; ∴ B · C

5. (A · B) · ~C; ∴ A · (B · ~C)

6. A ⊃ (B ⊃ C); A · ~C; ∴ ~B

7. ~B · (C ⊃ ~A); C; ∴ ~A · ~B

8. A ⊃ (B ⊃ ~C); B · C; A; ∴ C · ~C

9. A ⊃ B; ~A ⊃ ~C; ~C ⊃ ~D; ~B; ∴ ~D

10. ~A ⊃ (B · (C ⊃ ~D)); ~A · C; ∴ ~D · B

Use deduction to show that the following arguments are valid.

11. If the Democrats obstruct the president's legislative program, the market will lose confidence. The Democrats will obstruct that program only if they can gain politically by so doing. The Democrats will obstruct the president's legislative initiatives. Therefore, the Democrats will gain politically, but the market will lose confidence.

12. Georgia will lose the case if the Court decides to base its decision on *Davis*. Although the composition of the Court is more conservative than it was a few years ago, it will base its decision on *Davis*. So, Georgia will lose.

13. If many students come to the party, they'll eat lots of food and the party will be rowdy. If students eat lots of food, the party will be expensive. Many students will come to the party, although only a few have been invited. Thus, the party will be rowdy and expensive.

14. If Madsen gets some money from the deal, then Hendricks will too. If Hendricks gets some money, then Madsen will; moreover, if Hendricks gets some, then so will Franklin. If Franklin gets some money, Hendricks will. So, if Madsen gets some money, so will Franklin, and if Franklin gets some, so will Madsen.

15. This product will fail only if it isn't given adequate advertising support. If it fails, then the company won't show a profit for the

year. It won't be advertised adequately only if the vice president believes that another product has more potential. The product is great, but it will not succeed. It follows that the vice president thinks another product has more potential, and the company won't show a profit this year.

Fill in the missing justifications in the following proofs.

16. 1. A · (B ⊃ C)
    2. A ⊃ B   / ∴ C
    3. A
    4. B ⊃ C
    5. B
    6. C

17. 1. A · (B ⊃ C)
    2. A ⊃ B   / ∴ C
    3. A
    4. B ⊃ C
    5. A ⊃ C
    6. C

18. 1. (A ⊃ B) · (~A ⊃ C)
    2. ~B   / ∴ C
    3. A ⊃ B
    4. ~A ⊃ C
    5. ~A
    6. C

19. 1. A · (B · C)
    2. A ⊃ D
    3. B ⊃ E   / ∴ (C · D) · E
    4. A
    5. B · C
    6. B
    7. C
    8. D
    9. E
    10. C · D
    11. (C · D) · E

20. 1. ~A · (~B ⊃ C)
    2. (B ⊃ A) · (C ⊃ ~D)
    3. E ⊃ D
    4. ((~A · ~B) · ~E) ⊃ F   / ∴ F
    5. ~A
    6. ~B ⊃ C
    7. B ⊃ A

8. C ⊃ ~D
9. ~B
10. C
11. ~D
12. ~E
13. ~A · ~B
14. (~A · ~B) · ~E
15. F

 # 8.3 RULES OF IMPLICATION II: DISJUNCTIONS

The **addition** rule says that we may introduce a disjunction into a proof if we already have either disjunct. It underlies such arguments as

(9) Rachel is in London.

∴ Rachel is in London or East Grinstead.

It reflects the truth table definition of disjunction: p ∨ q is true if either p or q is true.

ADDITION (ADD)

n.	__p__ (or q)	
n+p.	p ∨ q	Add, n

We can use this rule to show that the argument

(10) Terry invests cautiously.

If she turns the account over to a professional manager or invests cautiously herself, then the funds will be secure.

∴ The funds will be secure.

is valid. We begin by stating the premises. We then obtain the antecedent of the conditional by addition and apply *modus ponens* to get the conclusion.

1. C		
2. (M ∨ C) ⊃ F	/ ∴ F	
3. M ∨ C	Add, 1	
4. F	MP, 2, 3	

How can we use a disjunction to derive a conclusion? The answer lies in two rules. The first is called **disjunctive syllogism.** It reflects the reasoning of this argument:

(11) Either Yvonne or Ursula arrived in town last night.

Yvonne didn't arrive last night.

∴ Ursula did.

If at least one of p or q is true and p is false, q must be true. Similarly, if p or q is true, p must be true if q is false.

DISJUNCTIVE SYLLOGISM (DS)

n.	p ∨ q		
m.	~p	(or ~q)	
p.	q	(or p)	DS, n, m

To illustrate this rule, let's show that this argument is valid:

(12) Iggy got the letter unless someone intercepted it.

If our agents supervised the delivery, then nobody intercepted it.

Our agents did supervise.

∴ Iggy got the letter.

1.	L ∨ I	
2.	S ⊃ ~I	
3.	S	/ ∴ L
4.	~I	MP, 2, 3
5.	L	DS, 1, 4

Disjunctive syllogism, incidentally, allows us to show that contradictions imply anything. If we have both p and ~p, then any arbitrary q follows.

1. p	
2. ~p	/ ∴ q
3. p ∨ q	Add, 1
4. q	DS, 3, 2

Another rule for using disjunctions amounts to "proof by cases." It underlies such arguments as

(13) Paul will win the event unless Ann scores perfectly.

   If Paul wins, Ann will get the silver medal.

   If Ann scores perfectly, she'll get the gold medal.

   ∴ Ann will win the silver or the gold medal.

CONSTRUCTIVE DILEMMA (CD)

n.	p ∨ r	
m.	p ⊃ q	
p.	r ⊃ s	
q.	q ∨ s	CD, n, m, p

CD lets us demonstrate the validity of

(14) Serge will write on the Russian Revolution or on the Versailles conference.

   If he writes on the Russian Revolution, he'll do well.

   If he writes on Versailles, he will get lost in details.

   If he gets lost in details, his paper will be too long.

   ∴ Serge will do well unless his paper is too long.

1. R ∨ C		
2. R ⊃ W		
3. C ⊃ D		
4. D ⊃ L	/ ∴ W ∨ L	
5. C ⊃ L	HS, 3, 4	
6. W ∨ L	DS, 1, 2, 5	

PROBLEMS

Show that the conclusion of each argument form is provable from the premises.

1. p · ~q; ∴ p ∨ q

2. p ∨ q; ~r; p ⊃ r; ∴ q

3. r ∨ p; ~r · s; ∴ p ∨ q

4. ~p ∨ q; ~q; ~p ⊃ r; ∴ r

5. p ⊃ q; ~r · s; p ∨ r; ∴ s · q

6. ~p ∨ q; ~q ∨ r; ~r; ∴ ~p

7. p ⊃ q; ~r ∨ q; ~q; ∴ ~p · ~r

8. p · ~s; r ⊃ s; p ⊃ (q ∨ r); ∴ q

9. p ⊃ r; ~s; p ∨ ~q; ~q ⊃ s; ∴ r

10. (p ∨ q) ⊃ r; s ⊃ p; (r ⊃ t) · s; ∴ t

11. p ∨ ~q; p ⊃ r; ~q ⊃ s; ~r · t; ∴ s

12. p ∨ q; r ∨ s; ~p · ~s; ∴ (q · r) ∨ t

13. (p ⊃ q) ∨ (r ⊃ s); ~(p ⊃ q); ~s; ∴ ~r

14. p · q; r · ~s; q ⊃ (p ⊃ k); k ⊃ (r ⊃ (s ∨ m)); ∴ m

15. p ∨ (r ∨ q); (r ⊃ m) · (q ⊃ n); (m ∨ n) ⊃ (p ∨ q); ~p; ∴ q

16. p ⊃ q; q ⊃ (r · s); p ∨ ~t; ~t ⊃ ~u; ~u ⊃ v; ∴ (r · s) ∨ v

Use deduction to show that the following arguments are valid.

17. The house contains either a fireplace or a deck; it contains either a fourth bedroom or a den. Evidently it doesn't have a fourth bedroom and doesn't have a deck. Hence, it has a den and a fireplace.

18. If management remains resolute, it can stop the hostile takeover attempt. But unless the company can improve its cash flow position, there will be no way to stop the takeover. Management will remain resolute only if it understands the consequences of the acquisition for its own position. Management understands the consequences, but the company can't improve its cash flow position. Therefore, management will not stop the hostile takeover.

19. Patricia is clever but won't work very hard. If she's clever, the boss will like her and either promote her or give her a bonus. If

Patricia is promoted, she'll work hard. Therefore, the boss will give Patricia a bonus.

20. No one knows where Rick was last night. Unless he has an alibi, he's the prime suspect in the murder. He has an alibi only if somebody knows where he was last night. Therefore, unless new evidence points to someone else, Rick is the prime suspect.

21. Peter will doubt me unless you come along and explain things to him. Margaret won't tell him anything about our idea. If you explain things to him, he'll understand. Margaret will tell him about our idea if he doubts me, but what she tells him will just make him angry. Therefore, Peter will understand unless you chicken out.

22. If John attends the meeting, he'll be able to convince at least a few people to vote against the development proposal. If we can get most of the neighborhood to sign a petition, then the developers will change their proposal. Either we'll get most of the neighborhood to sign a petition or John will attend the meeting. So, unless the developers change their proposal, John will be able to convince at least a few people to vote against it.

23. The dollar will fall if foreign banks sterilize their intervention in the currency markets or if the Fed does nothing to defend it. Germany and Japan are eager to keep their currencies strong, but they'll intervene in the markets if the dollar drops. If Germany and Japan want to maintain strong currencies, foreign banks will sterilize their interventions. It follows that they will intervene in the markets.

24. If the party maintains its current economic policy, there will be a flight of capital to other countries. But the party will maintain its policy if it tightens its control over the economy. If the party maintains its policy and capital flees to other countries, then it will tighten its grip on the economy. If the party maintains its current policy, however, the nation will have to pay large amounts of foreign debt in hard currency, and this it won't do. So, the party won't tighten its grip on the economy.

Fill in the missing justifications in the following proofs.

25. 1. ~A · (B ∨ C)
    2. B ⊃ D
    3. C ⊃ A   / ∴ D
    4. ~A
    5. ~C

6. B ∨ C
7. B
8. D

26. 1. A ∨ B
    2. A ⊃ C
    3. B ⊃ D
    4. D ⊃ E   / ∴ C ∨ E
    5. B ⊃ E
    6. C ∨ E

27. 1. A ⊃ B
    2. ~C ⊃ ~D
    3. A ∨ ~C
    4. (B ∨ ~D) ⊃ ~A   / ∴ ~C
    5. B ∨ ~D
    6. ~A
    7. ~C

28. 1. A ∨ B
    2. ~B ⊃ D
    3. ~A ⊃ C
    4. B ⊃ C
    5. A ⊃ E
    6. ~A ∨ ~B   / ∴ (C ∨ D) · (E ∨ C)
    7. E ∨ C
    8. C ∨ D
    9. (C ∨ D) · (E ∨ C)

29. 1. (A · B) ⊃ C
    2. (A · D) ∨ (A · B)
    3. ~C ∨ ~D
    4. (~D ⊃ E) · ~E   / ∴ D
    5. ~D ⊃ E
    6. ~~D
    7. ~E
    8. ~C
    9. ~(A · B)
    10. A · D
    11. D

30. 1. (A ⊃ B) ⊃ (A ⊃ C)
    2. A ∨ C
    3. ~E ⊃ (A ⊃ B)
    4. ~D · ~E
    5. C ⊃ D   / ∴ C · ~C
    6. ~D

> 7. ~C
> 8. A
> 9. ~E
> 10. A ⊃ B
> 11. B
> 12. A ⊃ C
> 13. C
> 14 C · ~C

 # 8.4   RULES OF REPLACEMENT: CONNECTIVES

The rules considered so far are sound: In any application of those rules, the premises imply the conclusion. But the reverse is generally not true. The formula p · q, for example, implies p, but p does not imply p · q. For that reason, the rules work in only one direction. We can apply them only to formulas with the appropriate main connectives. Simplification applies only to conjunctions—formulas with dots as their main connectives. Thus, rules of implication apply only to formulas as a whole. If we tried using them on parts of formulas, we would produce some howlers:

> (15) If you take the exam and get an A, you'll pass.   (T · A) ⊃ P
>
>   ∴ If you take the exam, you'll pass.   T ⊃ P   Simp
>   (WRONG)
>
> (16) If you ask her, she'll go out with you.   A ⊃ G
>
>   If you ask her or insult her, she'll go out with you.   (A ∨ I) ⊃ G   Add
>   (WRONG)

The rules to be presented in this section, however, *do* work in both directions. They express logical equivalences. For that reason, we will write them with the symbol ⇔, which is to be read as "is logically equivalent to." It is not a symbol of our logical language, on the same level as the tilde, dot, or wedge, but a symbol of the metalanguage in which we discuss that language.

Because these rules express logical equivalences, they apply to parts of formulas—more precisely, subformulas—as well as to entire formulas. If we substitute, for any subformula in a formula, a logically equivalent subformula, the result is equivalent to the original formula. This is called the **principle of replacement.** These rules are thus called rules of replacement.

DOUBLE NEGATION

The double negation rule allows us to add or delete two consecutive negation signs. Recall that negation turns a true statement into a false one and vice versa. If there are two consecutive negations, then, a true statement is turned into a false one, then back into a true one. Similarly, a false statement is turned true and then false again. So, the result is always equivalent to the original.

DOUBLE NEGATION (DN)

p	$\Leftrightarrow$	$\sim\sim$p	DN

Double negation is especially useful in combination with disjunctive syllogism, as this proof shows:

1. $\sim$p $\lor$ $\sim$q		
2. $\sim$r $\supset$ q		
3. r $\supset$ s		
4. $\sim$s		$/ \therefore \sim$p
5. $\sim$r		MT, 3, 4
6. q		MP, 2, 5
7. $\sim\sim$q		DN, 6
8. $\sim$p		DS, 1, 7

DEMORGAN'S LAWS

DeMorgan's laws, named for the British logician Augustus DeMorgan (1806–1871), allow us to simplify negated formulas. They give equivalents for negated conjunctions and disjunctions. The negation of a conjunction is equivalent to a disjunction:

(17) a. Don's mother and father won't both go to the wedding.
  b. Either Don's mother or his father won't go the wedding.
(18) a. Either Israel or Syria will disapprove of the plan.
  b. Not both Israel and Syria will approve of the plan.

If p and q are not both true, then one or the other is false. The second rule says that the negation of a disjunction is equivalent to a conjunction.

(19) a.   Neither Brazil nor Argentina have nuclear weapons.
     b.   Brazil doesn't have nuclear weapons, and Argentina doesn't either.

(20) a.   Kermit doesn't live in a garbage can, and Miss Piggy doesn't either.
     b.   Neither Kermit nor Miss Piggy lives in a garbage can.

If neither p nor q is true, then p and q are both false.

DeMorgan's Laws (DM)

$\sim(p \cdot q)$	$\Leftrightarrow$	$(\sim p \vee \sim q)$	DM
$\sim(p \vee q)$	$\Leftrightarrow$	$(\sim p \cdot \sim q)$	DM

DeMorgan's laws capture simple facts about the truth table definitions of conjunction and disjunction. A conjunction $p \cdot q$ is true if and only if both p and q are true. It is false if and only if either p or q is false. That is the content of the first law. A disjunction $p \vee q$ is true if and only if either p or q is true. So, it is false if and only if p and q are both false. That is the content of the second law.

These laws let us demonstrate the validity of (21), as this proof shows. (Note the use of both laws.)

(21) If the chair or the dean agreed, the grade was dropped from Jim's record, and he may take the course again.

      But the grade was not dropped.

      ∴ The dean did not agree.

1. $(C \vee D) \supset (G \cdot T)$	
2. $\sim G$	$/ \therefore \sim D$
3. $\sim G \vee \sim T$	Add, 2
4. $\sim(G \cdot T)$	DM, 3
5. $\sim(C \vee D)$	MT, 1, 4
6. $\sim C \cdot \sim D$	DM, 5
7. $\sim D$	Simp, 6

MATERIAL IMPLICATION

The next rule allows us to transform disjunctions into conditionals and vice versa. It reflects the equivalence we noted in Chapter 6 of

(22) a. The patient will die unless we operate.
b. If we don't operate, the patient will die.

MATERIAL IMPLICATION (IMPL)

$p \supset q$	$\Leftrightarrow$	$\sim p \vee q$	Impl

Material implication rests on the truth table definition of the conditional as being true if the consequent is true or the antecedent is false. It permits us to derive a simplified form of constructive dilemma:

(23) The United States will reject negotiations unless Libya renounces terrorism.

Libya will do so only if Gadhafi is no longer its leader.

∴ The United States will reject negotiations unless Gadhafi is no longer Libya's leader.

1. $N \vee R$	
2. $R \supset \sim Q$	$/ \therefore N \vee \sim Q$
3. $\sim\sim N \vee R$	DN, 1
4. $\sim N \supset R$	Impl, 3
5. $\sim N \supset \sim Q$	HS, 4, 2
6. $\sim\sim N \vee \sim Q$	Impl, 5
7. $N \vee \sim Q$	DN, 6

This rule, together with a rule of distribution to be presented in the next section, also permits a proof of a useful principle sometimes called *consequent conjunction*

(24) If the president vetoes, then the House will vote to override.

If the president vetoes, then the Senate will vote to override.

∴ If the president vetoes, then the House and Senate will vote to override.

1. V ⊃ H	
2. V ⊃ S	/ ∴ V ⊃ (H · S)
3. ~V ∨ H	Impl, 1
4. ~V ∨ S	Impl, 2
5. (~V ∨ H) · (~V ∨ S)	Conj, 3, 4
6. ~V ∨ (H · S)	Dist, 5
7. V ⊃ (H · S)	Impl, 6

TRANSPOSITION

The transposition rule (mentioned earlier, in Chapter 6) relates two equivalent conditionals; for example,

(25) a. If nobody called to tell him not to come, Ned will be here soon.
     b. If Ned isn't here soon, then somebody called to tell him not to come.

It is easy to see from the truth tables that p ⊃ q and ~q ⊃ ~p have the same truth conditions. The first is true if and only if p is false or q is true. The second is true if and only if ~q is false or ~p is true—that is, if and only if q is true or p is false.

TRANSPOSITION (TRANS)

p ⊃ q	⇔	~q ⊃ ~p	Trans

This rule allows a simple proof of the validity of

(26) If we advertise more, we will sell more.
      If we sell more, we make more money.
      If we do not advertise more, the company will not grow.
      ∴ If we don't make more money, the company will not grow.

1. A ⊃ S	
2. S ⊃ M	
3. ~A ⊃ ~G	/ ∴ ~M ⊃ ~G
4. A ⊃ M	HS, 1, 2

5. ~M ⊃ ~A	Trans, 4
6. ~M ⊃ ~G	HS, 5, 3

ABSORPTION

Another rule reflects the equivalence of

(27) a. If you promise, so will I.
    b. If you promise, we both will.

It too reflects the truth table definition of the conditional: p ⊃ q means that q is true if p is. But, if p is true, p is true. So, p and q are true if p is.

ABSORPTION (ABS)

p ⊃ q	⇔	p ⊃ (p · q)	Abs

This rule allows us to prove this argument valid:

(28) The Israelis will agree to the settlement only if the United States gives its approval.
The United States will give approval only if Saudi Arabia agrees.
But Israel and Saudi Arabia will not both agree to the settlement.
∴ The Israelis will not agree.

1. I ⊃ U	
2. U ⊃ S	
3. ~(I · S)	/ ∴ ~I
4. I ⊃ S	HS, 1, 2
5. I ⊃ (I · S)	Abs, 4
6. ~I	MT, 5, 3

EXPORTATION

These conditionals are equivalent:

(29) a. If you get the check and we sell the stock, we can afford it.
    b. If you get the check, then, if we sell the stock, we can afford it.

The equivalence can be checked by truth table: $(p \cdot q) \supset r$ is false only when p and q are true and r is false. Similarly, $p \supset (q \supset r)$ is false only when p is true and $q \supset r$ is false, and $q \supset r$ is false only when q is true and r is false.

EXPORTATION

$(p \cdot q) \supset r$	$\Leftrightarrow$	$p \supset (q \supset r)$	Exp

This rule permits an easy proof of the validity of

(30) If the House and Senate vote to override, the president will be embarrassed.
The House will vote to override.
∴ The president will avoid embarrassment only if the Senate does not override.

1. $(H \cdot S) \supset P$		
2. H	$/ \therefore {\sim}P \supset {\sim}S$	
3. $H \supset (S \supset P)$	Exp, 1	
4. $S \supset P$	MP, 3, 2	
5. ${\sim}P \supset {\sim}S$	Trans, 4	

THE BICONDITIONAL

How do we prove a biconditional? A biconditional is so called because it amounts to two conditionals. Mathematicians, for example, often prove biconditionals in two steps. They prove the "left-to-right" and "right-to-left" directions separately. In other words, they prove two conditionals. To use biconditionals, we can simply reverse this procedure. If we have established a biconditional, we can deduce either or both conditionals.

Alternatively, we can think of the meanings of biconditionals as given by truth tables. A biconditional is true if and only if the formulas on both sides agree in truth value. That is, they must both be true or both be false. The rule of material equivalence, then, has two forms:

MATERIAL EQUIVALENCE (EQUIV)

$p \equiv q$	$\Leftrightarrow$	$(p \supset q) \cdot (q \supset p)$	Equiv
$p \equiv q$	$\Leftrightarrow$	$(p \cdot q) \vee ({\sim}p \cdot {\sim}q)$	Equiv

We can use this rule to show the validity of

(31) Ed will run if and only if he thinks he can win.

Ed thinks he can win.

∴ Ed will run.

by showing that R ≡ W and W imply R.

---

1. R ≡ W	
2. W	/ ∴ R
3. (R ⊃ W) · (W ⊃ R)	B, 1
4. W ⊃ R	Simp, 3
5. R	MP, 4, 2

PROBLEMS

Theophrastus (371–286 B.C.), a pupil of Aristotle, cited these principles as hypothetical syllogisms. Show that their conclusions are provable from their premises.

1. r ⊃ p; p ⊃ q; ∴ r ⊃ q.

2. r ⊃ p; p ⊃ q; ∴ ~q ⊃ ~r.

3. r ⊃ p; ~r ⊃ q; ∴ ~q ⊃ p.

4. r ⊃ q; p ⊃ ~q; ∴ r ⊃ ~p.

5. r ⊃ p; ~r ⊃ q; ∴ ~p ⊃ q.

Show that the conclusion formulas of these argument forms can be proved from the premise formulas.

6. A ⊃ B; ~(A · B); ∴ ~A

7. (C · A) ∨ (~C · ~A); ∴ C ≡ A

8. A ⊃ B; (A · B) ⊃ ~C; ∴ A ⊃ ~C

9. ~(A ⊃ B); ∴ A · ~B

10. A ⊃ (B ⊃ C); ~C; ∴ A ⊃ ~B

11. C ⊃ A; B ⊃ C; ∴ ~B ∨ A

12. ~C ∨ ~B; ∴ ~(C · ~~B)

13. C ∨ A ; ~C ∨ ~A; ∴ C ≡ ~A

14. $C \lor A$; $C \supset B$; $A \supset D$; $\therefore$ $\sim B \supset D$

15. $B \supset C$; $\sim B \supset A$; $A \supset D$; $\therefore$ $C \lor D$

16. $\sim D \lor (D \cdot C)$; $(D \supset C) \supset B$; $\therefore$ B

17. $(C \cdot \sim A) \supset \sim B$; B; $\therefore$ $\sim C \lor (C \cdot A)$

18. $C \supset \sim A$; B; $B \supset (A \lor \sim D)$; $\therefore$ $D \supset \sim C$

19. $C \supset \sim A$; $\sim C \lor B$; A; $(A \cdot B) \supset C$; $\therefore$ $\sim B$

20. $C \supset \sim A$; $\sim C \supset \sim B$; $B \lor \sim D$; $\therefore$ $\sim A \lor \sim D$

21. $C \equiv A$; $\therefore$ $\sim C \equiv \sim A$

22. $C \cdot A$; $C \supset (D \lor B)$; $\sim (B \cdot A)$; $\therefore$ D

23. $D \supset C$; $(D \cdot C) \supset A$; $B \supset E$; $B \lor D$; $\therefore$ $A \lor E$

24. $C \lor (A \lor D)$; $E \cdot \sim G$; $\sim (\sim E \lor G) \supset \sim C$; $(D \supset B) \cdot \sim B$; $\therefore$ A

25. $\sim (A \lor \sim D)$; $\sim A \supset (B \lor C)$; $\sim C \lor \sim D$; $(B \lor E) \supset (F \cdot (G \supset \sim D))$;
$\therefore$ $\sim (F \supset G)$

Use deduction to show that the following arguments are valid.

26. If you are ambitious, you'll never achieve all your goals. But life has meaning only if you have ambition. Thus, if you achieve all your goals, life has no meaning.

27. If I'm right, then you're a fool. If I'm a fool, I'm not right. If you're a fool, I am right. Therefore, we're not both fools.

28. If Socrates died, he died either while he was living or while he was dead. But he did not die while living; moreover, he surely did not die while he was already dead. Hence, Socrates did not die. (Sextus Empiricus)

29. If you have a cake, just looking at it will make you hungry; if looking at it makes you hungry, you will eat it. So you can't both have your cake and fail to eat it.

30. A man cannot serve both God and Mammon. But if a man does not serve Mammon, he starves; if he starves, he can't serve God. Therefore a man cannot serve God.

31. Sarah knows that the company is unlikely to promote her. She will look for another job if and only if she's unhappy. She's bound to be unhappy if she knows that she's unlikely to be promoted. So, unless she has some other reason for staying, Sarah will look for another job.

32. If we can avoid terrorism only by taking strong retaliatory measures, then we have no choice but to risk innocent lives. But if we don't take strong retaliatory measures, we'll certainly

fall prey to attacks by terrorists. Nevertheless, we refuse to risk innocent lives. Consequently, we will fall prey to terrorism.

33. Nothing can be conceived as greater than God. If God existed in our imaginations, but not in reality, then something would be conceivable as greater than God (namely, the same thing, except conceived as existing in reality). Therefore, if God exists in our imaginations, He exists in reality. (Anselm)

34. If the objects of mathematics are material things, then mathematics can't consist entirely of necessary truths. Mathematical objects are immaterial only if the mind has access to a realm beyond the reach of the senses. Mathematics does consist of necessary truths, although the mind has no access to any realm beyond the reach of the senses. Therefore, the objects of mathematics are neither material nor immaterial.

35. We cannot both maintain high educational standards and accept almost every high school graduate unless we fail large numbers of students when (and only when) many students do poorly. We will continue to maintain high standards; furthermore, we will placate the legislature and admit almost all high school graduates. Of course, we can't both placate the legislature and fail large numbers of students. Therefore, not many students will do poorly.

Fill in the missing justifications in the following proofs.

36. 1. $A \equiv (A \vee B)$
    2. $\sim A$ $/ \therefore \sim B$
    3. $(A \supset (A \vee B)) \cdot ((A \vee B) \supset A)$
    4. $(A \vee B) \supset A$
    5. $\sim(A \vee B)$
    6. $\sim A \cdot \sim B$
    7. $\sim B$

37. 1. $A \supset B$
    2. $(A \cdot B) \supset C$
    3. $\sim A \supset D$
    4. $C \supset E$ $/ \therefore D \vee E$
    5. $A \supset (A \cdot B)$
    6. $A \supset C$
    7. $\sim A \vee C$
    8. $D \vee E$

38. 1. $(A \cdot B) \supset C$
    2. $(B \supset C) \supset (B \supset D)$
    3. $\sim(D \vee E) \cdot B$ $/ \therefore \sim A$
    4. $\sim(D \vee E)$

    5. A ⊃ (B ⊃ C)
    6. A ⊃ (B ⊃ D)
    7. ~D · ~E
    8. ~D
    9. B
  10. B · ~D
  11. ~~B · ~D
  12. ~(~B ∨ D)
  13. ~(B ⊃ D)
  14. ~A

39. 1. ~C ⊃ (~A ∨ ~B)
    2. (A · D) ∨ (A · B)
    3. ~(C · D)
    4. (D ∨ E) · ~E   / ∴ D
    5. ~E
    6. D ∨ E
    7. D
    8. ~C ∨ ~D
    9. ~~D
  10. ~C
  11. ~A ∨ ~B
  12. ~(A · B)
  13. A · D
  14. D

40. 1. (A ∨ B) ⊃ (A ∨ C)
    2. A ⊃ C
    3. E ⊃ (~B ⊃ A)
    4. D · E
    5. ~(C · D)   / ∴ ~(C ∨ ~C)
    6. D
    7. E
    8. ~C ∨ ~D
    9. ~~D
  10. ~C
  11. ~A
  12. ~B ⊃ A
  13. ~~B
  14. B
  15. A ∨ B
  16. A ∨ C
  17. C
  18. ~~C
  19. ~C · ~~C
  20. ~(C ∨ ~C)

■ 8.5 RULES OF REPLACEMENT: ALGEBRA

Several rules illustrate important algebraic properties of conjunction and disjunction. The rules for commutativity (or commutation) and associativity (or association) say that order and grouping in a conjunction or disjunction make no difference to truth values. This is obvious from the truth tables. A continued conjunction is true if and only if all conjuncts are true, no matter how they are ordered or grouped. Similarly, a continued disjunction is true if and only if at least one disjunct is true. Changes in order or grouping yield an equivalent statement.

(32) a. Either Joe will call, or Kate will write you a letter.
  b. Either Kate will write you a letter, or Joe will call.

(33) a. Sandra and Dave worked on the project.
  b. Dave and Sandra worked on the project.

(34) a. Either Willy will walk, or Willy or Ron will bunt.
  b. Either Willy will walk or bunt, or Ron will bunt.

(35) a. France and Germany have sent representatives, and Hungary has, too.
  b. France has sent representatives, and Germany and Hungary have, too.

COMMUTATIVITY (COM)

$p \vee q$	$\Leftrightarrow$	$q \vee p$	Com
$p \cdot q$	$\Leftrightarrow$	$q \cdot p$	Com

ASSOCIATIVITY (ASSOC)

$(p \vee q) \vee r$	$\Leftrightarrow$	$p \vee (q \vee r)$	Assoc
$p \cdot (q \cdot r)$	$\Leftrightarrow$	$p \cdot (q \cdot r)$	Assoc

Another rule says that a conjunction or disjunction with identical components is redundant. We can eliminate repetition: Both $p \cdot p$ and $p \vee p$ are equivalent to p.

TAUTOLOGY (TAUT)

p	⇔	p ∨ p	Taut
p	⇔	p · p	Taut

This rule is surprisingly useful. It permits, for example, a simple proof of a special form of constructive dilemma:

1. p ∨ q		
2. p ⊃ r		
3. q ⊃ r	/ ∴ r	
4. r ∨ r	CD, 1, 2, 3	
5. r	I, 4	

DISTRIBUTION

Another rule says that conjunctions of disjunctions are equivalent to disjunctions of conjunctions. It has two forms. The first underlies the equivalence of

(36) a. Bob and either Vanna or Carol will star in the series.
     b. Either Bob and Vanna or Bob and Carol will star in the series.

while the second underlies the equivalence of

(37) a. Either Zeke or Al and Sam are bringing a dog.
     b. Zeke or Al is bringing a dog, and Zeke or Sam is bringing a dog.

DISTRIBUTION (DIST)

p · (q ∨ r)	⇔	(p · q) ∨ (p · r)	Dist
p ∨ (q · r)	⇔	(p ∨ q) · (p ∨ r)	Dist

This rule allows an easy proof of the validity of

(38) I will go with you unless Zeke will be there.

I will go with you unless Yvonne will be there.

Zeke and Yvonne will not both be there.

∴ I will go with you.

---

1. I ∨ Z		
2. I ∨ Y		
3. ~(Z · Y)		/ ∴ I
4. (I ∨ Z) · (I ∨ Y)		Conj, 1, 2
5. I ∨ (Z · Y)		Dist, 4
6. I		DS, 5, 3

P R O B L E M S

Construct a deduction to show that each of the following arguments is valid.

1. If Adam comes to the party, then Barbara will come, too. Furthermore, if Barbara comes to the party, so will Adam. Carlos and Adam will come unless Barbara does not come. So, Barbara will come to the party only if Carlos does.

2. Interest rates will rise unless Congress enacts a tax increase. But rates will rise even if Congress raises taxes. If interest rates rise, the unemployment level will rise. Either unemployment won't increase or the budget deficit will. It follows that there will be an increase in the budget deficit.

3. God is omnipotent if and only if He can do everything. If He can't make a stone so heavy that He can't lift it, then He can't do everything. But if He can make a stone so heavy that He can't lift it, He can't do everything. Therefore, either God is not omnipotent or God does not exist.

4. If the president pursues arms limitation talks, then, if he gets the foreign policy mechanism working more harmoniously, the European left will acquiesce to the placement of additional nuclear weapons in Europe. But the European left will never acquiesce to that. So, either the president won't get the foreign policy mechanism working more harmoniously or he won't pursue arms limitation talks.

5. If we introduce a new product line or give an existing line a new advertising image, then we'll be taking a risk, and we may

lose market share. If we don't introduce a new product line, we won't have to make large expenditures on advertising. So, if we don't take risks, we won't have to make large expenditures on advertising.

6. My cat does not sing opera unless all the lights are out. If I am very insistent, then my cat sings opera; but if I either turn out all the lights or howl at the moon, I am very insistent indeed. I always howl at the moon if I am not very insistent. Therefore, my lights are out, I am very insistent, and my cat is singing opera.

7. If we continue to run a large trade deficit, then the government will yield to calls for protectionism. We won't continue to run a large deficit only if our economy slows down or foreign economies recover. So, if foreign economies don't recover, then the government will resist calls for protectionism only if our economy slows down.

8. If companies continue to invest money here, then the government will sustain its policies. If they don't invest here, those suffering will be even worse off than they are now. But if the government sustains its policies, those suffering will be worse off. Thus, no matter what happens, the suffering will be worse off.

9. There will be scorpions in the house this week if there's any building going on in the neighborhood. There will be building if the depression in the housing market ends. If there are scorpions in the house, the cats will kill them and bring them to me as trophies. Therefore, if the housing depression ends, my cats will bring me scorpions as trophies.

10. If you profit from our mutual arrangement, then so will I. If I profit, then so will you. If you profit, of course, you can pay back some of the money you owe. If I profit from the arrangement and you can pay back some of the money, I'll be able to help you. Hence, I'll be able to help you unless one of us fails to profit from our mutual arrangement.

11. If people don't conserve energy, or let the environment suffer, utilities will have no choice but to rely on nuclear power. They'll avoid relying on atomic energy, moreover, just in case it's prohibitively expensive and risky. Nuclear power will remain expensive. If it also remains risky, therefore, the environment will suffer unless people conserve energy.

12. Applications to universities will fall unless universities do something to counteract demographic trends. If the number of

applications declines, then universities will teach fewer students. If they teach fewer students, they will have to shrink and become even more expensive. Therefore, unless universities somehow counteract demographic trends, they will become more expensive.

13. If Congress erects trade barriers and grants further subsidies to American farmers, then other countries will retaliate. If that happens, however, the economy will slow down; if there is a slowdown, then, unless Congress finds new sources of revenue, it won't give farmers additional subsidies. If Congress doesn't erect trade barriers, it won't give farmers added subsidies, but at least other countries won't retaliate. Therefore, unless Congress finds new revenue sources, farmers will get no additional subsidies from Congress.

14. If advertisers continue to use Saturday morning cartoons to peddle toys by creating shows about them, pressure will build for regulation, and ratings will decline if parents understand what's going on. If ratings decline, then advertisers will stop using cartoons to sell toys by having toys as their main characters. If pressure builds for regulation, then either advertisers will stop or Congress will act. If Congress acts, then they will stop. Hence, if advertisers continue, parents won't know what's going on, children will become obsessed with material goods, and American culture will irrevocably decline.

15. If the earth has been visited recently by extraterrestrial beings, then the government will have kept the information silent although the visits occurred. If some reported UFO sightings have been authentic, then the earth has been visited recently by extraterrestrials. But if those sightings have been authentic, then our current understanding of our place in the universe is seriously mistaken. Therefore, either our understanding of our place in the universe is seriously mistaken or, even if there have been recent extraterrestrial visits to the earth and the government has kept information about them silent, no reported UFO sightings are authentic.

Construct deductions to demonstrate the validity of the following:

16. $\sim C \vee \sim A$; $\therefore A \supset \sim C$

17. $C \equiv (C \cdot A)$; $\therefore C \supset A$

18. $A \vee (B \vee C)$; $\sim B$; $\therefore A \vee C$

19. $A \supset B$; $\sim(B \cdot A)$; $\therefore \sim A$

20. A ≡ (B ∨ D); ~A; ∴ ~B

21. A ≡ B; ~(H ⊃ B); ∴ ~A

22. ~(C · ~A) ∨ ~C; ∴ C ⊃ A

23. C ⊃ (A ∨ B); ∴ (C ⊃ A) ∨ B

24. A · B; ∴ ~(~B · ~C) · A

25. ((A · B) · C) · ~D; ∴ A · (B · ~D)

26. A ∨ (B ∨ C); ∴ B ∨ (A ∨ C)

27. (~A ⊃ B) ∨ C; ∴ ~A ⊃ (B ∨ C)

28. (C ∨ A) ∨ ~B; ∴ ~C ⊃ (~A ⊃ ~B)

29. (B · ~C) ∨ (A · B); ∴ (C ⊃ A) · B

30. A ≡ B; A ⊃ C; ∴ A ⊃ (B · C)

31. (A ⊃ (A · B)) ⊃ (C ⊃ D); C · H; A ≡ B; ∴ D · H

32. ~A ≡ ~B; ~B ⊃ ~C; ~A ⊃ ~D; ~A · E; ∴ ~D · ~C

33. A ⊃ (B ⊃ ~C); ~~B ⊃ ~A; C · A; ∴ B ⊃ (~A · ~C)

34. ~C ≡ ~B; A ∨ ~B; A ≡ D; ∴ D ∨ ~C

35. A · B; A ⊃ ~~C; B ⊃ D; ~C ∨ H; A ⊃ E; ∴ (H · D) · E

36. A ∨ B; C ≡ B; ∴ C ∨ A

37. A ⊃ (B ∨ C); (~B · M) ∨ (D ⊃ A); ~(~C ⊃ ~A); ∴ ~D ∨ B

38. ~D ⊃ ~K; (D · E) ⊃ (A ≡ B); ~(~A ∨ B); ∴ E ⊃ ~K

39. A · D; B ∨ C; A ≡ ~B; ∴ C · D

40. (C ∨ A) · B; A ⊃ D; ∴ ~C ⊃ (B ⊃ D)

Other authors have adopted rules for sentential logic that correspond to these argument forms. Show that each argument form is valid in our system.

41. p ≡ ~q; ∴ ~(p ≡ q)   (Negated Biconditional)

42. ~(p ≡ q); ∴ p ≡ ~q   (Negated Biconditional)

43. p ≡ q; ∴ ~q ≡ ~p   (Biconditional Transposition)

44. ~q ≡ ~p; ∴ p ≡ q   (Biconditional Transposition)

45. p ≡ q; ∴ q ≡ p   (Biconditional Commutation)

46. p ⊃ q; p ⊃ r; ∴ p ⊃ (q · r)   (Consequent Conjunction)

47. p ⊃ q; r ⊃ s; ~q ∨ ~s; ∴ ~p ∨ ~r   (Destructive Dilemma)

48. p ⊃ q; p ⊃ r; ~q ∨ ~r; ∴ ~p   (Destructive Dilemma)

49. ~(p · q); p; ∴ ~q   (Conjunctive Argument)

50. p ∨ q; ~q ∨ r; ∴ p ∨ r   (Disjunctive Transitivity)

Construct proofs to demonstrate the validity of the following arguments. (Warning: These are challenging!)

51. ~(~C · A) · (C ≡ ~A); ∴ C ≡ (A ⊃ B)

52. C · (~A · ~E); C ⊃ (D ⊃ B); D ⊃ (B ⊃ (E ∨ A)); ∴ ~D

53. C ⊃ (D · B); (B ∨ ~D) ⊃ (A · H); H ≡ J; ∴ C ⊃ J

54. (A ≡ (B · A)) ≡ (A ≡ C); A ⊃ B; C · D; ∴ A · (B · (C · D))

55. D ⊃ (B · E); ~H ⊃ A; (A ⊃ C) · ~G; A ⊃ D; (B · C) ⊃ ~H;
    ∴ A ≡ (B · C)

56. B · G; (D · ~C) ⊃ ~A; (B ⊃ ~E) ≡ ~A; B ⊃ D; (B · D) ⊃ ~C;
    ∴ G · ~E

57. ~(A ⊃ ~B); C ⊃ (~A ∨ ~B); (C ∨ D) ≡ E; ∴ E ≡ D

58. A ≡ B; B ∨ A; B ⊃ C; ∴ A · C

■ 8.6 CATEGORICAL PROOFS

So far, our system allows us to construct only hypothetical proofs. Our system is **complete:** Given some premises, we can deduce any conclusion that follows from them. But we cannot yet deduce a theorem from no premises at all.

To do that, however, we need to add only one rule. It differs from our other rules, for it allows us to introduce something into a proof rather than derive something from a previous line. It is also very simple. Every formula of the form p ⊃ p is a tautology. English sentences like

(39) a. If you're wrong, you're wrong.
      b. If this is Tuesday, this is Tuesday.

and so on are logical truths. At any point in a proof, therefore, we may write down a conditional whose antecedent is the same as its consequent:

SELF-IMPLICATION (SI)

n.	p ⊃ p	SI

This rule has no premises and refers to no previous lines. No matter what has happened earlier in the proof, p ⊃ p must be true. So, we can write such a formula at any point.

We can use self-implication to show that the formula ~~p ⊃ p is a tautology.

---

1. p ⊃ p	SI
2. ~~p ⊃ p	DN, 1

This is a categorical proof; it has no assumptions. It shows that ~~p ⊃ p is a theorem of our system.

More interesting, we can prove the tautology traditionally called the law of excluded middle, p ∨ ~p:

---

1. p ⊃ p	SI
2. ~p ∨ p	Impl, 1
3. p ∨ ~p	Com, 2

We can also prove the law of noncontradiction, ~(p · ~p):

---

1. p ⊃ p	SI
2. ~p ∨ p	Impl, 1
3. ~p ∨ ~~p	DN, 2
4. ~(p · ~p)	DM, 3

In fact, any tautology can be proved using the rules considered so far.

PROBLEMS

Show that these formulas are theorems of our natural deduction system.

1. (p ⊃ p) ∨ (p ⊃ q)

2. (p ⊃ q) ⊃ (~q ⊃ ~p)

3. (p ∨ p) ⊃ p

4. $(p \lor q) \supset (q \lor p)$

5. $p \equiv \sim\sim p$

6. $((p \lor q) \lor \sim r) \supset \sim\sim(p \lor (\sim\sim q \lor \sim r))$

7. $(p \lor (q \lor r)) \equiv ((p \lor q) \lor r)$

8. $(p \cdot q) \equiv (q \cdot p)$

9. $q \supset (p \lor q)$

10. $(p \lor (q \lor r)) \supset (q \lor (p \lor r))$

11. $(\sim p \supset p) \supset p$

12. $p \supset (\sim p \supset q)$

13. $(p \cdot q) \supset p$

14. $(p \supset (p \supset q)) \supset (p \supset q)$

15. $(p \supset q) \supset ((p \supset r) \supset (p \supset (q \cdot r)))$

16. $(p \supset r) \supset ((q \supset r) \supset ((p \lor q) \supset r))$

17. $(p \supset (q \supset r)) \supset (q \supset (p \supset r))$

18. $p \supset (q \supset p)$

19. $(p \cdot q) \supset (q \lor r)$

20. $(p \supset q) \lor (q \supset p)$

21. $(((p \cdot q) \lor (p \cdot \sim q)) \lor (\sim p \cdot q)) \lor (\sim p \cdot \sim q)$

Use SI to demonstrate the validity of the following:

22. $A \lor B$; $B \supset C$; $\therefore A \lor C$

23. $A \supset B$; $\sim A \supset C$; $\therefore B \lor C$

24. $A \supset B$; $\sim D \supset \sim B$; $\sim A \supset C$; $\sim D \lor E$; $\therefore C \lor E$

25. $A \supset C$; $\therefore (A \cdot B) \supset (B \supset C)$

■ 8.7 INDIRECT PROOFS

Although our rules allow us to establish the validity of every valid argument form, another method can make proofs easier: the method of **indirect proof.** As we saw in Chapter 5, Aristotle used this method for proving the validity of syllogisms. It involves assuming the premises and then, in addition, the contradictory of the conclusion.

Why do indirect proofs work? Why, that is, do they show arguments to be valid? Recall that an argument is valid if the truth of its premises guarantees the truth of its conclusion—if, equivalently, its premises can never be true while its conclusion is false. An indirect proof (like a semantic tableau; see Chapter 7) begins with the assumption that the premises are true and the conclusion is false and shows that this leads to a contradiction. If it does, the proof shows that it is impossible for the premises to be true while the conclusion is false. The proof shows, in other words, that the argument is valid. In effect, then, an indirect proof begins with an assumption that an argument is invalid and shows that the assumption is absurd.

In propositional logic, indirect proof is especially useful for proving negations and disjunctions. Consider, for example, this valid argument:

(40)   Pat and Terry aren't taking chemistry.

Rob will pass chemistry only if he gets help.

If Rob is to pass, he'll get help if and only if Terry or Pat takes chemistry, too.

∴ Rob won't pass chemistry.

Proving its validity in our system as it stands is difficult and unintuitive:

1. ~(P ∨ T)	
2. R ⊃ H	
3. R ⊃ ((P ∨ T) ≡ H)	/ ∴ ~R
4. R ⊃ (((P ∨ T) ⊃ H) · (H ⊃ (P ∨ T)))	Equiv, 3
5. ~R ∨ (((P ∨ T) ⊃ H) · (H ⊃ (P ∨ T)))	Impl, 4
6. (~R ∨ ((P ∨ T) ⊃ H)) · (~R ∨ (H ⊃ (P ∨ T)))	Dist, 5
7. ~R ∨ (H ⊃ (P ∨ T))	Simp, 6
8. ~R ∨ (~H ∨ (P ∨ T))	Impl, 7
9. (~R ∨ ~H) ∨ (P ∨ T)	Assoc, 8
10. ~R ∨ ~H	DS, 9, 1
11. ~H ∨ ~R	Com, 10
12. H ⊃ ~R	Impl, 11
13. R ⊃ ~R	HS, 2, 12
14. ~R ∨ ~R	Impl, 13
15. ~R	Taut, 14

This proof is not very natural at all. It would be much more straightforward to reason in the following way. "Suppose the premises were true but that ~R were false, so R were true. Then H would also be true, and so would (P ∨ T) ≡ H; thus, P ∨ T would be true, contradicting the first premise." This is an indirect proof. It begins by assuming the premises and the negation of the conclusion. The proof ends with a contradiction. For this reason, the method of indirect proof is traditionally called **reductio ad absurdum**—a reduction to absurdity of an assumption.

In general, indirect proofs have the form

INDIRECT PROOF

Premises

Negation of Conclusion   AIP

.

.

.

Contradiction

Every indirect proof begins with assumptions: first, any premises or other hypotheses on which the conclusion depends and, second, the assumption for indirect proof (AIP), which is the negation of the desired conclusion. The proof succeeds when we reach a contradiction of the form p · ~p for any formula p.

Proving the validity of (40) is much easier indirectly:

---

1. ~(P ∨ T)	
2. R ⊃ H	
3. R ⊃ ((P ∨ T) ≡ H)	/ ∴ ~R
4. ~~R	AIP
5. R	DN, 4
6. H	MP, 2, 5
7. (P ∨ T) ≡ H	MP, 3, 5
8. ((P ∨ T) ⊃ H) · (H ⊃ (P ∨ T))	Equiv, 7
9. H ⊃ (P ∨ T)	Simp, 8
10. P ∨ T	MP, 9, 6
11. (P ∨ T) · ~(P ∨ T)	Conj, 10, 1

Another example of the usefulness of indirect proof is in showing the validity of this admittedly peculiar argument form, here proved directly:

1. $(p \supset q) \supset p$	$/ \therefore p$
2. $\sim(p \supset q) \vee p$	Impl, 1
3. $\sim(\sim p \vee q) \vee p$	Impl, 2
4. $(\sim\sim p \cdot \sim q) \vee p$	DM, 3
5. $p \vee (\sim\sim p \cdot \sim q)$	Com, 4
6. $(p \vee \sim\sim p) \cdot (p \vee \sim q)$	Dist, 5
7. $p \vee \sim\sim p$	Simp, 6
8. $p \vee p$	DN, 7
9. p	Taut, 8

Again, this proof seems roundabout and unilluminating. Indirect proof is oxymoronically more straightforward:

1. $(p \supset q) \supset p$	$/ \therefore p$
2. $\sim p$	AIP
3. $\sim(p \supset q)$	MT, 1, 2
4. $\sim(\sim p \vee q)$	Impl, 3
5. $\sim\sim p \cdot \sim q$	DM, 4
6. $\sim\sim p$	Simp, 5
7. $\sim p \cdot \sim\sim p$	Conj, 2, 6

Indirect proofs allow us to demonstrate the validity of tautologies as well as arguments with premises, even without the additional rule of section 8.6. Without that rule, there is no way to begin a proof, for all the other rules allow us to derive formulas from previous lines; none allows us to introduce a line out of nowhere. Indirect proofs, however, introduce the negation of the conclusion as a hypothesis. That allows a proof even without any additional premises.

Consider, for example, the traditional law of excluded middle, $p \vee \sim p$:

1.	~(p ∨ ~p)	AIP
2.	~p · ~~p	DM, 1

The law of contradiction, ~(p · ~p), allows a similarly simple proof:

1.	~~(p · ~p)	AIP
2.	p · ~p	DN, 1

To take a more complex example, consider (p · q) ⊃ p:

1.	~((p · q) ⊃ p)	AIP
2.	~(~(p · q) ∨ p)	Impl, 1
3.	~~(p · q) · ~p	DM, 2
4.	(p · q) · ~p	DN, 3
5.	p · q	Simp, 4
6.	p	Simp, 5
7.	~p	Simp, 4
8.	p · ~p	Conj, 6, 7

P R O B L E M S

On the way to the barber shop (adapted from Lewis Carroll): You are trying to decide which of three barbers—Allen, Baker, and Carr—will be in today. You know Allen has been sick and so reason that (1) if Allen is out of the shop, his good friend Baker must be out with him. But, since they never leave the shop untended, (2) if Carr is out of the shop, then, if Allen is out with him, Baker must be in. Show that (1) and (2) imply

1. that not all three are out;

2. that Allen and Carr are not both out;

3. that, if Carr and Baker are in, so is Allen.

Do the following problems using indirect proof:

4. A ⊃ B; ~A ∨ ~B; ∴ ~A

5. A ≡ B; ∴ ~(~A ∨ ~C) ⊃ B

6. A ≡ B; ∴ ~(G ⊃ ~B) ⊃ A

7. A ≡ ~B; A ∨ B; C; ∴ A

8. C ⊃ (~B ∨ D); C ⊃ ~D; ∴ C ⊃ ~B

9. D ⊃ (B · C); A ⊃ (~B · ~E); ∴ (A · D) ⊃ F

10. (C ⊃ A) ⊃ (B ⊃ D); ~(C · A) ⊃ D; ∴ ~D ⊃ ~B

11. ~C ∨ (A · B); (B ∨ ~A) ⊃ (D · H); (H · J) ∨ ~(H ∨ J); ∴ ~C ∨ J

12. ~(~A · B); (A ∨ C) ⊃ ~(M · D); B ≡ D; ∴ M ⊃ ~(B ∨ D)

13. A ⊃ (B ⊃ C); (~B ≡ D) ⊃ ~A; A ∨ M; ∴ (C · D) ⊃ M

14. (A ≡ B) ≡ C; ~(A ≡ ~C); (B ∨ ~D) ⊃ E; ∴ E

15. (B ≡ ~A) ⊃ ~C; (~B · D) ∨ (A · M); (D ∨ M) ⊃ C; ∴ A ⊃ B

PREDICATE LOGIC

The theory of categorical syllogisms treats arguments having a very restricted form. Every statement must have the structure *All F are G, No F are G, Some F are G,* or *Some F are not G,* where F and G are terms applying to individual objects. Every syllogistic argument must have two such statements as premises and one as conclusion, with the terms meshing in just the right way. As a result, syllogistic logic covers a rather limited domain. It has no resources for dealing with the ancient skeptical argument analyzed in Chapters 1 and 6:

(1) If Socrates died, he died either when he was living or when he was dead.

But he did not die while living; for assuredly he was living, and as living he had not died.

Nor when he died; for then he would be twice dead.

∴ Socrates did not die.

Sentential logic, by taking statements as basic units of analysis, oversees a broader realm. It can handle arguments with any number of premises and statements of any length and with any degree of complexity. But it, too, suffers from narrow horizons. Just as the theory of the syllogism cannot solve problems characteristic of sentential logic, so sentential logic fails to solve syllogistic problems. Consider even a simple example:

(2) All humans are animals.

Some humans are philosophers.

∴ Some animals are philosophers.

This argument is surely valid. Yet sentential logic cannot explain why. It has no choice but to construe (2) as

(3) H

P

∴ A

which is not valid. The same happens with any syllogism. The validity of arguments such as (2) depends on the structure within statements, not on the structure relating distinct statements. No theory that declines to analyze atomic statements can hope to account for syllogistic reasoning.

The split between syllogistic and propositional logic, which began in Greece in the third or fourth century B.C., persisted for more than two thousand years. Neither theory could account for arguments that the other took as paradigms of correct reasoning. In the nineteenth century, logic became symbolic, which paved the way to a unification of the propositional and syllogistic realms.

Around 1879, two logicians working independently—Gottlob Frege (1848–1925), a German, and Charles Sanders Peirce (1839–1914), an American—overcame the ancient divergence between syllogistic and propositional logic. They introduced symbols representing **determiners,** expressions such as *all, some, no, every, any,* and so on. Frege and Peirce used two symbols: the universal quantifier, which we will write with parentheses and is often written ∀ (Peirce's work has Π), and the existential quantifier, ∃ (in Peirce, Σ). The universal quantifier corresponds roughly to the English *all, every,* and *each;* the existential quantifier, to the English *some, a,* and *an.*

To see how quantifiers unify propositional and syllogistic logic, notice that the theories use different basic units. Propositional logic treats statements lacking connectives as unanalyzed, while syllogistic logic treats terms as basic. Frege and Peirce combined these schemes by making terms behave, in essence, as if they were statements. That enabled propositional tools to work on syllogistic and, as it turned out, far more complex arguments. Modern **predicate logic** has the power to express and evaluate a very large group of sentences and arguments. This chapter presents a guide to representing English sentences in a fragment of the theory and develops tableau and deduction methods for evaluating them.[1]

---

[1]We do not cover relational predicates *(loves, admires, gives)* or quantifiers that overlap in scope *(There are things that are so ugly that if they are beautiful, anything is)* in this chapter. Our semantic tableau method is a decision procedure for our fragment, though not for predicate logic as a whole. There is no decision procedure for the whole of predicate logic, as the American logician Alonzo Church showed in 1936.

■ 9.1 QUANTIFIERS

Propositional logic takes statements as basic. To expand it to include syllogistic arguments, we must see how atomic statements are put together. The theory of syllogisms gives us the crucial clue: Among the basic building blocks of statements are *terms*. Terms are true or false of individual objects. *Man* and *eats heartily,* for example, are terms. Pick any object you like; it will either be a man or not be a man. It will either eat heartily or not eat heartily. Alone, terms are not true or false. They are true or false *of objects.* Objects of which a term are true **satisfy** it; the term **applies to** them. We can think of terms as classifying objects into two categories: those of which they are true and those of which they are false. The set of objects of which a term is true is called its **extension.**

Just as in syllogistic logic, we will introduce capital letters to stand for terms, or, as they are called in modern logic, **predicates.**

English Predicate	Symbolic Predicate
(is a) man	M
eats	E
(is) red	R
(is) somewhat strange	S

CONSTANTS

Terms are not statements; taken alone, they are not true or false. Instead, they are true or false of objects. One way of making a term into a statement, then, is to apply it to an object. In English, we do this by attaching it to a name or pronoun. We might make each predicate above, for example, into a statement in this way:

English Predicate	Statement
(is a) man	Socrates is a man.
eats heartily	Fido eats heartily.
(is) red	This is red.
(is) somewhat strange	That is somewhat strange.

To symbolize such statements, we need symbols that stand for objects in the way that names or pronouns do. We will use lowercase letters (except *x, y,* and *z*) to stand for objects. These symbolic names are **individual constants.**

Armed with symbols corresponding to terms and names or pronouns, we can symbolize the statements above as

English Predicate	Symbolic Predicate	Statement	Formula
(is a) man	M	Socrates is a man.	Ms
eats heartily	E	Fido eats heartily.	Ef
(is) red	R	This is red.	Rt
(is) somewhat strange	S	That is somewhat strange.	Sa

One way of building formulas of predicate logic, then, is to combine individual constants with predicates. We will always write predicates before the individual constants to which they apply. So, Sa is a formula, but aS is not. The formulas in the table—Ms, Ef, Rt, and Sa—are **atomic,** for they contain no connectives or quantifiers.

The language of predicate logic includes that of propositional logic. So, the connectives ~, ·, ∨, ⊃, and ≡ are available for linking statements and formulas together.

Statement	Symbolization
Pittsburgh is beautiful.	Bp
Albuquerque is not damp.	~Da
Beth or Carlota will come.	Cb ∨ Cc
If you go, I will fill in.	Gu ⊃ Fi
Max will play if and only if Al and Bob don't.	Pm ≡ (~Pa · ~Pb)

Variables

The other components of the language of predicate logic are **quantifiers** and **individual variables.** Consider the statement

(4) Something is missing.

To symbolize it, we cannot use an individual constant for *something*. We do not want to say that any object in particular is missing. The statement says that *missing* applies to some object or other. Quantifiers and variables allow us to say, in effect, that *x is missing* is true for some object x. We write the symbolic equivalent of

(5) (for some x)(x is missing),

which is

(6) (∃x)Mx.

The symbol ∃ is the **existential quantifier.** (∃x)Mx means that at least one thing is missing.

Individual variables such as x link quantifiers to the predicates they accompany. Variables—lowercase letters from the end of the alphabet (x, y, and z)—act much as variables for numbers act in algebra. They stand for no objects in particular; they mark places where names of particular objects could go and *range over* the set of objects under discussion—the **universe of discourse.** The context usually supplies the universe of discourse. If people report that something is missing, for example, we are unlikely to take them to mean that there is at least one thing on earth, or in the entire universe, that is missing. More likely, they mean that something in the room or that they own or that they need is missing.

We may read (6) as saying

(7) a.   for some x, x is missing
     b.   some x is such that x is missing
     c.   there is an x such that x is missing
     d.   an x is such that x is missing

In better English, these become

(7) a.   something is missing
     b.   there is something missing
     c.   an object is missing

The words *thing* and *object* serve in English much as variables such as x serve in predicate logic.

To take another example, suppose we want to say, recalling the Ray Stevens song, that everything is beautiful. To say that Pittsburgh is beautiful and that Ingrid Bergmann is beautiful, we can introduce the predicate B and the individual constants p and b and write Bp and Bb. To say that everything is beautiful, however, we need to say that *x is beautiful* is true for every object x. The universal quantifier does this. We can prefix the quantifier to Bx, writing the symbolic equivalent of

(8) (for every x)(x is beautiful)

or

(9) (x)Bx

This says

(10) a.   for every (all, each, any) x, x is beautiful
      e.   every (all, each, any) x is such that x is beautiful

or, in more acceptable English,

(11) a.   everything is beautiful
      b.   all things are beautiful
      c.   each object is beautiful
      d.   any object is beautiful

Variables have no meanings independent of their symbolic context. In (6) and (9), M and B represent *missing* and *beautiful,* respectively. They cannot be interchanged without changing the translation manual relating English statements to their symbolic representations. Variables, in contrast, can be interchanged, with very few restrictions, without altering meaning. *Something is missing* could just as well be symbolized by

(12) (∃y)My

as by (9). This is true because the variables x, y, and z themselves have no significance. They link quantifiers to predicates and mark places where constants could be placed. Any sort of mark could do the same job.

The expressions Mx, Bx, and By are not themselves statements. Neither are such expressions as Mx · Bx or Bx ⊃ By. They are called **statement functions** (or **propositional functions** or **open sentences**). They are not true or false. But they become statements when all their variables are *bound* by a matching quantifier. Like connectives, quantifiers have scope: the quantifier together with the statement function immediately following it. Variables are **bound** when in the scope of a quantifier on the same variable, and **free** otherwise. All variables in statements are bound.

The expressions on the left below are statements. Those in the center are statement functions; their free variables are listed to the right.

STATEMENT	STATEMENT FUNCTION	FREE VARIABLES	BOUND VARIABLES
(∃x)Fx	Fx	x	
(y)Gy	Gy	y	
(∃x)(y)(Gx · Gy)	(y)(Gx · Gy)	x	y
(∃x)(y)(Hx ∨ Jy)	(∃x)(Hx ∨ Jy)	y	x
(∃x)(y)(Lx ⊃ My)	(Lx ⊃ My)	x, y	
(∃x)Fx ⊃ (x)Hx	(∃x)Fx ⊃ Hx	x (in Hx)	x (in (∃x)Fx)

P R O B L E M S

Symbolize the following statements in predicate logic with the predicate letters given in parentheses.

1. Alfred is angry. (A)

2. Barbara is not a senior. (S)

3. Something's wrong. (W)

4. Everything's fine and dandy. (F, D)

5. There are no witches. (W)

6. If Curly is not happy, someone will pay. (H, P)

7. If Ed can do the job, anyone can. (D)

8. Everything makes Don look guilty, but he's innocent, and I can prove it. (M, I, P)

9. If something has dignity, then not everything has a price. (D, P)

10. Fran got a raise, and Gladis will too if someone recommends her and no one objects. (G, R, O)

■ 9.2 CATEGORICAL STATEMENT FORMS

Most of the time, we want to say something about, for example, some people or every frog, not about something or everything. If we could do this, it would be easy to represent any categorical proposition in predicate logic.

Universal statements assert something about all members of a class. They have the forms

(13) Universal affirmative: All F are G

Universal negative: No F are G

We might want to represent, for instance,

(14) All frogs swim. But no frogs do the backstroke.

In essence, predicate logic points out that *All frogs swim* says that if an object is a frog, it swims. Similarly, *No frogs do the backstroke* says that if an object is a frog, it does not do the backstroke. The Stoics recognized the equivalence between universal and conditional statements, but not until Frege and Peirce did anyone see its significance. To symbolize (14) we want to say something like

(15) for all x, if x is a frog, then x swims; and

for all x, if x is a frog, then x does not do the backstroke.

Using quantifiers and connectives, this becomes

(16) $(x)(Fx \supset Sx) \cdot (x)(Fx \supset {\sim}Bx)$

The conditional is ideal for representing universal statements. Consider a truth table reflecting part of the symbolization of *All frogs swim:*

X IS A FROG	X SWIMS	X IS A FROG $\supset$ X SWIMS
T	T	T
T	F	F
F	T	T
F	F	T

All frogs swim is false only if there is a frog that does not swim. The same is true of the conditional; it is false only when x is a frog that does not swim. The formula $(x)(Fx \cdot Sx)$ thus comes out true exactly when *All frogs swim* is true.

UNIVERSAL STATEMENT FORM	SYMBOLIZATION
All F are G	$(x)(Fx \supset Gx)$
No F are G	$(x)(Fx \supset {\sim}Gx)$

Particular statements have the structures

(17) a. Particular affirmative: Some F are G
      b. Particular negative: Some F are not G.

To take an example, consider

(18) Some bacteria are immune to antibiotics.

Once again, we can represent *This E. coli is immune to antibiotics* as Ie and *Some things are immune to antibiotics* as ∃xIx. But (18) expresses a relationship between the terms *bacteria* and *are immune to antibiotics*. With the existential quantifier, conjunction expresses the right relationship: To say that some bacteria are immune to antibiotics is to say that for some object x, x is a bacterium and x is immune to antibiotics. So we can symbolize (18) as

(19) $\exists x(Bx \cdot Ix)$

Again, consider a truth table:

X IS A BACTERIUM	X IS IMMUNE TO ANTIBIOTICS	X IS A BACTERIUM · X IS IMMUNE TO ANTIBIOTICS
T	T	T
T	F	F
F	T	F
F	F	F

For *Some bacteria are immune to antibiotics* to be true, there must be some x such that x is both a bacterium and is immune to antibiotics. A bacterium that is vulnerable to antibiotics (row two), a virus that is immune (row three), or a more complex organism that is vulnerable (row four) are no help at all; they do nothing to make the statement true. (*Note:* using a conditional instead of a conjunction, by giving us the value T on the last two rows, would allow immune viruses and vulnerable complex organisms to make *Some bacteria are immune to antibiotics* true!)

Particular negative statements are very similar to their affirmative counterparts. To say that some F are not G is to say that for some object x, x is F but not G. Their symbolizations are just like those of particular affirmatives, but with an added negation.

PARTICULAR STATEMENT FORM	SYMBOLIZATION
Some F are G	$(\exists x)(Fx \cdot Gx)$
Some F are not G	$(\exists x)(Fx \cdot {\sim}Gx)$

Universal negative and particular affirmative statements are contradictories. This is one logical relation on which the traditional and modern squares of opposition agree. (See Chapter 4.) *No F are G* and *Some F are G* always have opposite truth values. Each is equivalent to the negation of the other. So, we can symbolize *No F are G* as ${\sim}(\exists x)(Fx \cdot Gx)$ as well as $(x)(Fx \supset {\sim}Gx)$. *No frogs do the backstroke*, then, we can represent in two ways:

(20) Every frog does not do the backstroke. $(x)(Fx \supset {\sim}Bx)$

There is no frog that does the backstroke. ${\sim}(\exists x)(Fx \cdot Bx)$

Any categorical proposition can be symbolized in predicate logic. The logic of this chapter has the power to cover syllogistic reasoning—and much more besides.

PROBLEMS

Symbolize these statements in predicate logic, using the predicates in parentheses.

1. All books have pages. (B: books; H: have pages)

2. Some relatives are nosy. (R: relative; N: nosy)

3. No employers like to be regulated. (E: employers; L: like to be regulated)

4. Some answers are not correct. (A: answers; C: correct)

5. Everyone who can do these problems is smart. (P: person who can do these problems; S: smart)

6. Nobody Jill invited came to the party. (J: person Jill invited; C: came to the party)

7. Not every politician is honest. (P: politician; H: honest)

8. If some workers complain, everyone will be better off. (W: worker; C: complain; P: person; B: better off)

9. Some statements are easy to symbolize, but some aren't. (S: statement; E: easy to symbolize)

10. All incumbents will be reelected if and only if some challengers falter. (I: incumbent; R: reelected; C: challenger; F: falters)

 # 9.3   SYMBOLIZATION

Proper names and singular pronouns translate into predicate logic as individual constants; terms translate as predicates. This seems simple, but natural languages such as English present many complications. This section is a quick guide to some of them.

DETERMINERS

The determiners *all, each, any,* and *every* generally translate as universal quantifiers, while *some* and *a(n)* generally translate as existentials. But even these rules have exceptions. First, *a* and *an* have a **generic** or **general** use, where they seem to talk about typical members of a kind. Thus

(21) A whale is a mammal.

is not saying that some whales are mammals but (roughly) that all are. Here, the indefinite article corresponds roughly to a universal quantifier. (For further details on recognizing generic uses, see Chapter 4, page 153.)

Second, *a, an, any,* and *some* all interact with conditionals when they are part of the antecedent. We can treat

(22) If someone is hurt, you'll get into trouble.

as a conditional with the quantified antecedent

(23) Someone is hurt

which we can symbolize as

(24) $(\exists x)Hx$

As a whole, then, we can symbolize (22) as

(25) $(\exists x)Hx \supset Ta$

But, when the consequent contains a word that refers to something in the antecedent, this strategy fails. We could try to translate

(26) If someone is hurt, he/she will cry.

as

(27) $(\exists x)Hx \supset Cx$

but this is not a statement: The x in Cx is not in the scope of the quantifier $(\exists x)$. Changing the parentheses to give the quantifier scope over the entire formula does no good.

(28) $(\exists x)(Hx \supset Cx)$

is a statement, but it says the wrong thing. By the definition of the conditional, it is equivalent to

(29) $(\exists x)(\sim Hx \vee Cx)$

But this says that someone is either unhurt or crying. And this is true so long as someone is crying. But that is not enough to make *If someone is hurt, he/she will cry* true.

To represent (26), we must use a universal quantifier with the entire formula as its scope:

(30) $(x)(Hx \supset Cx)$

This says that everyone who is hurt cries, which is equivalent to (25). So, when the determiners *a, an,* and *some* appear in the antecedent of a conditional and a pronoun in the consequent refers to them, they correspond to universal quantifiers.

Notice that (26) is equivalent to

(31) If anyone is hurt, he/she will cry.

Any translates as a universal quantifier, but with the widest possible scope. This explains why *any* often seems similar to *some* or *a*.

(32) John didn't see any deer.

is equivalent to

(33) John didn't see a deer.

not to

(34) John didn't see every deer.

because *any* translates as a universal quantifier to the left of the negation sign:

(35) $(x) \sim (Sx \cdot Dx)$

Only functions much like *all*, except that it reverses the order of the relevant expressions. *Only F are G* amounts to *All G are F.*

Everything, anything, something, and so on, all act like the corresponding determiners; *thing* functions, more or less, as a variable. *Everybody, anybody, somebody, everyone, anyone,* and *someone* all act like *every person, any person,* and so forth. They often force the use of a quantifier together with a predicate P (for *person*) linked appropriately to the rest of the statement.

But an important part of interpreting formulas is assigning them a *universe of discourse* or *domain.* This is a set, determined by context, containing the objects being talked about. The quantifiers and variables *range over* the domain: We interpret the universal quantifier as saying that something is true for all elements of the domain and similarly interpret the existential quantifier as saying that something is true for at least one element of the domain.

We can often simplify symbolization by assigning an appropriate universe of discourse. If, in the context of an argument, we are speaking of nothing but people, we can limit the domain to the set of people. If we do, then a universal quantifier will have the effect of *for all x in the set of people,* that is, *for all people x.* The universal quantifier, in such a case, represents the English expressions *anybody* and *everybody* without using a predicate meaning *person.*

Argument (36), for example, can be symbolized without the predicate P, but (37) requires it.

(36) Everybody can learn logic. So, everybody can learn logic or rock climbing. $(x)Lx; \therefore (x)(Lx \lor Rx)$

(37) Everybody can learn logic. Some animals cannot learn logic. So, some animals are not people. $(x)(Px \supset Lx); (\exists x)(Ax \cdot \sim Lx);$ $\therefore (\exists x)(Ax \cdot \sim Px)$

ADJECTIVES

Adjectives—words such as *good, red, friendly,* and *logical*—modify nouns. With a few exceptions, they translate as predicates, linked by conjunctions to the predicates representing the nouns they modify. Adjectival phrases, consisting of an adjective modified by an adverb, for example, function in the same way. They must be treated as a single unit. Thus:

STATEMENT	SYMBOLIZATION
All friendly cats purr.	$(x)((Fx \cdot Cx) \supset Px)$
Some artists are unhappy people.	$(\exists x)(Ax \cdot (Ux \cdot Px))$
John is an amazingly wealthy logician.	$Wj \cdot Lj$
Some very clever students aren't very industrious.	$(\exists x)((Cx \cdot Sx) \cdot {\sim}Ix)$

Most adjectives and adjectival phrases thus translate as conjunctions. The set of colorless gases is the set of things that are both colorless and gases. It is the intersection of the set of colorless things and the set of gases. Adjectives for which this is true are called **intersective.**

Certain adjectives are not intersective. They do not translate directly. They have meanings that relate somehow to the nouns they modify, so they and their nouns must translate as a single unit. Luckily, identifying these *nonintersective* adjectives is easy. Wealthy logicians are both wealthy and logicians; red Chevrolets are both red and Chevrolets. But alleged criminals are not alleged and criminals. Good pianists are not simply pianists and good. Former congressmen are not former and congressmen; large mice are not large and mice. *Good pianist* means something like *good as a pianist; large mouse,* something like *large for a mouse.* This is why the following arguments fail:

(38) a. Every pianist is a lover.
          ∴ Every good pianist is a good lover.

      b. All mice are animals.
          ∴ All large mice are large animals.

      c. Dan, a former congressman, is an inmate.
          ∴ Dan, a former inmate, is a congressman.

      d. Charles is an alleged criminal.
          ∴ Charles is a criminal.

These examples point out an important difference among nonintersective adjectives. The set of large mice is a subset of the set of mice; the set of good pianists is similarly a subset of the set of pianists. Other nonintersective adjectives, such as *alleged, fake,* and *former,* are not like this. The set of fake diamonds is not a subset of the set of diamonds. Likewise, the set of former congressmen is not a subset of the set of congressmen. These adjectives must translate, together with the nouns they modify, as a single predicate. Adjectives like *large* and *good,* in contrast, may translate as conjunctions, but with an important difference from intersective adjectives. We can render *large mouse* as Lx · Mx, where M represents *mouse,* but only if we construe L as translating, not *large,* but *large for a mouse.*

RELATIVE CLAUSES

Relative clauses are English expressions formed from statements. They begin, generally, with *that* or a word starting with *wh,* such as *who, which, when,* or *where,* though these words are often omitted. Relative clauses frequently act like adjectives, modifying nouns or noun phrases. Like intersective adjectives, we may think of them as conjoined to the nouns they modify.

STATEMENT	SYMBOLIZATION
There is a nice apartment that is close to downtown.	$(\exists x)((Nx \cdot Ax) \cdot Cx)$
Everybody who has served as ambassador is here.	$(x)((Px \cdot Sx) \supset Hx)$

Some relative clauses restrict the group of things to which the noun phrase they modify applies. If I tell you that everyone I know prefers Mexican to Chinese food, then I am speaking, not of everyone, but just of everyone I know. The examples above contain such **restrictive** relative clauses. Other clauses, however, make almost parenthetical comments about their nouns or noun phrases.

Both restrictives and nonrestrictives, in symbolic representations, connect to the remainder of the formula by conjunction. Most of the time, therefore, it makes no difference to the translation whether a given clause is restrictive. When universal quantifiers are involved, however, and the clause modifies the subject noun phrase, it does matter. Consider these statements:

STATEMENT	SYMBOLIZATION
All the Democratic candidates for president, who are already campaigning, support labor unions.	$(x)(Dx \supset Lx) \cdot (x)(Dx \supset Cx)$ or $(x)(Dx \supset (Lx \cdot Cx))$
All the Democratic candidates for president who are already campaigning support labor unions.	$(x)((Dx \cdot Cx) \supset Lx)$

The only difference is the pair of commas setting off the relative clause in the first statement. There the clause is clearly nonrestrictive. It asserts that all the Democratic candidates for president support labor unions and remarks, on the side, as it were, that all those candidates are already campaigning. The second, in contrast, does not claim that all the Democratic candidates support labor unions; it asserts only that all those who are already campaigning do so.

Most relative clauses in actual discourse are restrictive. To tell whether a given clause is restrictive, ask whether the clause is helping to specify what the statement is talking about or providing additional information concerning an already determinate topic. English offers two linguistic hints. First, *that* often signals that a clause is restrictive; *which,* with some exceptions (for example, the phrases *in which* and *with which*), often signals that it is nonrestrictive. Relative clauses often begin with other *wh* words, however, or with no special word at all. In these cases, there are no signals. Furthermore, *that* and *which* are often misused. Second, and more reliably, commas often do, and always can, set off nonrestrictive clauses from the rest of the statement. Restrictives, on the other hand, reject commas in this role. Hence, virtually all relative clauses set off by commas are nonrestrictive. For those not set off by commas, there is a simple test: Try inserting commas. If the result sounds acceptable, the clause is probably nonrestrictive. Otherwise, it is restrictive.

ADVERBIAL MODIFIERS

Adverbs, such as *quickly, well, anytime,* and *somewhere,* modify verbs. They specify how, when, or where a certain condition holds or a certain activity occurs. Unfortunately, most adverbs have no direct symbolizations in predicate logic. We must represent them, together with the verbs they modify, as we do nonintersective adjectives. Some adverbs are like *large* and *good.* Anyone who is walking slowly, for example, is walking. To express this, we can write Wx · Sx, where W symbolizes *walks* and S symbolizes *is slow for a walker.* But beware of such adverbs as *allegedly:* That John allegedly stole the money does not mean that he stole the money. *Allegedly stole* must translate as a single predicate.

Some adverbs, however, translate into predicate logic differently. We will call *always, anytime, whenever, wherever, anywhere, sometimes, never,* and so on, **adverbs of quantification.** They may be symbolized with quantifiers—often, together with a predicate T for *time* or P for *place.*

STATEMENT	SYMBOLIZATION
Children are always cruel.	(x)(Cx ⊃ Rx)
Democrats never oppose government spending.	(x)(Dx ⊃ ~Ox)
Mosquitos are everywhere!	(x)(Px ⊃ Mx)
Sometimes you win; sometimes you lose.	(∃x)(Tx · Wx) · (∃x)(Tx · Lx)

Connectives

Propositional connectives can link quantified statements together, but they can also link noun and verb phrases together within statements. When possible, it is best to split such phrases. Connectives linking noun or verb phrases should be replaced by connectives linking statements:

Original	Interpret as
Abraham Lincoln and Calvin Coolidge were Republican presidents.	Abraham Lincoln was a Republican president, and Calvin Coolidge was a Republican president.
Fred likes hot dogs and hamburgers.	Fred likes hot dogs and Fred likes hamburgers.
All lions and tigers are cats.	All lions are cats and all tigers are cats.

Note especially the last example. The conjoined noun phrase *lions and tigers* can tempt us into a translation

(39) $(x)((Lx \cdot Tx) \supset Cx)$

which says that everything that is both a lion and a tiger is a cat. Nothing, however, is both a lion and a tiger. Separating statements results in the formula

(40) $(x)(Lx \supset Cx) \cdot (x)(Tx \supset Cx)$

which captures the meaning of the original (and which is equivalent to $(x)((Lx \lor Tx) \rightarrow Cx)$—everything that is either a lion *or* a tiger is a cat). When existential quantifiers are involved or when the connectives are in the verb phrase, splitting makes little difference. In subject noun phrases, however, it is vital.

As in the case of propositional logic, however, we must take care to split only those statements for which the process preserves meaning.

(41) Harry loves chips and salsa.

may not be equivalent to

(42) Harry loves chips and Harry loves salsa.

He may love the combination without being very excited about the individual components or vice versa. *Mary and Susan own the entire company* is probably not saying that Mary owns the entire company and that Susan does, too, but that they own the entire company between them.

P R O B L E M S

Are the following adjectives intersective? If not, give examples to show why.

1. blue

2. round

3. fast

4. great

5. gregarious

6. interesting

7. supposed

8. bad

9. intelligent

10. little

11. famous

12. spherical

13. remarkable

14. experienced

15. so-called

Symbolize the following statements in predicate logic, using the predicates listed.

16. Nobody who's ever been to Gulag is a pacifist. (William F. Buckley Jr.) (G, P)

17. He that cannot be angry is no man. (Thomas Dekker) (A, M)

18. God is with those who patiently persevere. (Arab proverb) (P, G)

19. Only the shallow know themselves. (Oscar Wilde) (S, K)

20. Everything that looks to the future elevates human nature. (Walter Savage Landor) (L, E)

21. He who does kind deeds becomes rich. (Hindu proverb) (K, R)

22. Every good thing that comes is accompanied by trouble. (Maxwell Perkins) (G, A)

23. They sicken of the calm, who know the storm. (Dorothy Parker) (S, K)

24. . . . and now nothing will be restrained from them, which they have imagined to do. (*Genesis* 11:6) (R, I)

25. The secret of success in life is known only to those who have not succeeded. (John Collins) (S, K)

26. We receive only what we give. (Samuel Taylor Coleridge) (R, G)

27. Light grows the burden which is well borne. (Ovid) (L, B, W)

28. Any man who has a job has a chance. (Elbert Hubbard) (M, J, C)

29. We believe easily what we fear or what we desire. (La Fontaine) (B, F, D)

30. No man who wanted to be a great man ever was a great man. (John Hunter) (M, W, G)

31. . . . the things which are seen are temporal; but the things which are not seen are eternal. (*II Corinthians* 4:18) (S, T, E)

32. They also live who swerve and vanish in the river. (Archibald MacLeish) (L, S, V)

33. A man who cannot tolerate small ills can never accomplish great things. (Chinese proverb) (M, T, A)

34. The man who lives for himself is a failure. (Norman Vincent Peale) (M, L, F)

35. Only madmen and fools are pleased with themselves. . . . (Benjamin Whichcote) (M, F, P)

36. . . . nothing great in the world has ever been accomplished without passion. (G. F. W. Hegel) (G, W, A, P)

37. If the only tool you have is a hammer, you tend to see every problem as a nail. (Abraham Maslow) (T, H, P, N)

38. Socialists and totalitarian governments are doomed to support the past. (George Gilder) (S, T, G, D)

39. Meditation is a gift confined to unknown philosophers and cows. (Finley Peter Dunne) (U, P, C, M)

40. Only little boys and old men sneer at love. (Louis Auchincloss) (L, B, O, M, S)

41. Blessed are those who give without remembering and take without forgetting. (Elizabeth Bibesco) (B, G, R, T, F)

42. A white lie is always pardonable. But he who tells the truth without compulsion merits no leniency. (Karl Kraus) (W, L, P, T, C, M)

The following statements contain both quantifiers and connectives. Symbolize them in predicate logic, exposing as much structure as possible, using the predicates listed.

43. If any would not work, neither should he eat. (*II Thessalonians* 3:10) (W, E)

44. It is the dull man who is always sure, and the sure man who is always dull. (H. L. Mencken) (D, S)

45. We all learn by experience but some of us have to go to summer school. (Peter de Vries) (L, S)

46. Every luxury must be paid for, and everything is a luxury. . . . (Cesare Pavese) (L, P)

47. What we obtain too cheap we esteem too lightly. . . . (Thomas Paine) (C, L)

48. One either meets or one works. One cannot do both at the same time. (Peter Drucker) (M, W)

49. No rule for success will work if you won't. (Elmer G. Leterman) (R, W)

50. Nothing that is proved is obvious; for what is obvious shows itself and cannot be proved. (Joseph Joubert) (P, O, S)

51. If an electrical engineer has a pulse, he'll get a job. (Vicki Lynn) (E, P, G)

52. . . . it is not poetry, if it make no appeal to our passions or our imagination. (Samuel Taylor Coleridge) (P, A, I)

53. Any job is meaningful if you need money. (Millicent Fenwick) (J, M, N)

54. I distrust all systematizers, and avoid them. (Friedrich Nietzsche) (D, S, A)

55. He that is never suspected is either very much esteemed or very much despised. (Lord Halifax) (S, E, D)

56. The wretched reflect either too much or too little. (Publilius Syrus) (W, M, L)

57. Life does not agree with philosophy: there is no happiness that is not idleness, and only what is useless is pleasurable. (Anton Chekhov) (H, I, U, P)

58. All change is not growth, as all movement is not forward. (Ellen Glasgow) (C, G, M, F)

59. Every social injustice is not only cruel, but it is economic waste. (William Feather) (S, I, C, W)

60. When a man is wrong and won't admit it, he always gets angry. (Thomas Haliburton) (M, W, A, G)

61. Any mental activity is easy if it need not take reality into account. (Marcel Proust) (M, A, E, R)

62. It is only the cynicism that is born of success that is penetrating and valid. (George Jean Nathan) (C, S, P, V)

63. There are well-dressed foolish ideas just as there are well-dressed fools. (Nicolas Chamfort) (W, F, I, P)

64. To be beloved is all I need, and whom I love, I love indeed. (Samuel Taylor Coleridge) (B, N, L, I)

65. All men have aimed at, found, and lost. . . . (William Butler Yeats) (M, A, F, L)

66. Nothing will ever be attempted if all possible objections must be first overcome. (Jules W. Lederer) (A, P, O, F)

67. If you build a castle in the air, you won't need a mortgage. (Philip Lazarus) (B, C, A, M, N)

68. Some people with great virtues are disagreeable while others with great vices are delightful. (François La Rochefoucauld) (P, V, D, C, L)

69. Every gift which is given, even though it be small, is in reality great, if it is given with affection. (Pindar) (G, V, S, R, A)

70. Shallow men believe in luck, wise and strong men in cause and effect. (Ralph Waldo Emerson) (S, M, L, W, T, C)

71. When all is summed up, a man never speaks of himself without loss; his accusations of himself are always believed; his praises never. (Michel Eyquem de Montaigne) (M, S, L, A, B, P)

9.4 QUANTIFIED TABLEAUX

The method of semantic tableaux extends easily to incorporate quantifiers. As in propositional logic, semantic tableaux determine whether argument forms are valid. Semantic tableaux for predicate logic use all the propositional tableau rules, together with four new rules for the quantifiers. Each quantifier has two rules: one for its occurrences on the left side of tableau branches, the other for its occurrences on the right.

RULES

The quantifier rules all use the notion of an *instance*. Say that $\mathfrak{J}(c)$ is the result of substituting constant c for every occurrence of variable v throughout the statement function $\mathfrak{J}(v)$. If $(v)\mathfrak{J}(v)$ and $(\exists v)\mathfrak{J}(v)$ are statements, then $\mathfrak{J}(c)$ is an *instance* of them. Conversely, $(v)\mathfrak{J}(v)$ and $(\exists v)\mathfrak{J}(v)$ are *generalizations* of $\mathfrak{J}(c)$.

The left side of a tableau represents truth, while the right represents falsity. "Left" rules thus tell us what we can infer from the supposition that a quantified formula is true; "right" rules tell us what we can infer from the assumption that such a formula is false.

Existential Left (∃L)

$\checkmark \quad (\exists v)\mathfrak{J}(v) \; |$

$\mathfrak{J}(c) \; |$

Here c must be a constant new to the tableau.

This rule says that we may replace an existential formula on the left by one of its instances, if the constant we substitute for the variable is new to the tableau. To see why the rule has this form, assume that an existential statement—say,

(43) Someone in this room is a spy.

is true.[2] This tells us nothing about who the spy is. Nevertheless, we need a way of referring to the spy so that we can record additional information about the same person. In English, we might just use *the spy* in this role, as we have been doing in the preceding two statements. Or, we might use a pronoun. In our system of tableaux, we achieve the same effect by introducing a name for the spy—as we might also in English by saying,

---

[2]*Someone in this room is a spy* asserts that there is at least one spy in the room. What follows in the text assumes, for the sake of simplicity, that there is only a single spy. If there is more than one, then what is said about *the spy* applies to any of the spies; we cannot distinguish them with the information at hand.

"call the spy *Karla*"—in the guise of a constant that has not appeared anywhere on the tableau before. We must use a new constant, because we cannot say who the spy is. It is hardly fair to say to Fred, "Someone in this room is a spy. Let's call him *Fred*." Similarly, it would be outrageous for a mathematician to say, "So, there's a point on the interval at which the derivative is zero. Let's call this point, say, 9.3." If we do not know which objects make the existential statement true, we need a new constant to avoid making illicit identifications.

To take an example, suppose that we want to learn whether (∃x)(Fx · ~Fx)—*something is both F and not F*—is satisfiable or contradictory. As in propositional logic, we place the formula on the left of a tableau, assuming that it is true, and see whether a contradiction results.

$$
\begin{array}{ll}
\checkmark & (\exists x)(Fx \cdot \sim Fx) \\
\checkmark & Fa \cdot \sim Fa \\
& Fa \\
\checkmark & \sim Fa \\
& \qquad Fa \\
& \quad Cl
\end{array}
$$

The tableau closes, so the formula is contradictory; nothing can be both F and not F at the same time.

Existential Right (∃R)

$$
\begin{array}{l}
(\exists v)\Im(v) \; * \\
\Im(c)
\end{array}
$$

Here c may be any constant.

This rule says that we may dispatch an existential formula on the right by writing, also on the right, any instance of that formula. The constant we use in ∃R need not be new, though it may be; any constant whatever will do. Furthermore, we may apply this rule more than once. The *star* (*) is a **temporary dispatch mark**; it indicates that, though we have already applied ∃R to this formula, we may come back and apply it again. To see why we may repeat this rule, suppose that we know it is false that someone in this room is a spy. Then we can infer that it is false that Al is a spy, that Beth is a spy, that Carl is a spy, that Dorothy is a spy, and so on for each person in the room.

To take an example, suppose we want to find out whether (∃x)(Fx ∨ ~Fx)—*something is either F or not F*—is a tautology. We place it on the right side of a tableau, assuming that it can be false.

$$(\exists x)(Fx \vee \sim Fx) \text{ *}$$
$$Fa \vee \sim Fa \quad \checkmark$$
$$Fa$$
$$\sim Fa \quad \checkmark$$
$$Fa$$
$$Cl$$

Since the tableau closes, the formula is valid. The instance here contains a new constant. It must, not because the rule demands it, but because no constants had yet appeared on the branch. In general, we will try to avoid introducing new constants. We will introduce them under only two circumstances: (1) when a rule (existential left or universal right) requires it and (2) when the branch contains no constants.

Universal Left (UL)

$$\text{* } (v)\Im(v) \mid$$
$$\Im(c) \mid$$

Here c may be any constant.

This rule says that a universal formula on the left allows us to write any instance of it on the left. Again, the constant we use does not have to be new. In fact, any instance of the formula will do. Furthermore, we can repeat this rule, too. If everyone here knows logic, then it follows that I know logic, you know logic, Fred knows logic, Samantha knows logic, and so on for each person here.

Suppose that we want to discover whether (x)Fx implies (∃x)Fx. We place these formulas on the left and right side of a tableau, respectively, and find out whether there is any circumstance making (x)Fx true but (∃x)Fx false.

* (x)Fx	
	(∃x)Fx *
	Fa
Fa	
	Cl

As this tableau shows, there is not; (x)Fx does imply (∃x)Fx. Neither quantifier rule used here requires the use of new constants. We can use a single constant for both instances.

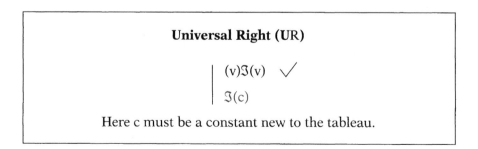

Universal Right (UR)

(v)ℑ(v) ✓
ℑ(c)

Here c must be a constant new to the tableau.

This rule says that we can replace a universal formula on the right by an instance of it on the right, using a constant that has not appeared before on the tableau. Suppose it is false that everyone in the department has a

Ph.D. It follows that at least one person in the department does not have a Ph.D. We cannot infer anything about who these persons are, but we want to reason about them, so we need some way of referring to them. To correspond roughly to the English *the person or persons in the department without a Ph.D,* we introduce a new constant. If the constant were not new, we would be illicitly assuming something about the identity of those without Ph.Ds.

To take an example, suppose that we want to know whether (x)(Fx ⊃ Fx)—*all Fs are Fs*—is a tautology.

$$
\begin{array}{c|l}
 & (x)(Fx \supset Fx) \quad \checkmark \\
 & Fa \supset Fa \quad \checkmark \\
Fa & \\
 & Fa \\
\hline
Cl &
\end{array}
$$

This tableau tells us that it is. Our search for an interpretation making the formula false results in contradiction.

STRATEGIES

In propositional logic, any way of applying tableau rules produces the same result. Nevertheless, some ways are more efficient than others. We thus adopted two strategies for simplifying tableaux. The first urged us to close branches as soon as possible. Once some formula has appeared in both sides of a branch, applying further rules to the branch can make no difference; the branch—or branches, if further applications split the original—will still have a formula that appears on both sides and so will close. The second strategy urged us to avoid splitting tableaux as long as possible. Once a branch splits, applying rules to the formulas on it above the split forces us to write the result on each resulting branch.

These strategies remain useful in predicate logic. Furthermore, two other strategies help to simplify tableaux. Our third rule of strategy will be to

3. Apply ∃L and UR before applying ∃R and UL. That is, apply the quantifier rules introducing new constants as soon as possible.

Observing this principle minimizes the number of constants in a tableau.

To see this, consider an example. Suppose that we want to evaluate the argument

(44) Some rulers are dictators.

All dictators have absolute power.

∴ Some rulers have absolute power.

We can symbolize this as

(45) (∃x)(Rx · Dx)

(x)(Dx ⊃ Px)

∴ (∃x)(Rx · Px)

Following the rule of strategy results in the first tableau, while introducing the new constant after applying the other quantifier rules results in the longer second tableau.

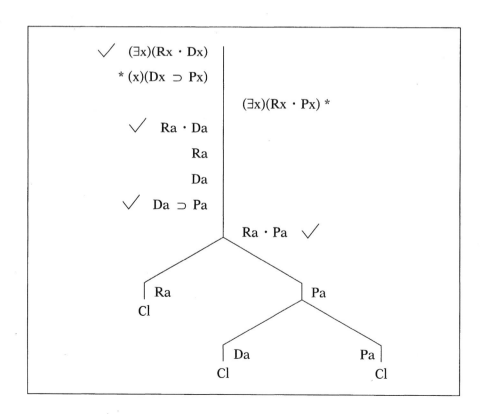

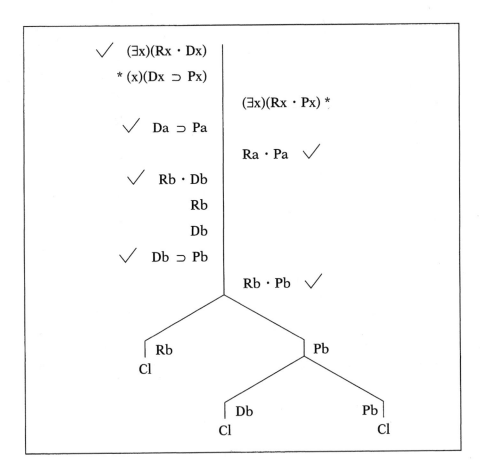

Our fourth rule of strategy will be

4. Do not introduce new constants unless you must.

Two quantifier rules, ∃R and UL, do not require a new constant. They accept any constant, whether or not it is already on the tableau. Consequently, when applying these rules, use any available constants. Introduce new ones only if a rule requires it or no constants are available on the tableau.

To see why, again consider an example. The argument

(46) Some conservative southerners are Democrats

∴ Some Democrats are conservative

has a form

(47) (∃x)((Cx · Sx) · Dx)

∴ (∃x)(Dx · Cx)

Following the fourth principle produces the first tableau, while ignoring it results in the second tableau.

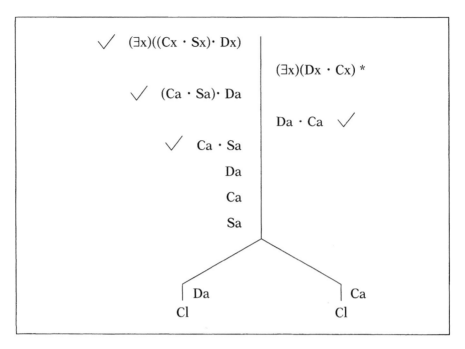

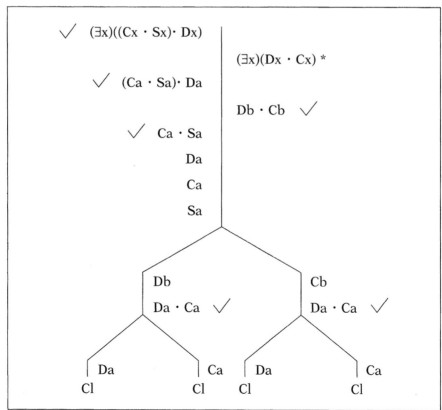

The second tableau closes only because we returned to instantiate the existential formula on the right a second time using the original constant.

This points out an important fact: Some tableaux will close only if we use appropriate constants in applying quantifier rules. The last tableau would not have closed if we had given up too early. Consequently, it is important to follow one more policy:

5. Continue instantiating as far as possible using the constants already on the branch.

∃R and UL do not require new constants and can be repeated. Indeed they must be repeated, using the constants available on the tableau, until (1) the tableau closes or (2) the available constants are exhausted. In predicate logic as a whole, there is another possibility, for tableaux do not constitute a decision procedure for predicate logic. (In fact, there is no decision procedure for predicate logic.) Sometimes applying the rules for tableau construction results in an infinite tableau: The tableau will never close, but the opportunities for further instantiation will never be exhausted. In the fragment of predicate logic developed in this chapter, however, tableaux are a decision procedure. Tableaux, in this chapter, always come to an end.

P R O B L E M S

Evaluate the following arguments for validity by using tableaux.

1. All tableau problems are easy. Everything that is not easy gives me a headache. Therefore, some things that give me a headache are not tableau problems.

2. Something about the way she moves excites me. Some songs are about the way she moves. Thus, some songs excite me.

3. All flying horses are quick and clever; all flying horses live forever. And sad but true, all horses die. It follows that no horses fly.

4. Some illegal acts go unpunished. All blatantly wrong acts are punished. Therefore, some illegal acts are not blatantly wrong.

5. All who do not remember the past are condemned to repeat it. No one condemned to repeat the past looks forward to the future with eagerness. So, everyone who eagerly looks forward to the future remembers the past.

6. John respects no one who insults him. Everyone who does not like John insults him. Thus, John respects everyone who likes him.

7. I like any bread that is not too sweet. I dislike some breads that do not contain rye flour. So, some breads that do not contain rye flour are too sweet.

8. Some untreated illnesses do not become serious. Home remedies are fine for any illness that is not serious. Thus, home remedies are fine for some untreated illnesses.

9. Lori is unhappy with some people who didn't write thank-you notes. Lori will send presents next year to everyone with whom she's happy. Therefore, some people who didn't write thank-you notes won't get presents from Lori next year.

10. There are people without jobs who are not well nourished. Nobody who is not absent from the welfare rolls fails to be well nourished. Therefore, some people with no jobs are absent from the welfare rolls.

11. Anyone who is not an idiot can see that Jake is lying. Some people in this room cannot tell that Jake is lying. Hence, some people in this room are idiots.

12. Most people would consider anyone who shows no fear in the face of danger to be courageous. Anybody who would be considered courageous by most people deserves recognition. Therefore, only those who show fear in the face of danger deserve no recognition.

13. Genevieve befriends anyone who has been treated unfairly. But she doesn't befriend some really obnoxious people. So, some really obnoxious people receive fair treatment.

14. Not all who fail to graduate drop out. All who do not drop out eventually get their degree. Thus, not all who graduate eventually get their degree.

15. The president is willing to appoint anyone who didn't work for opposing candidates. Not only people who worked for opposing candidates will accept the job. So, it is not true that nobody the president is willing to appoint will take the job.

16. No Ivy League colleges are inexpensive. No state-affiliated college is expensive. A college is private if and only if it is not state affiliated. It follows that every Ivy League college is private.

17. No mammals but bats can fly. Every commonly kept house pet is a mammal, but none are bats. So nothing that can fly is a commonly kept house pet.

18. Nothing stupid is difficult. Everything you can do is stupid; anything that is not difficult, I can do better than you. So anything you can do, I can do better.

19. There are no good books that do not require their readers to think. Every book that has inspired acts of terror has been inflam-

matory. No inflammatory books require their readers to think. Therefore, all books that have inspired acts of terror are no good.

20. Anyone with some brains can do logic. Nobody who has no brains is fit to program computers. No one who reads this book can do logic. So no one who reads this book is fit to program computers.

Evaluate these arguments, taken from Lewis Carroll, by symbolizing them in predicate logic and using semantic tableaux.

21. No one who means to go by the train, cannot get a conveyance, and has not enough time to walk to the station can do without running. This party of tourists mean to go by the train and cannot get a conveyance, but they have plenty of time to walk to the station. Thus, this party of tourists need not run.

22. Some healthy people are fat. No unhealthy people are strong. So, some fat people are not strong.

23. All uneducated people are shallow. Students are all educated. So, no students are shallow.

24. All young lambs jump. No young animals are healthy unless they jump. So, all young lambs are healthy.

25. A prudent man shuns hyenas. No banker is imprudent. Thus, no banker fails to shun hyenas.

26. Some unauthorized reports are false. All authorized reports are trustworthy. Thus, some false reports are not trustworthy.

27. Improbable stories are not easily believed. None of his stories are probable. So, none of his stories are easily believed.

28. All wise men walk on their feet; all unwise men walk on their hands. Therefore, no man walks on both.

29. John never orders anything I ought to do; Peter never orders anything I ought not to do. So, John and Peter never give the same order.

30. None of my boys are clever; none but a clever boy could solve this problem. So, none of my boys could solve this problem.

Evaluate these statements as tautologous, contingent, or contradictory.

31. Something is either physical or nonphysical.

32. Something is both physical and nonphysical.

33. Everything nonphysical is physical.

34. If nothing is physical, then something is nonphysical.

35. If nothing is physical, then nothing is nonphysical.

Use tableaux to evaluate the following argument forms for validity.

36. $(\exists x)(Gx \cdot \sim Fx)$; $(x)(Gx \supset Hx)$; $\therefore (\exists x)(Hx \cdot \sim Fx)$

37. $(\exists x)(Gx \cdot Fx)$; $(x)(Fx \supset \sim Hx)$; $\therefore (\exists x)\sim Hx$

38. $(x)(Gx \supset (Hx \cdot Jx))$; $(x)((Fx \vee \sim Jx) \supset Gx)$; $\therefore (x)(Fx \supset Hx)$

39. $(x)((Gx \vee Kx) \supset Hx)$; $\sim(\exists x)(Fx \cdot Gx)$; $\therefore (x)\sim(Fx \cdot Hx)$

40. $(x)(\sim Gx \vee \sim Hx)$; $(x)((Jx \supset Fx) \supset Hx)$; $\therefore \sim(\exists x)(Fx \cdot Hx)$

41. $\sim(\exists x)(Gx \cdot \sim Hx)$; $(x)(Fx \supset \sim Gx)$; $\therefore (x)((Fx \cdot \sim Gx) \supset \sim Hx)$

42. $(\exists x)(Gx \cdot \sim Hx)$; $\sim(\exists x)(Fx \cdot \sim Gx)$; $\therefore \sim(x)(Fx \supset Hx)$

43. $(\exists x)Fx \vee (\exists x)Gx$; $\therefore (\exists x)(Fx \vee Gx)$

44. $(\exists x)Fx \supset (y)(Gy \supset Hy)$; $(\exists x)Jx \supset (\exists x)Gx$; $\therefore (\exists x)(Fx \cdot Jx) \supset (\exists z)Hz$

45. $(\exists x)Fx \vee (\exists x)Gx$; $(x)(Fx \supset Gx)$; $\therefore (\exists x)Gx$

46. $(x)(Fx \supset (Gx \vee Hx))$; $(x)((Jx \cdot Fx) \supset \sim Gx)$; $(x)(\sim Fx \supset \sim Jx)$;
    $\therefore (x)(Jx \supset Hx)$

47. $(x)((Fx \cdot Gx) \supset Hx)$; $Ga \cdot (x)Fx$; $\therefore Fa \cdot Ha$

48. $(x)(Fx \equiv Gx)$; $\therefore (\exists x)Fx \equiv (\exists x)Gx$

49. $(\exists x)(Fx \equiv Gx)$; $\therefore (x)Fx \equiv (\exists x)Gx$

50. $(x)(Fx \equiv (y)Gy)$; $\therefore (x)Fx \vee (x)\sim Fx$

■ 9.5 QUANTIFIED PROOFS

Natural deduction extends easily to predicate logic. All rules of propositional deduction apply in quantification theory. To deal with quantifiers, the system adds several new rules. The deduction system that emerges is sound, for every conclusion provable from a set of premises follows from them. The system mirrors closely the processes of reasoning that people use in a wide variety of contexts.

DEDUCTION RULES FOR QUANTIFIERS

The deduction rules needed for predicate logic are straightforward. Say that $\mathfrak{F}(c)$ is the result of substituting c for every occurrence of v throughout the statement function $\mathfrak{F}(v)$. If $(v)\mathfrak{F}(v)$ and $(\exists v)\mathfrak{F}(v)$ are statements, then $\mathfrak{F}(c)$ is an *instance* of them. Conversely, $(v)\mathfrak{F}(v)$ and $(\exists v)\mathfrak{F}(v)$ are *generalizations* of $\mathfrak{F}(c)$.

Each quantifier has two rules. One introduces it into proofs. The other allows us to use it to derive other formulas. The introduction rule for the existential quantifier allows us to move to an existentially quantified formula from any instance of that formula. Called **existential generalization,** it takes the form

EXISTENTIAL GENERALIZATION (EG)

n. $\underline{\Im(c)}$

n+p. $(\exists v)\Im(v)$ EG, n

Here c may be any constant.

Existential generalization allows us to infer an existentially quantified statement from any instance of it. It sanctions the step from an instance to its corresponding existential generalization. Suppose that our universe of discourse consists entirely of people. If Jones, for example, is a spy, then we may conclude that someone is a spy.

The rule of **existential instantiation** allows us to move from an existentially quantified formula to an instance of it.[3] It is almost exactly the reverse, then, of the existential generalization rule. But it imposes a restriction: The instance must involve a constant new to the proof. The rule says that we may drop an existential quantifier serving as a main connective in a formula and substitute for the quantified variable a constant that has not appeared earlier in the proof or in the conclusion. The constant must have appeared nowhere in the deduction. And it must not appear in the deduction's conclusion.

EXISTENTIAL INSTANTIATION (EI)

n. $\underline{(\exists v)\Im(v)}$

n+p. $\Im(c)$ EI, n

Here c must be a constant new to
the proof and must not appear in the
proof's conclusion.

This rule reflects a very important feature of indefinite descriptions like *a house* or *a room with a view:* They not only assert existence but, like proper names, also introduce a constant that can figure in subsequent discourse. The existential quantifier itself plays only the first role. Existential instantiation, however, allows us to simulate the second. It allows us to refer to the object whose existence is asserted in the existentially quantified formula.

---

[3]The American philosopher W. V. Quine first formulated existential exploitation in this way in 1950 in his *Methods of Logic* (Cambridge: Harvard University Press, 1950, 1982).

The restriction on the rule—that we use a constant appearing neither in the proof above nor in the conclusion—prevents us from using it
to derive invalid conclusions. To see why the constant must be new, consider this argument:

(48) Some people are nice, and some aren't.

   ∴ Some people are both nice and not nice.

Plainly, the argument is invalid. The premise is a truism; the conclusion
is a contradiction. But, if we were to ignore the requirement that we use
a new constant in applying EI, we could show the corresponding argument form to be valid:

---

1. (∃x)Nx · (∃x)~Nx   /∴ (∃x)(Nx · ~Nx)

2. (∃x)Nx   Simp, 1

3. (∃x)~Nx   Simp, 1

4. Na   EI, 2

5. ~Na   EI, 3   (Wrong!)

6. Na · ~Na   Conj, 4, 5

7. (∃x)(Nx · ~Nx)   EG, 6

To see that we must not use a constant appearing in the conclusion, consider this argument:

(49) Some people are crazy.

   ∴ You're crazy.

This, too, is a terrible argument. But, if we were to ignore the requirement that EI not introduce a constant appearing in the conclusion, we
could show it to have a valid form (where *a* symbolizes *you*):

---

1. (∃x)Cx   /∴ Ca

2. Ca   EI, 1   (Wrong!)

This proof is flawed. But there would be nothing wrong with step 2 if the
proof did not end at that line and did not conclude with a formula containing *a*.[4]

---

[4]The system of rules in this chapter is sound in the sense that the rules never lead us
astray; they never allow us to prove a formula that is not valid or permit us to establish the

The third rule for quantifiers is **universal instantiation.** If we know that something is true about every object, then we can conclude that it is true for each particular object we consider. If God loves everyone, then God loves you, me, and the Earl of Roxburgh. If Jane likes everyone she meets, then she likes you, if she has met you; she likes me, if she has met me; and so on. The rule of universal instantiation says that from a universally quantified statement, we may infer any of its instances.

UNIVERSAL INSTANTIATION (UI)

n.	(v)$\mathfrak{F}$(v)	
n+p.	$\mathfrak{F}$(c)	UI, n

Here c may be any constant.

This rule does not require a new constant. In fact, it is silly to use a new constant in applying UI unless no constants at all appear in the proof up to this line. If constants a and b appear earlier, then, from (x)Fx, we can infer Fa or Fb or both. If we also have a formula (x)Gx, we can infer Ga or Gb. And, if we have (x)(y)(Hx ⊃ Jy), then we can obtain (y)(Ha ⊃ Jy) or (y)(Hb ⊃ Jy) and, in another step, any of Ha ⊃ Ja, Ha ⊃ Jb, Hb ⊃ Ja, and Hb ⊃ Jb. We could also infer similar formulas with other constants. Unless we are forced to introduce those constants in other ways, however, using them to exploit a universal formula will serve no purpose.

---

validity of an invalid argument form. Nevertheless, the existential instantiation rule is unsound; it justifies the inference from (∃v)$\mathfrak{F}$(v) to $\mathfrak{F}$(c), so long as c is new to the proof. To preserve the system's soundness, we need to add restrictions to our proof system.

Most of our rules are **truth preserving:** The truth of the rules' premises guarantees the truth of their conclusions. Existential instantiation, however, is not truth preserving. The same is true of universal generalization; the inference from *Gorbachev is Russian* to *Everybody's Russian* fails. But, although EI and UG are not truth preserving, they never lead us to an illegitimate conclusion. The rules are **conservative** in that any formula without the constant introduced or generalized on that follows from the conclusion of the rule also follows from the rule's premise.

To put this another way: Our system of rules as a whole is sound, but proofs are not sound line by line. We can never prove a conclusion that does not follow from the premises, but we may deduce intermediate statements in the proof that do not follow from them. The restrictions make sure that these deviations do not affect the outcome of the proof.

Some logicians believe that proofs must be sound line by line and, so, formulate EI and UG differently. There is a way of interpreting our system as sound line by line, even with the rules as they are. Doing that requires that we reinterpret constants—in particular, constants used in EI and UG. So far, we have thought of constants, in a given context, as standing for objects. We could choose to think of them as marking places for objects, but without necessarily standing for anything in particular. Or, we could think of them as standing for a different kind of object. Developing these strategies goes beyond the scope of this book. But both lead to serious and interesting analyses of logic. For the first, see Hans Kamp, "A Theory of Truth and Semantic Representation," in Irene Heim, *The Syntax and Semantics of Definite and Indefinite Noun Phrases* (Ph.D. diss., University of Massachusetts at Amherst, 1983), and Hans Kamp and Uwe Reyle, *From Discourse to Logic* (Dordrecht, Netherlands: Kluwer, 1995). For the second, see Kit Fine, *Reasoning with Arbitrary Objects* (Oxford: Basil Blackwell, 1985).

To see how these rules work, consider

(50) Something is rotten in the state of Denmark.
     Whatever is in the state of Denmark is in Europe.
     ∴ Something is rotten in Europe.

---

1. $(\exists x)(Rx \cdot Dx)$		
2. $(y)(Dy \supset Ey)$	$/\therefore (\exists z)(Rz \cdot Ez)$	
3. $Ra \cdot Da$	EI, 1	
4. $Da \supset Ea$	UI, 2	
5. Da	Simp, 3	
6. Ea	MP, 4, 5	
7. Ra	Simp, 3	
8. $Ra \cdot Ea$	Conj, 7, 6	
9. $(\exists z)(Rz \cdot Ez)$	EG, 8	

This proof makes use of important guidelines for doing quantified proofs:

1. Whenever possible, use EI before UI.
2. Do not use new constants unless you must.

P R O B L E M S

Use deduction to show that these arguments are valid.

1. God created everything. So, God created Texas.

2. God created everything. So, God created something.

3. God created everything. Some things are evil. So, God created something evil.

4. Nothing logical frustrates me. This book frustrates me, so this book is illogical.

5. Any player who bats over .300 could be a hot item in the free-agent market. There is a Pirates player who bats over .300. So, some Pirate could be a hot free agent.

6. There are new houses that are very large but that are not very well built. Every very large house is impressive to visitors. Therefore, some houses that impress visitors are not very well built.

7. All insects in this house are large and hostile. Some insects in this house are impervious to pesticides. Thus, some large, hostile insects are impervious to pesticides.

8. Some students cannot succeed at the university. All students who are bright and mature can succeed. It follows that some students are either not bright or immature.

9. Nothing written by committee is easy to write or easy to read. Some documents written by committee are nevertheless extremely insightful. So some extremely insightful documents are not easy to read.

10. Some of the cleverest people I know are clearly insane. Any of the cleverest people I know could prove that this argument is valid. Hence, some people who could prove this argument valid are clearly insane.

11. All managers who put their own welfare above that of their company endanger the interests of stockholders. Some CEOs are managers who put their own welfare above that of their company. Thus, some managers who endanger stockholder interests are CEOs.

12. Some business professors spend their time on highly esoteric, theoretical questions. Nobody who spends his or her time on such matters does research that has much to do with the real world. Consequently, some business professors do not do research that has much to do with the real world.

13. Some utility companies are predicting brownouts in their service regions during this summer. No utilities that can easily and affordably purchase power from other utilities are predicting brownouts for this summer. Hence, some utility companies cannot easily and affordably purchase power from other utilities.

14. Some analysts insist that we are in the middle of a historic bull market, but others say that the market will soon collapse. Nobody who is expecting the market to collapse is recommending anything but utility stocks. None who believe that M1 controls the direction of the economy contend that we are in the midst of a historic bull market. Thus, some analysts are recommending only utility stocks, but some do not believe that M1 controls the economy's direction.

These are some of the syllogisms that Aristotle considered valid, together with their medieval names. Show that each is valid in predicate logic. (Some require extra assumptions; they are listed in parentheses. Translate universal negative statement forms with a universal quantifier.)

15. Darii: Every M is L; Some S is M; ∴ Some S is L.

16. Ferio: No M is L; Some S is M; ∴ Some S is not L.

17. Festino: No L is M; Some S is M; ∴ Some S is not L.

18. Baroco: Every L is M; Some S is not M; ∴ Some S is not L.

19. Darapti: Every M is L; Every M is S; (There are Ms;) ∴ Some S is L.

20. Felapton: No M is L; Every M is S; (There are Ms;) ∴ Some S is not L.

21. Disamis: Some M is L; Every M is S; ∴ Some S is L.

22. Datisi: Every M is L; Some M is S; ∴ Some S is L.

23. Bocardo: Some M is not L; Every M is S; ∴ Some S is not L.

24. Ferison: No M is L; Some M is S; ∴ Some S is not L.

Medieval logicians added other syllogistic patterns to those Aristotle explicitly held valid. Show that these "subaltern moods" are valid, at least with the added assumptions in parentheses.

25. Barbari: Every M is L; Every S is M; (There are Ss;) ∴ Some S is L.

26. Celaront: No M is L; Every S is M; (There are Ss;) ∴ Some S is not L.

27. Cesaro: No L is M; Every S is M; (There are Ss;) ∴ Some S is not L.

28. Camestros: Every L is M; No S is M; (There are Ss;) ∴ Some S is not L.

Theophrastus, who succeeded Aristotle as head of the Lyceum, also added additional syllogistic principles. Show that these, too, are valid, with the added assumptions in parentheses.

29. Baralipton: Every M is L; Every S is M; (There are Ss;) ∴ Some L is S.

30. Dabitis: Every M is L; Some S is M; ∴ Some L is S.

31. Fapesmo: Every M is L; No S is M; (There are Ms;) ∴ Some L is not S.

32. Frisesomorum: Some M is L; No S is M; ∴ Some L is not S.

Use deduction to establish the validity of the following argument forms.

33. (x)Fx; ∴ (∃x)Fx

34. (∃x)Fx; (x)Gx /∴ (∃x)(Fx · Gx)

35. (∃x)Fx; (x)(Fx ⊃ ~Gx); ∴ (∃x)~Gx

36. (x)(Gx ⊃ Hx); (∃x)(Fx · Gx); ∴ (∃x)(Fx · Hx)

37. (x)~(Gx · Hx); (∃x)(Fx · Gx); ∴ (∃x)(Fx · ~Hx)

38. $(x)(Hx \supset Gx)$; $(\exists x)(\sim Gx \cdot Fx)$; $\therefore (\exists x)(Fx \cdot \sim Hx)$

39. $(x)(\sim Gx \supset \sim Hx)$; $(\exists x)(Fx \cdot \sim Gx)$; $\therefore (\exists x)(Fx \cdot \sim Hx)$

40. $(x)(Gx \supset Hx)$; $(\exists x)(Gx \cdot Fx)$; $\therefore (\exists x)(Fx \cdot Hx)$

41. $(x)(Gx \supset \sim Hx)$; $(\exists x)(Gx \cdot \sim Fx)$; $\therefore (\exists x)(\sim Fx \cdot \sim Hx)$

42. $(\exists x)(Gx \cdot Hx)$; $(x)(Gx \supset Fx)$; $\therefore (\exists x)(Fx \cdot Hx)$

43. $(\exists x)(Gx \cdot \sim Hx)$; $(x)(Gx \supset Fx)$; $\therefore (\exists x)(Fx \cdot \sim Hx)$

44. $(x)(Hx \supset \sim Gx)$; $(\exists x)(Gx \cdot Fx)$; $\therefore (\exists x)(Fx \cdot \sim Hx)$

45. $(\exists x)(Hx \cdot Gx)$; $(x)(Gx \supset Fx)$; $\therefore (\exists x)(Fx \cdot Hx)$

46. $(\exists x)(\sim Fx \cdot \sim Gx)$; $(x)(\sim Gx \supset \sim Hx)$; $\therefore (\exists x)(\sim Hx \cdot \sim Fx)$

47. $(x)(Fx \supset (Gx \supset \sim Hx))$; $(\exists y)(Hy \cdot Fy)$; $\therefore (\exists x)\sim Gx$

48. $(x)(Fx \vee \sim Gx)$; $(x)(Fx \supset Hx)$; $(x)(\sim Gx \supset Jx)$; $(\exists x)\sim Jx$; $\therefore (\exists x)Hx$

49. $(x)(Fx \equiv Gx)$; $(x)(Gx \supset Hx)$; $(\exists x)(Fx \vee Gx)$; $\therefore (\exists x)(Fx \cdot Hx)$

50. $(z)(\sim Hz \equiv (x)(Fx \cdot Gz))$; $(x)Fx$; $(\exists y)Gy$; $\therefore (\exists x)\sim Hx$

■ 9.6 UNIVERSAL GENERALIZATION

Introducing a universal formula requires a new rule.

UNIVERSAL GENERALIZATION (UG)

---

n.	$\mathfrak{F}(c)$	
n+p.	$(v)\mathfrak{F}(v)$	UG, n

Here (1) c must not occur in $(v)\mathfrak{F}(v)$;
     (2) c must not occur in the assumptions or conclusion of the proof; and
     (3) c must not have been introduced by EI.[5]

---

[5]To handle predicate logic in its full generality, with relational predicates and quantifiers that overlap in scope, we need to add a fourth restriction on UG:

    4. No term remaining in $(v)\mathfrak{F}(v)$ may depend on c.

This provides a way to distinguish, in a proof, *something or other* from *some one thing* or *something in particular*. To understand the restriction, we need to define **dependence** among constants.

A constant c *immediately depends on* a constant d in a proof if and only if c is introduced into the proof by applying EI to a formula containing d.

A constant c *depends on* a constant d if and only if there is a chain of constants $c_1, \ldots, c_n$ such that c immediately depends on c_1; c_1 immediately depends on $c_2, \ldots$; and c_n immediately depends on d. (The chain may be empty; immediate dependence is a kind of dependence.)

UG is most often used near the end of a proof. To use UG to derive a universal conclusion, this rule says, derive an instance of it first. The idea behind UG is that we are generalizing on an arbitrarily chosen object. The restrictions on UG require that the object be indeed arbitrary. If the proof works for c, it would work for any object. Consequently, though we prove something about c, we have shown how to prove it about anything. And this justifies drawing a universal conclusion.

To take an example of a successful universal generalization, consider this inference:

(51) Everything created by God is good.

   Everything is a creation of God.

   ∴ Everything is good.

To establish its validity, we construct a universal proof:

1. $(x)(Cx \supset Gx)$		
2. $(x)Cx$	$/ \therefore (x)Gx$ [Ga]	
3. Ca	UI, 2	
4. $Ca \supset Ga$	UI, 1	
5. Ga	MP, 4, 3	
6. $(x)Gx$	UG, 5	

To show that everything is G, we show that some arbitrarily chosen object a is G. UG succeeds because a occurs in neither assumptions (lines 1 and 2) nor in the universal conclusion (line 6) and was not introduced by EI.

UG forces us to instantiate with a new constant—not in applying the rule itself, but earlier, when we use UI. (Here, in step 3.) To set up a later UG, we must use a constant that has not appeared before and will not appear in the conclusion. That constant stands for an arbitrarily chosen object. By proving something about this arbitrary object, we justify drawing a universal conclusion.

Above, beside our desired universal conclusion (x)Gx (on the right side of line 2), we wrote [Ga]. In general, when seeking to show a universal statement of the form $(v)\mathfrak{F}(v)$, it is useful to write, bracketed off to the right as an unofficial part of the proof, an instance $\mathfrak{F}(c)$ with a new constant c—that is, a constant that has appeared nowhere above, in the proof or in the desired conclusion (also written unofficially on the right). For purposes of further setups of UG or uses of EI, treat that new constant as if it were actually part of the proof; do not use it, for example, in applying EI. The bracketed instance is guaranteed to satisfy the restrictions on UG. Proving it thus sets up the use of UG that yields the universal conclusion.

To see why this is useful, consider a more complicated proof:

(52) Alvin is the teacher, and Betty is a student who has an after-school job. All the students want good grades. People are lazy if and only if they are not willing to work hard. All who want good grades are willing to work hard. Therefore, Betty is willing to work hard, and none of the students are lazy.

---

1. Ta · (Sb · Jb)

2. (x)(Sx ⊃ Gx)

3. (x)(Lx ≡ ~Wx)

4. (x)(Gx ⊃ Wx)   / ∴ Wb · (x)(Sx ⊃ ~Lx) [Sc ⊃ ~Lc]

We write [Sc ⊃ ~Lc] in the right margin, planning the eventual application of UG that will produce (x)(Sx ⊃ ~Lx). We use a new constant, c, since a and b appear in the premises and b appears in the conclusion. We then proceed to simplify premise 1:

---

5. Ta   Simp, 1

6. Sb · Jb   Simp, 1

7. Sb   Simp, 8

At this point, it is tempting to begin instantiating the universal quantifications on lines 2–4 using the constant b. But that would prevent us from using UG later; b appears in the premises. That restriction matters, moreover; if we were to ignore it here, we could show that all the students have after-school jobs! We must resist temptation and use the new constant c, as planned.

---

8. Sc ⊃ Gc   UI, 2

9. Lc ≡ ~Wc   UI, 3

10. Gc ⊃ Wc   UI, 4

Now, we apply propositional rules, aiming to derive Sc ⊃ ~Lc, according to plan.

11. Sc ⊃ Wc		HS, 8, 10
12. (Lc ⊃ ~Wc) · (~Wc ⊃ Lc)		Equiv, 9
13. Lc ⊃ ~Wc		Simp, 12
14. ~~Wc ⊃ ~Lc		Trans, 13
15. Wc ⊃ ~Lc		DN, 14
16. Sc ⊃ ~Lc		HS, 11, 15

Now, we can use UG. Our use of c guarantees that the restrictions are satisfied.

17. (x)(Sx ⊃ ~Lx)	UG, 16

Finally, we must show Wb. We could go back to lines 2–4 and instantiate again with b. But it is easier to note that we showed, on line 11, that an arbitrary student is willing to work hard. We can apply UG (again, with satisfaction of restrictions guaranteed) to line 11 and then instantiate to b:

18. (x)(Sx ⊃ Wx)	UG, 11
19. Sb ⊃ Wb	UI, 18
20. Wb	MP, 19, 7
21. Wb · (x)(Sx ⊃ ~Lx)	Conj, 20, 17

The restrictions on UG prevent us from proving very silly arguments valid. Consider the first restriction, that c not appear in (v)ℑ(v). To see why this is necessary, suppose we tried to argue

(53) All liberals are Democrats.

∴ If anyone is a liberal, everyone is a Democrat.

Clearly, the argument fails. We might attempt this:

1. (x)(Lx ⊃ Dx)	/∴ (y)(x)(Lx ⊃ Dy) [Lb ⊃ Da]
2. La ⊃ Da	UI, 1
3. (x)(Lx ⊃ Da)	UG, 2 (Wrong!)
4. (y)(x)(Lx ⊃ Dy)	UG, 3

(The symbolization of the conclusion may seem peculiar, but it is equivalent to the more intuitive translation (∃x)Lx ⊃ (y)Dy.) Since a occurs in (x)(Lx ⊃ Da), line 3 violates the first restriction on UG.

The second restriction prohibits c from appearing in the assumptions or conclusion of the proof. If we allowed c to appear in the assumptions, we could prove valid such poor arguments as

(54) Seinfeld is a comic.

∴ Everybody's a comic.

Ignoring restriction (2), we could attempt the proof

1. Cs	/ ∴ (x)Cx [Ca]
2. (x)Cx	UG, 1 (Wrong!)

The constant s, on which line 2 tries to generalize, appears in an assumption. So, UG cannot apply.

A similar argument shows the point of restriction (3), that c must not have been introduced by EI. Consider this invalid argument:

(55) There is a pope.

∴ Everybody is pope.

We might try to prove it valid as follows:

1. (∃x)Px	/∴ (x)Px [Pa]
2. Pb	EI, 1
3. (x)Px	UG, 2 (Wrong)

We used existential instantiation to introduce b, so we cannot later universally generalize on it.

QUANTIFIER NEGATION

Two **quantifier negation** rules relate quantified formulas to their negations. In the process, these rules relate the universal and existential quantifiers. In fact, they show how to define each quantifier in terms of the other. They underlie the equivalence of the following pairs of statements:

(56) a.   Not everyone will show up.
    b.   Somebody will fail to show up.

(57) a.   Jerry couldn't see a thing.
    b.   Everything was such that Jerry couldn't see it.

QUANTIFIER NEGATION (QN)

$\sim(\exists v)\mathfrak{J}(v)$	$\Leftrightarrow$	$(v)\sim\mathfrak{J}(v)$	QN, n
$\sim(v)\mathfrak{J}(v)$	$\Leftrightarrow$	$(\exists v)\sim\mathfrak{J}(v)$	QN, n

We can use quantifier negation to demonstrate the validity of

(58)   Everyone who came to the party was arrested.

Not all my friends were arrested.

∴ Not all my friends came to the party.

After symbolizing, we can construct this proof:

1.	$(x)(Cx \supset Ax)$	
2.	$\sim(x)(Fx \supset Ax)$	$/ \therefore \sim(x)(Fx \supset Cx)$
3.	$(\exists x)\sim(Fx \supset Ax)$	QN, 2
4.	$\sim(Fa \supset Aa)$	EI, 3
5.	$\sim(\sim Fa \vee Aa)$	MC, 4
6.	$\sim\sim Fa \cdot \sim Aa$	DM, 5
7.	$Fa \cdot \sim Aa$	DN, 6
8.	Fa	Simp, 7
9.	$\sim Aa$	Simp, 7
10.	$Ca \supset Aa$	UI, 1
11.	$\sim Ca$	MT, 10, 9
12.	$Fa \cdot \sim Ca$	Conj, 8, 11
13.	$\sim\sim Fa \cdot \sim Ca$	DN, 12

14. ~(~Fa ∨ Ca)	DM, 13
15. ~(Fa ⊃ Ca)	MC, 14
16. (∃x)~(Fx ⊃ Cx) ·	EG, 11
17. ~(x)(Fx ⊃ Cx)	QN, 12

PROBLEMS

Aristotle considered the first four of the following syllogisms valid; Theophrastus added the last. Show that each is valid in quantification theory. (Some require extra assumptions; they are listed in parentheses. Symbolize universal negative statement forms with universal quantifiers.)

1. Barbara: Every M is L; Every S is M; ∴ Every S is L.

2. Celarent: No M is L; Every S is M; ∴ No S is L.

3. Cesare: No L is M; Every S is M; ∴ No S is L.

4. Camestres: Every L is M; No S is M; ∴ No S is L.

5. Celantes: No M is L; Every S is M; ∴ No L is S.

The following arguments are adapted from Lewis Carroll's *Symbolic Logic*. Use deduction to show that each is valid.

6. All diligent students are successful; all ignorant students are unsuccessful. So, no diligent students are ignorant.

7. All selfish men are unpopular. All obliging men are popular. So, all obliging men are unselfish.

8. Ill-managed business is unprofitable. Universities are invariably ill managed. Therefore, all universities are unprofitable.

9. No professors are ignorant. All vain people are ignorant. So, no professors are vain.

10. Bores are dreaded. No one who is dreaded is ever begged to prolong his visit. So, no bore is ever begged to prolong his visit.

11. All wise men walk on their feet; all unwise men walk on their hands. Therefore, every man walks on either his hands or his feet.

12. No experienced person is incompetent. Everyone on our staff is always blundering; no competent person is always blundering. Therefore, everyone on our staff is inexperienced.

13. No terriers wander among the signs of the zodiac. Nothing that does not wander among the signs of the zodiac is a comet. Nothing but a terrier has a curly tail. Therefore, comets do not have curly tails.

14. All members of the House of Commons have perfect self-command; no member of parliament who wears a coronet should ride in a donkey race. All members of the House of Lords wear coronets. All members of parliament who are not members of the House of Commons are members of the House of Lords. Therefore, no members of parliament should ride in a donkey race, unless they have perfect self-command.

15. Showy talkers think too much of themselves. No really well-informed people are bad company. People who think too much of themselves are not good company. Therefore, showy talkers are not really well informed.

16. Things sold in the street are of no great value. Nothing but rubbish can be had for a song. Eggs of the Great Auk are very valuable. It is only what is sold in the street that is really rubbish. Therefore, eggs of the Great Auk cannot be had for a song.

17. None of the unnoticed things met at sea are mermaids. Things entered in the log, as met with at sea, are sure to be worth remembering. I have never met with anything worth remembering, when on a voyage. Things met with at sea, that are noticed, are sure to be recorded in the log. Therefore, I have never met a mermaid.

18. There is no box of mine here that I dare open. My writing desk is made of rosewood. All my boxes are painted, except what are here. There is no box of mine I dare not open, unless it is full of live scorpions. All my rosewood boxes are unpainted. Therefore, my desk is full of live scorpions.

19. Every idea of mine that cannot be expressed in a syllogism is really ridiculous; none of my ideas about bath-buns are worth writing down. No idea of mine that fails to come true can be expressed as a syllogism. I never have any really ridiculous idea that I do not at once refer to my solicitor. My dreams are all about bath-buns. I never refer any idea of mine to my solicitor unless it is worth writing down. Therefore, all my dreams come true.

20. The only animals in this house are cats. Every animal is suitable for a pet that loves to gaze at the moon. When I detest an animal, I avoid it. No animals are carnivorous, unless they prowl

at night. No cat fails to kill mice. No animals ever take to me, except what are in this house. Kangaroos are not suitable for pets. None but carnivora kill mice. I detest animals that do not take to me. Animals that prowl at night always love to gaze at the moon. Therefore, I always avoid a kangaroo.

Show, using deduction, that these argument forms are valid.

21. $(x)(Fx \cdot Gx)$; $\therefore$ $(x)Fx \cdot (x)Gx$

22. $(x)Fx \cdot (x)Gx$; $\therefore$ $(x)(Fx \cdot Gx)$

23. $(x)(Fx \equiv Gx)$; $(x)Fx$; $\therefore$ $(x)Gx$

24. $(x)(Fx \equiv (Gx \cdot Hx))$; $\therefore$ $(x)(Fx \supset Gx) \cdot (x)(Fx \supset Hx)$

25. $(x)((Fx \lor Gx) \supset Hx)$; $\therefore$ $(x)(Fx \supset Hx) \cdot (x)(Gx \supset Hx)$

26. $(x)(Fx \supset Hx)$; $(x)(Gx \supset Hx)$; $\therefore$ $(x)((Fx \lor Gx) \supset Hx)$

27. $(x)(Fx \equiv (\sim Gx \cdot \sim Fx))$; $\therefore$ $(x)Gx$

28. $(x)(Fx \equiv (\sim Gx \supset \sim Fx))$; $\therefore$ $(x)Gx$

29. $(x)((Fx \cdot Gx) \supset Hx)$; $(x)Gx \cdot (x)Fx$; $\therefore$ $(x)(Fx \cdot Hx)$

30. $(x)(Fx \supset (Gx \lor Hx))$; $(x)((Jx \cdot Fx) \supset \sim Gx)$; $(x)(\sim Fx \supset \sim Jx)$; $\therefore$ $(x)(Jx \supset Hx)$

PART IV

INDUCTION

10

GENERALIZATIONS
AND ANALOGIES

P art III of this book discusses deductive validity. An argument is deductively valid if and only if the truth of its premises guarantees the truth of its conclusion. Many successful, commonly advanced arguments, however, are not deductively valid. Most of our general knowledge about the world rests on **induction**—reasoning "from the known to the unknown," in the words of John Stuart Mill (1806–1873). Such reasoning is reliable but not valid deductively. Deduction dominates mathematics and a few other abstract disciplines. Everywhere else, induction rules. For this reason, Mill called induction "the main question of the science of logic."[1]

Consider a very simple argument:

(1) Every cat I have ever known liked to eat fish.

∴ All cats like to eat fish.

The premise of this argument could be true even if the conclusion were false. Perhaps some cats I have never met hate fish. Nevertheless, the premise offers evidence for the conclusion. There is a logical relation between the two, but it is not implication. If I have known a wide variety of cats, moreover, the argument is fairly reliable.

An argument that is reliable without being deductively valid is **inductively strong.** The premises of such an argument support the conclusion but do not establish it conclusively. That is why Mill says one reasons to a conclusion one does not know (for certain) to be true. This chapter discusses the notion of inductive strength and probes some common types of inductive arguments. The next two chapters discuss progressively more complex topics on the logic of induction.

---

[1]John Stuart Mill, *A System of Logic* (New York: Harper and Brothers 1874), p. 224.

■ 10.1 INDUCTIVE STRENGTH

Inductively strong arguments are successful without being valid deductively. The mark of validity is that the premises imply the conclusion; the conclusion cannot be false while the premises are all true. The truth of the premises guarantees the truth of the conclusion. In an inductively reliable argument, there is no guarantee. But the truth of the premises does provide evidence for the truth of the conclusion. The truth of the premises makes the truth of the conclusion probable.

> An argument is **inductively strong** if and only if *(a)* it is not deductively valid and *(b)* given the truth of the premises, the truth of the conclusion is probable.

This argument, for instance, is inductively strong:

(2) All the crows that have ever been observed have been black.

∴ All crows are black.

People have observed many crows, of many different shapes and sizes, in many different places, and at many different times. If all have been black, it seems probable that all crows are black. Drawing this conclusion from the collected observations of crow-watchers through the ages seems reasonable. Nevertheless, the conclusion's truth is not guaranteed; a purple crow may somewhere lurk unobserved. The argument is strong, but not valid.

Another, more complex example:

(3) 43 percent of voters surveyed leaving the polls said that they voted Democratic.

∴ 43 percent of all voters voted Democratic.

If the poll was taken correctly, then the premise of this argument supports its conclusion. The survey results indicate that it is probable that 43 percent of all voters voted Democratic. They do not, however, guarantee this conclusion. In 1988, two major news networks declared, on the basis of exit surveys, that Michael Dukakis won Illinois. The survey methods were sophisticated; the surveys were conducted properly. But, when all votes were counted, George Bush had won the state. Similarly, in 1996, the major networks declared that Bob Dole finished third in the New Hampshire primary, behind Pat Buchanan and Lamar Alexander. When the votes were tallied, however, Dole finished second.

Just as an argument can be deductively valid even though one or more of its premises is in fact false, an argument with a false premise can be inductively strong. As an example, consider

(4) All the crows that have ever been observed have been purple.

∴ All crows are purple.

This argument, like (2), is inductively strong: It is not deductively valid; and although its premise is in fact false, its conclusion is nevertheless probable conditioned on the assumption that its premise is true.

Inductively strong and deductively valid arguments do differ in two important ways. First, adding a premise can make an inductively strong argument weak. In a deductively valid argument, in contrast, the truth of the premises guarantees the truth of the conclusion. So, no matter what information is added to the premises, the conclusion still follows. For example, this argument is valid:

(5) Sam and Dave love baseball.

∴ Dave loves baseball.

It remains valid, no matter what premises we add:

(6) Sam and Dave love baseball.

Jill prefers basketball, and Susan likes squash.

Dave has seen the Pirates play dozens of times.

Sam has seen baseball only on television.

∴ Dave loves baseball.

In fact, it remains valid even if we add information that directly contradicts our premise:

(7) Sam and Dave love baseball.

Dave hates baseball.

Sam hates baseball.

∴ Dave loves baseball.

Any argument with inconsistent premises is trivially valid: If an argument has inconsistent premises, it cannot be the case that the premises are all true and the conclusion false—simply because it cannot be the case that the premises are all true.

But adding premises to an inductively strong argument may make it weak. The premises of an inductively strong argument support the conclusion but do not establish it conclusively. So, additional information may undermine the support offered by the original premises. For example, this argument is plausible:

(8) Sam and Dave love baseball.

∴ Many of Dave's friends love baseball.

Dave's love for baseball may suggest that many of his friends love baseball, too. People tend to become friends with people who share their interests. So, the premise offers some support for the conclusion. Adding premises, however, may destroy whatever reliability the argument has:

(9) Sam and Dave love baseball.

Dave's other friends are Sam, Susan, Jill, Andy, and Pat.

Jill, Andy, and Pat think baseball is stupid.

∴ Many of Dave's friends love baseball.

So, adding premises may make an inductively strong argument weak, although adding premises never makes a valid argument invalid.

The second major difference between inductive and deductive arguments is that deductive validity is an all-or-nothing affair, while inductive strength is a matter of degree. An argument is valid if and only if the truth of its premises guarantees the truth of its conclusion. All valid arguments are equally valid. (A guarantee is a guarantee.) It makes no sense to make comparative judgments of validity. Nor does it make sense to speak of an argument as "somewhat valid," "moderately valid," or "very valid."

Inductive strength, in contrast, comes in degrees. Given the truth of the premises of an inductively strong argument, the conclusion is probable, and probability is a matter of degree. The premises may offer more or less support for the conclusion. So, some inductively strong arguments may be stronger than others. We can speak sensibly of arguments being "somewhat strong," "moderately strong," "very strong," and the like.

Suppose that we are investigating a murder. We find a witness who says that she heard exactly three gunshots. We may conclude, provisionally, that there were exactly three gunshots:

(10) Lucy says she heard exactly three gunshots.

∴ There were exactly three gunshots.

How strong is this argument? That depends on how reliable Lucy is. If we find additional witnesses who also report three shots, we can construct a stronger argument:

(11) Lucy, Ricky, Fred, and Ethel all say they heard exactly three gunshots.

∴ There were exactly three gunshots.

But further information may lead to a weaker argument:

(12) Four witnesses say they heard exactly three gunshots.

Six witnesses say they heard exactly four gunshots.

One witness reports hearing exactly five gunshots.

∴ There were exactly three gunshots.

This argument seems unreliable. The witness reports strongly support the assertion that there were at least three gunshots but offer weak support for the conclusion that there were exactly three.

P R O B L E M S

Say whether each argument in Chapter 1, section 1.2, is deductively valid, inductively reliable, or neither.

■ 10.2   ENUMERATION

The simplest kind of inductively strong argument is argument by **enumeration**. Arguments by enumeration try to justify a general conclusion by listing instances of that conclusion. Suppose we want to argue that all cats like fish. We might take as premises sentences such as

(13) Jessica is a cat, and Jessica likes fish.

Gwen is a cat, and Gwen likes fish.

Snowball is a cat, and Snowball likes fish.

No matter how many such sentences we accumulate, the conclusion "All cats like fish" will not follow deductively. (Even if the list of premises mentions every cat that exists, "All cats like fish" will not be a deductive consequence, since the conclusion would be false if there were more cats than there in fact are and at least one of them disliked fish; thus the joint truth of such premises is logically compatible with the falsity of the conclusion.) Nevertheless, these sentences do support that conclusion. The more premises of this form we acquire, the stronger that support becomes. The argument with the sentences in (13) as premises is not very strong; it mentions only three of the millions of cats covered by the conclusion. If we add such premises as

(14) Tina is a cat, and Tina likes fish.

Leela is a cat, and Leela likes fish.

Penelope is a cat, and Penelope likes fish.

the argument becomes slightly stronger. The conclusion follows with a greater degree of probability.

This example suggests a pattern. Arguments by enumeration go from instances to generalizations. The instances used to support a conclusion of the form "All Fs are G"—"All cats like fish," for example—have the form "x is both F and G"; for example, "x is a cat and likes fish." They are often called **supporting** or **confirming instances** of the conclusion. Arguments by enumeration thus have the basic structure

ARGUMENT BY ENUMERATION

SUPPORTING INSTANCES	a_1 is both F and G
	a_2 is both F and G
	.
	.
	.
	a_n is both F and G
GENERALIZATION	∴ All Fs are G

If the conclusion has the form "No Fs are G," the supporting instances follow the pattern "x is F and not G." To show that no cats eat fruit, for example, we can enumerate premises of the form "x is a cat and does not eat fruit."

Arguments by enumeration can adopt forms closely related to, but not identical with, this basic structure. Instead of listing the supporting instances one by one, we could group instances together, presenting, in effect, subgeneralizations:

(15) All tabbies that have been observed like fish.

All calicos that have been observed like fish.

∴ All cats like fish.

Or, we could combine all the known supporting instances together into a single group:

(16) All cats that have been observed like fish.

∴ All cats like fish.

This displays a central feature of arguments by enumeration: They argue from observed—or, more generally, known—instances to generalizations. The known instances may be listed, presented in groups, or bunched together into a single group in the premises. The key move is from facts about known instances in the premises to known and unknown cases in the conclusion.

EVALUATING ENUMERATIONS

Arguments by enumeration generalize from known instances. They are never deductively valid. Even if we were to enumerate all instances of a certain kind, our list of premises would not by itself imply the generalized conclusion. As Bertrand Russell (1872–1970) noted, we would also need the premise that we have listed all the instances of the relevant kind. Suppose that Ann has two brothers, Jake and Harry. The argument

(17) Jake and Harry are both dentists.

∴ All Ann's brothers are dentists.

is not valid. We would need to supplement (17) with an additional premise to obtain a valid argument:

(18) Jake and Harry are both dentists.

Jake and Harry are Ann's only brothers.

∴ All Ann's brothers are dentists.

Even though they are not valid, some arguments by enumeration are very strong. Dogs of all known breeds, for example, have noses. It seems reasonable to conclude that dogs of all breeds, known and unknown, observed and unobserved, have noses. There could be a breed of dog, never before encountered, that features a very unusual head with nothing that we would call a nose. But it seems very unlikely. That all known dog breeds have a certain feature provides strong evidence that all dog breeds have that feature.

Other arguments by enumeration, however, are weak. A child who has seen only one dog that happens to be gentle might infer that all dogs are gentle. The evidence presented by observation of one dog offers weak support for any generalization about all dogs.

What distinguishes strong from weak arguments by enumeration? Several factors help to determine the reliability of these arguments. These factors relate to characteristics of the **sample,** the class of objects appealed to in the premises.

1. Sample Size   How many objects of the appropriate kind have been observed? What percentage of the total population is in the sample? Other things being equal, the more objects that have been observed, the more reliable the argument from them to the conclusion. Inferring things about all federal agencies on the basis of observations about just one agency, for example, is not very reliable. Generalizing from observations of two agencies is somewhat better; generalizing from observations of six agencies is much more reliable. Percentages, as well as absolute numbers, make a difference. Six agencies may constitute a reasonable percentage of all federal agencies to survey, but six people would constitute a much smaller percentage of the total number of people. A conclusion based on an enumeration involving six agencies, other things being equal, would be far more reliable than a conclusion based on an enumeration involving six people.

Generalizations obtained from inadequate enumeration are called **hasty.** Concluding that all government employees are underpaid because a few observed ones are, or that welfare cheating is widespread because of one or two highly publicized cases, is to make a hasty generalization.

2. SAMPLE VARIATION   How varied are the observed instances? How homogeneous is the sample? Other things being equal, the more varied the sample, the more reliable the argument. If we make a generalization about all government agencies on the basis of our experience with six agencies that all were involved in military affairs, our argument is less reliable than one generalizing from experience with six very different agencies. Similarly, our generalizations about people are more reliable if we have observed a wide variety of people. Suppose we want to learn what American citizens think about crime. If we ask just Republicans, we will get one set of answers. If we ask just Democrats, we will get another. If we ask just schoolchildren or just farmworkers or just city dwellers, we will get still different results. To generalize accurately, we need to ask people of different ages, in different parts of the country, and of different ethnic backgrounds, political persuasions, and professions.

In saying that it is desirable to have, for example, people of different age groups among the instances over which we generalize, the point is not that we want equal numbers of people from each age group—to have asked the same number of people between the ages of 80 and 90 as between 30 and 40, and so forth. What we aim for, rather, is that, among the instances over which we generalize, the proportion of people from any particular age group is close to that group's proportion in the population as a whole. Some kinds of difference matter more than others. We would not appreciably increase the strength of our generalization if we made sure to include, among those we asked for their views on crime, people with various different favorite flavors of ice cream. Ideally, we want our sample to be *representative*—to mirror the total population in its relevant characteristics. (Unrepresentative samples are *biased*.)

As the philosopher Ludwig Wittgenstein (1889–1951) pointed out, looking at just one kind of example is a common source of philosophical error. The same holds in all fields. Unless a sample is varied enough, it is likely to be biased. The bias may arise from limitations of the method used to gather the sample. It may result from habits of thought that are hard to break. And, it may arise from a tendency, conscious or unconscious, to ignore contrary evidence.

PROBLEMS

For each of the following generalizations, say whether each additional premise listed makes the inference stronger or weaker, and explain. (Consider each added premise separately.)

1. All my cats like turkey. So, all cats like turkey.
   (a) I have two cats.
   (b) I have fifteen cats.
   (c) All my cats are calicos.

(d) One of my cats is a calico; several are tabbies; one is Persian; and one is Siamese.

(e) All my cats are former strays.

2. Werner has taken several philosophy courses with professors who smoke pipes. He concludes that all philosophy professors smoke pipes.
   (a) Werner takes several more philosophy courses, and all those professors also smoke pipes.
   (b) All Werner's philosophy courses have been taken at the same university.
   (c) All Werner's philosophy professors studied at Frankfurt.
   (d) Almost all professors on Werner's campus smoke pipes.
   (e) Werner takes another philosophy course and never sees the professor smoke.

3. Abigail meets some people from the Maritime Provinces; all are named "Leblanc." She concludes that everyone in the Maritimes is named "Leblanc."
   (a) Everyone Abigail met belongs to the same family.
   (b) Everyone Abigail met is a teacher at a college or university.
   (c) Abigail met the Leblancs at different times and in different places.
   (d) The Leblancs Abigail met all know each other.
   (e) Abigail met the Leblancs many years ago.

4. Orlando knows several people who spent some of their child-hood in France. All are very fond of wine. He concludes that everyone who lived for a while in France as a child likes wine.
   (a) One person is an accountant, one a museum curator, and one a lawyer.
   (b) Everyone else Orlando knows likes wine, too.
   (c) Orlando meets people who spent part of their childhood in Spain; they are also very fond of wine.
   (d) Most of Orlando's other acquaintances prefer beer to wine.
   (e) Orlando visits France and finds that most people drink wine with meals.

5. Millie knows several friends who own Fords and like them; she concludes that everyone who owns a Ford likes it.
   (a) Millie's friends all live in New York.
   (b) Some of Millie's friends take good care of their cars; others are negligent about maintenance.
   (c) Some of Millie's neighbors also drive Fords; all like them.
   (d) One of Millie's friends bought her car in Dallas; the others bought theirs in Austin or San Antonio.
   (e) Most of Millie's friends drive only around town; one drives to Houston and back each week.

◼ 10.3   STATISTICAL GENERALIZATIONS

Rare before the twentieth century, statistical arguments are now commonplace. In a typical newspaper, statistics fill the financial pages, the sports pages, and some of the advertising. The national news pages may discuss inflation rates and poll results; the local pages may discuss a survey of citizens' attitudes. Each of the following headlines, for example, would accompany an article prominently featuring statistical reasoning:

Low P/E Stocks Outperform Market, Study Finds

Inflation Slows in August; Trade Gap Narrows

Race Too Close to Call, Poll Shows

"Build It Somewhere Else": Residents Want New Airport, But Not Nearby

Because statistical reasoning is so common, no study of logic would be complete without some examination of it.

Evaluating statistical arguments, however, is difficult. Statistics is a significant and complex field of mathematics. Moreover, many frequently used English words pertaining to statistics, such as "average," have more specific uses in statistics than in everyday discourse. Finally, statistical arguments often omit important information. When the *Wall Street Journal* reports poll results, it publishes nearby a brief summary of how the poll was conducted. Few news sources report such information, however, and few readers know how to interpret it. Yet how the poll was conducted may have a large bearing on what it means.

Polls and surveys rely on a form of argument known as **statistical** or **inductive generalization.** They seek to learn what percentage of people prefer the Republican candidate for president, have incomes over $100,000 per year, or feel unsatisfied with their current toothpaste. To find out, they ask, not everyone, but a relatively small collection of people, called a **sample.** They use information about the sample to draw a conclusion about the total population under study.

Arguments by enumeration, considered earlier, draw the conclusion that all Fs are G based on the fact that all Fs that have been observed are G. But what if our observations find some Fs that are G and others that aren't G? For example, after testing 300 bumpers from a certain factory, we might find that 9 are defective and 291 perform properly. Clearly, this does not support the conclusion that all bumpers produced by the factory are defective. Nevertheless, our sample can support a conclusion about the total population of bumpers the factory manufactures: Namely, the percentage of defective bumpers in the total population is the same as the percentage observed in the sample, in this case 3%.[2]

---

[2]That is, the number of defective bumpers in the sample (9) divided by the number of bumpers in the sample (300) equals 3 divided by 100.

Still the argument that the percentage of defective bumpers in the total population exactly matches the percentage in the sample might not be very strong. Its strength will depend on such factors as the total number of bumpers produced by the factory, the number of different models of bumper, and how accurately the sample reflects this product line. One way to create a stronger argument would be to make a more modest claim as the conclusion. In this case, for example, rather than concluding that 3% of the total population of bumpers are defective, we might conclude that from 1% to 5% of bumpers manufactured by the factory are defective. Another way to express the same conclusion is to say that 3%, ±2%, of bumpers manufactured by the factory are defective.

The general idea is that if we use a sample as the basis for a claim about the distribution of a certain characteristic in some larger population, we do not always have to argue that the percentage of the total population possessing the characteristic exactly matches the percentage of the sample observed to have the characteristic. We can argue instead that the percentage of the total population possessing the characteristic falls within some range of values centered on the percentage observed in the sample. Roughly speaking then, a statistical generalization has the following form:

Statistical Generalization

n% of observed Fs are G

∴ n%, ± e%, of Fs are G

That is, on the basis of a sample of Fs, n% of which are observed to be G, it is argued that between n−e% and n+e% of the total population of Fs are G, for some number e.[3] The percentage e is known as the **margin of error** of the generalization. Also, the range from n−e% to n+e% is called the generalization's **confidence interval;** and the generalization is said to have an **accuracy** of 1−e%.

In statements of statistical generalizations, the margin of error e is often presented apart from the percentage n, especially when several conclusions are drawn on the basis of a given sample. For example, the results of a poll assessing the preferences of registered voters for major party candidates may be given in this format:

---

[3]This presentation of the form of a statistical generalization is intended to indicate the main features of such arguments. It is a simplified representation in that the premises make no mention of such factors as the size of the sample, the size of the total population of Fs, and the method by which the sample was selected. As will be discussed, these and other factors constitute part of the basis for the claim made by the conclusion and are relevant to the evaluation of the argument.

DEMOCRATIC	REPUBLICAN	OTHER	UNDECIDED
42%	43%	4%	11%

(Margin of error: 3%)

Here the columns report the preferences actually observed among the sample. On the basis of these observations, the poll claims that between 39% and 45% of registered voters prefer the Democratic candidate, that between 40% and 46% of registered voters prefer the Republican, and so forth. The margin of error (and, indirectly, the confidence interval) is often reported in poll results in such phrases as "within 3 points," "plus or minus 5 percent," and so on. The poll is said to be accurate to within the margin of error.

In the case of a statistical generalization purporting to show that n%, ±e%, of Fs are G, if the premises are true, there is still no complete *guarantee* that the distribution of things that are G in the total population of Fs falls within the confidence interval. And, as with inductive arguments in general, the strength of a statistical generalization is something that admits of degrees. Some such generalizations are very weak; others very strong; and still others somewhere in between. Given enough information about the particular circumstances in which a statistical generalization has been or might be made—such characteristics as the size of the sample, the way in which the members of the sample are selected, and the size and diversity of the total population—the strength of the generalization can actually be expressed as a numerical value, called the **reliability** or **degree of confidence** of the generalization. For example, a generalization asserting the defectiveness of 3%, ±2%, of bumpers manufactured by a particular factory might have a reliability of 80%. This means that, in the case of generalizations with the same characteristics, four times out of five the percentage of defective bumpers in the total population will fall within the specified confidence interval. The degree of confidence associated with a statistical generalization is sometimes also referred to as the **confidence level** of the generalization.

An analogy may clarify the relationship between accuracy (or margin of error) and reliability: The difference between accuracy and reliability is like the difference between a point spread and odds in betting on a game. The odds indicate the level of confidence the parties have about the teams' performances; the point spread indicates something about the accuracy of their estimate of the score. Clearly, the odds and the point spread are interrelated. The same bettors would assign different odds to a bet, depending on the point spread. Similarly, the accuracy and reliability of a sample interrelate. Other things being equal, lowering the accuracy of a generalization (that is, increasing the margin of error) will tend to raise its reliability.

To represent these concepts visually:

$\quad$ 100%

$\quad$ n+e%   **Margin of error:** e%

$\quad$ n%   **Accuracy:** 1−e%

$\quad$ n−e%   **Confidence interval:** n−e% to n+e%

Confidence level: m% (m% of values for population fall within confidence interval)

SAMPLE SIZE

When we infer something about a large group from information gathered about a sample, we assume that the sample is **representative** of the large group: that the relevant characteristics of the sample are typical of the large group as a whole. An unrepresentative sample is **biased.** The reliability of a statistical generalization thus depends crucially on how well the sample represents the population being studied.

All other things being equal, large samples are more likely to be representative than smaller samples. The more cases observed, the stronger the generalization based on them. How large a sample must we consider to establish the strength of a generalization? There is no simple answer to this question. Too small a sample leads to weakness; too large a sample wastes resources. To determine an appropriate sample size, we need to specify two factors: the accuracy we want and the reliability we want. Only relative to these can we determine an appropriate sample size.

To illustrate how these concepts apply to statistical generalizations: Suppose we want to find out how many students at the university consider themselves Democrats. Say that the university has 50,000 students. How many do we need to survey to obtain an adequate sample? Before answering, we must specify a desired margin of error. Suppose we say that we want the answer to be correct to within 5%. That is, if our poll finds that 40% of the students sampled consider themselves Democrats, we will draw the conclusion that between 35% and 45% of the student body at large consider themselves Democrats. The range of 35% to 45% in this case is the confidence interval established for the survey.

Second, we must specify a desired degree of reliability or confidence. Just how reliable, in other words, do we want our generalization to be? We cannot guarantee that the sample's characteristics will be within 5%

of those of the entire student body without asking virtually everyone. So, we need to know what degree of reliability is acceptable. We might specify a 90% confidence level: Nine times out of ten, the percentage of, say, Democrats in the entire population will fall within the confidence interval we have specified. We could represent these decisions as follows. The form of the generalization will be

(19) n% of the sample are F

∴ n%, ±e%, of the population are F   Confidence: m%

where e% is the margin of error, n− e% to n+e% is the confidence interval, and m% is the confidence level, the measure of the reliability of the inference. In our example, choosing 5% for a margin of error (or, equivalently, 95% accuracy) and 90% reliability, the generalization will have the form

(20) n% of the sample consider themselves Democrats.

∴ n%, ±5%, of the students consider themselves Democrats.

Confidence: 90%

This means that the percentage of students considering themselves Democrats will lie between n−5% and n+5%, except for a one in ten chance. To obtain this level of confidence and margin of error, we would have to survey about 270 students. (This is true provided that the university has more than about 6,000 students. The sample of 270 will allow us to generalize to populations of 6,000 or 25,000 or 50,000 or 500,000 with almost exactly the same confidence and margin of error. In statistical generalizations, the percentage of the population sampled is usually irrelevant to reliability. This is surprising; explaining why it is true goes beyond the scope of this book. But it is what allows opinion polls to be economically and practically feasible.)

Supposing, again, that we find 40% of the sample calling themselves Democrats, our generalization will have the form

(21) 40% of the sample consider themselves Democrats.

∴ 40%, ±5%, of the students consider themselves Democrats.

Confidence: 90%

The smaller the margin of error we are willing to tolerate and the more reliability we demand, the larger the sample must be. If we require 95% reliability, for example, we will have to interview about 400 students. Shrinking the margin of error below 5% or increasing the reliability above 95% requires a sharper increase in the size of the sample. The following table indicates how large a sample is needed in a large population (that is, over about 330,000) to obtain various margins of error and levels of confidence.

		CONFIDENCE			
		99%	95%	90%	80%
	1%	16,650	9,600	6,725	4,100
	2%	4,163	2,400	1,681	1,025
MARGIN	3%	1,850	1,067	747	456
OF	4%	1,041	600	420	256
ERROR	5%	666	384	269	164
	10%	154	96	68	41

This assumes that about half the population will exhibit the trait we are investigating. Rare characteristics are much harder to estimate and pose special problems of their own. This also assumes that the population being studied is relatively *homogeneous*—that is, that it does not divide naturally into subpopulations with very different characteristics. We need to take the variation of the population into account to determine both how big and how varied the sample ought to be.

SAMPLE VARIATION

Groups are rarely completely homogeneous; their members typically display some degree of variation. The more varied the sample we take, the more likely it is to represent accurately the variation in the group as a whole. Increasing the variation in a sample, therefore, is one way to increase the reliability of a statistical generalization.

Suppose, for example, that we want to compare several supermarkets to see which has the lowest prices. Supermarkets sell many different products. If we compare the supermarkets by examining prices on a single item—rutabagas, say—any argument to their pricing across the board will be weak. Considering prices of lettuce, carrots, oranges, and jalapeño peppers will help some, but not much, since all these items are fresh produce; prices on them may not reflect prices on other kinds of goods. An adequate sample would have to include items of different kinds, capturing more of the variation in products. A sample consisting of rutabagas, pork chops, canned beets, cornflakes, shampoo, bread, cheese, and milk, for example, would allow for a much more reliable generalization than one consisting solely of eight fruits, eight breakfast cereals, or eight brands of dog food.

How do we ensure that our sample is varied enough for a strong generalization? The most common technique is to use a **random sample:** one drawn randomly from the population.

> A sample of a population is **random** if and only if every member
> of the population has an equal chance of being included in it.

Random sampling does not guarantee a representative or even widely varied sample, but it does make one likely.

Random sampling works well with populations that are fairly homogeneous or about which we have little further information. With populations that divide into distinct subgroups with different characteristics, we can do better with **stratified random sampling:** applying random sampling techniques separately to each subgroup. Stratified sampling also provides more detailed information. It permits generalizations about the various subgroups, or **strata,** as well as about the population as a whole. In surveying our university students for political affiliations, for example, we might want to poll students in business, liberal arts, and education colleges separately. In addition, we might want to distinguish graduate from undergraduate students. This would lead to a survey with six separate random samplings.

Obtaining a random sample, within strata or in the population at large, is difficult. An ideal method is to number all the population's members and then choose the sample using a random number sequence. But this faces two problems. First, many populations are difficult to number. Imagine facing Winnie-the-Pooh's task of counting the bees in even a single hive! Even already-numbered populations, such as the student body at a college or university, may present problems, since Social Security numbers, student numbers, and the like are not public information. A statistician cannot simply send in a fee and get a list of everyone in the United States by Social Security number; questions of privacy rule out the use of these already-available numbering systems. Second, there is no way to force people to respond to survey questions. Even if everyone selected for the sample can be found—and this is not a trivial matter—many may decline to answer. People may be pressed for time, be concerned about privacy, or hate surveys. This presents a problem to the statistician, for it may introduce bias.

Suppose that only 80% of those questioned in a randomly chosen sample respond. Is the responding sample still representative? Perhaps, if the unresponsive people were distributed randomly. In many cases, however, reasons why people respond, or fail to respond, are relevant to the survey's topic. People at the top and bottom of the socioeconomic ladder are notoriously hard to find and also notoriously unresponsive; few opinion polls reflect their views. Because social and economic standing are relevant to opinions on many topics, omission of these groups introduces bias into many surveys. To evaluate a survey based on even a carefully drawn random sample, we must look at the **response rate,** the proportion of people who responded as a fraction of all those asked to respond. In general, the greater the response rate, the less significant a source of bias unresponsive members of the sample constitute. If the response rate is low—and some surveys have response rates well below 50%—we should interpret the results with caution.

Because numbering a population can be difficult, most samples are selected in a less than ideal way. Sometimes, pollsters choose a system that numbers many but not all people: a voter registration list, for example, a driver's license list, or the telephone book. These can introduce bias. Many people do not register to vote; many people do not have licenses to drive; and many people have no telephone or an unlisted number. If any of these factors relate statistically to what is being studied, omitting such people creates bias.

Sometimes, pollsters choose samples in even less reliable ways. Someone may stand on a street corner or in a shopping mall and ask questions of passersby. Clearly, this is not a random sample—not everyone has an equal chance of passing by a given point on a given street during a given period of time. Kinsey's pioneering sex research in the 1940s found startlingly high rates of sexual deviancy among American men; it has since been discovered that his sample had a high percentage of prisoners. Not everyone, of course, has an equal chance of being in prison at a given time. Similarly, a magazine's survey of its readers may reveal something about the people who read it, but probably not about the population at large. The *Reader's Digest* confidently predicted that Alf Landon would swamp Franklin Delano Roosevelt in the 1936 presidential election, for example, because its readers strongly preferred Landon.

Perhaps the worst are surveys with self-selected samples. A television program, for example, may flash a number and ask people to call to register their views. The sample that results is self-selected: People choose whether to be a part of it or not. Such samples are almost always biased, for a person's decision to respond depends on his or her feelings about the issue in question. For this reason, the selection almost always relates to the issue at hand and distorts the result. Consequently, surveys with self-selected samples are usually so flawed as to be virtually meaningless. Yet newspapers or magazines still occasionally contain an insert to be mailed back expressing some opinion for a survey of readers. Several widely publicized studies of female sexuality published in the 1970s and 1980s used self-selected samples: One magazine asked readers to fill out a questionnaire, and another researcher distributed questionnaires through women's groups (mostly, chapters of the National Organization for Women). Although such studies may turn up interesting information, their samples are almost sure to be biased, and any statistical generalizations drawn from those samples are suspect.

PROBLEMS

For each of the following survey topics, evaluate the sampling methods proposed. Do they contain sources of bias? If so, how serious are they?

1. How many students are happy with the education they're receiving at State?
   (a) Pass out questionnaires in large introductory courses.
   (b) Pass out questionnaires at registration.
   (c) Ask students in lines at registration.
   (d) Interview students, chosen at random, from each floor of each dormitory.
   (e) Put a questionnaire in the student newspaper, to be mailed in.

2. Which of three factories produces the smallest percentage of flawed widgets?
   (a) Test every hundredth widget produced at each factory between 9 A.M. and 5 P.M. for a week.
   (b) Test randomly selected widgets produced at each factory between 9 A.M. and 5 P.M. for a week.
   (c) Test randomly selected widgets produced at each factory during any shift for a week.
   (d) Test randomly selected widgets produced at each factory during any shift for a month.
   (e) Test randomly selected widgets produced at each factory, separately for each shift, for a month.

3. How many students at Tech support raising the university's admissions standards?
   (a) Ask representatives in the student government.
   (b) Ask students at the Union at lunchtime.
   (c) Ask students in randomly selected classes.
   (d) Ask students in the library.
   (e) Put up posters asking students to write their comments.

4. What percentage of El Paso residents will vote Republican in the next election?
   (a) Survey people in each El Paso shopping center.
   (b) Call randomly selected telephone numbers with El Paso exchanges.
   (c) Survey randomly selected people from voter registration lists.
   (d) Run a TV ad, asking people to call one number if they plan to vote Democratic and another if they plan to vote Republican.
   (e) Divide the city into census tracts, group them by income and education level, and survey people at randomly selected addresses in each group.

The following are especially difficult topics to examine using surveys. Can you suggest any methods for each? Do the methods contain sources of bias?

5. How many residents of San Diego are illegal aliens?

6. How many husbands cheat on their wives?

7. How many people have seen or are seeing a psychiatrist?

8. How many people did the last census fail to count?

9. How many people are homeless?

10. How many people cheat on their taxes?

A common statistical fallacy results from making precise numerical assertions concerning things that cannot easily be quantified. Another results from making comparative judgments, such as "n% better," while leaving one term of the comparison unclear (n% better than what?). In each of the following, what is the numerical measure measuring? What are the terms of any comparisons made, or is this unclear? Are there any reasons to be suspicious of the claim?

11. New SoothTooth gets your teeth 50% brighter!

12. Now, 20% more bleach!

13. Two out of three doctors surveyed prefer Curital.

14. The group eating a low-fat diet experienced an average 30% drop in blood sugar.

15. Children in affluent schools learning 27% more, study finds.

16. New, improved Whitewash gets your laundry 42% whiter!

17. Married men 15% happier than bachelors, psychologist says.

18. Married men live 15% longer.

19. 10% more fiber than other leading brands.

20. Pitching is 75% of baseball.

The next two items are opinions concerning a presidential election as a survey of public opinion. Using the concepts of this chapter, analyze and evaluate each argument.

21. Only about half the eligible electorate went to the polls November 8 [1988] . . . the half-nation that votes is not what statisticians would call a random sampling. . . . [It] is heavily skewed toward the rich and well-educated, the white and the conservative. . . . Factoring out the gender gap and the minorities gap from the

turnout figures, the electoral equation brutally belies the myth that "the American people" gave Bush his victory. *(The Nation)*

22. Yet the *New York Times* and CBS News tested the impact of voter turnout by polling a single sample of voters both before and after the election to record the preferences of those who failed to vote. On the basis of this survey, the *Times* reported that if everyone had voted, Bush would have won by an even larger margin—about 11 percentage points. (Joshua Muravchik)

23. Consider teaching evaluations, in which students, near the end of a semester, fill out a questionnaire about their opinion of a course and its instructor. (The instructor, typically, does not see the surveys until well after grades are submitted.) Can you name any sources of bias in this procedure?

24. Suppose that you are writing an article for your school or town newspaper and want to include a survey of students' (or residents') attitudes about a particular question. How can you draw an adequate sample? *(a)* Describe a method for choosing your sample, and then *(b)* critique it, analyzing it for possible sources of bias. *(c)* If feasible, exchange your answer to *(a)* with other students, and critique their methods.

■ 10.4 ANALOGIES

Many inductive arguments are arguments by **analogy.** An analogy is a similarity, in certain respects, between distinct things. And an argument by analogy is an argument inferring a similarity from other similarities. Quite often, analogies fulfill nonargumentative roles. They help to describe, to illustrate, and to explain. In literature, similes and metaphors are kinds of analogy. Consider, for example, these speeches from Shakespeare:

(22) No, no, no no! Come, let's away to prison:
We two alone will sing like birds i' the cage:
When thou dost ask me blessing, I'll kneel down
And ask of thee forgiveness: so we'll live,
And pray, and sing, and tell old tales, and laugh
At gilded butterflies, and hear poor rogues
Talk of court news; and we'll talk with them too,
Who loses and who wins, who's in, who's out;
And take upon's the mystery of things,
As if we were God's spies: and we'll wear out,
In a wall'd prison, packs and sects of great ones
That ebb and flow by the moon. (*King Lear* V, iii, 8–19)

(23) Out, out, brief candle!
Life's but a walking shadow, a poor player
That struts and frets his hour upon the stage
And then is heard no more: it is a tale
Told by an idiot, full of sound and fury,
Signifying nothing. (*Macbeth* V, v, 23–28)

The first of these compares the speaker and listener, in prison, to birds in a cage, singing regardless of what happens outside the confines of their allotted space. It goes on to compare them to spies of God, capable of seeing and knowing anything without being a part of it. The second speech contains several metaphors for life: as a candle, about to be blown out; as a shadow, an ephemeral, elusive, dependent bit of reality; as an actor, on a stage only temporarily; as a tale told by an idiot, without meaning or design. Much of the power of great literature depends on the use of simile and metaphor to shed new light on familiar scenes and situations.

Analogy is also extremely useful in science. Modern science has made its most powerful advances by developing theories of entities beyond the reach of our senses: atoms, molecules, subatomic particles, waves, and fields. Mathematical equations define their character. But using analogies to relate them to more familiar things helps to make them more intelligible. Contemporary cosmology, for example, holds that the universe is expanding outward in all directions, without any center to the expansion. This concept is difficult to picture. But consider a balloon being inflated. Imagine that stars and galaxies are dots on the balloon's surface. No dot is in the center; yet, from the perspective of each dot, all other dots are moving away from it. The analogy helps to explain how an uncentered expansion is possible.[4] Perhaps the most dramatic example of analogy in science has been the conception of the atom as a minute planetary system, with the nucleus as sun and electrons as planets. Charles Darwin used a variety of metaphors for the process of natural selection. He described nature as a "tangled bank," the development of species as a "piece of seaweed endlessly branching," natural selection as "war," and variation as "wedging" and relied on an extended analogy between Malthus's economic portrait of scarcity and competition in the natural world. Indeed, some scientists hold that analogy is the key to scientific discovery: ". . . hypotheses," geologist G. K. Gilbert wrote, "are always suggested through analogy."[5]

---

[4]Alan P. Lightman discusses this example in "Magic on the Mind: Physicists' Use of Metaphor," *American Scholar* 58 (1989): 97–101.

[5]"The Origin of Hypotheses, Illustrated by the Discussion of a Topographic Problem," *Science* 3 (1896): 1–13.

How are analogies used in argument? An analogical argument infers a similarity from other similarities. Consider, for example, this argument from Henry Kissinger's *White House Years:*

(24) The superpowers often behave like two heavily armed blind men feeling their way around a room, each believing himself in mortal peril from the other, whom he assumes to have perfect vision. Each side should know that frequently uncertainty, compromise, and incoherence are the essence of policymaking. Yet each tends to ascribe to the other side a consistency, foresight and coherence that its own experience belies. Of course, over time even two armed blind men in a room can do enormous damage to each other, not to speak of the room.

This argument begins by comparing the superpowers to two heavily armed blind men. After listing several similarities, the argument points out that the two men present extreme danger to each other and to their surroundings. The unstated conclusion is that the superpowers also pose a serious threat to each other and to their surroundings—the world as a whole.

As this example suggests, arguments by analogy have these forms:

ARGUMENT BY ANALOGY

$a_1, \ldots, a_i$ and b are all $F_1, F_2, \ldots,$ and F_j.

$a_1, \ldots, a_i$ are G.

∴ b is G.

All Hs and all Js are $F_1, F_2, \ldots,$ and F_j.

All Hs are G.

∴ All Js are G.

We may write Kissinger's argument in this form, as follows:

(25) Both the superpowers and two heavily armed blind men feeling their way around a room

(a) believe themselves to be in mortal peril from the other;

(b) assume the other to have perfect vision;

(c) should know that frequently uncertainty, compromise, and incoherence are the essence of policy making;

(d) ascribe to the other side a consistency, foresight, and coherence that their own experiences belie.

> Over time even two armed blind men in a room can do enormous damage to each other, not to speak of the room.
>
> ∴ Over time the superpowers can do enormous damage to each other, not to mention their surroundings (the world).

In this argument, a pair of heavily armed blind men plays the role of a_1 (in this case i = 1). a_1 is compared to the pair of superpowers, in the role of b. *(a)* through *(d)* list characteristics $F_1, \ldots, F_4$ that these two pairs a_1 and b have in common. The last premise attributes to a_1 the ability to do enormous damage over time—this characteristic corresponds to G. The conclusion then attributes the same ability to the pair of superpowers, b.

Many analogical arguments are, in effect, arguments by modeling. Using them, we can reason about one situation by reasoning about another, similar situation. The situation we analyze serves as a model for the situation about which we wish to draw a conclusion. Modeling a situation offers many advantages. The model is usually clearer, better understood, and more familiar than the circumstance being modeled. Consequently, analyzing the model—seeing how its parts relate and interact, seeing how it evolves, reacts, and causes changes in other circumstances—is easier than reasoning directly about the less familiar and less comprehensible situation we want to understand. Kissinger, for example, reasons about the superpowers indirectly by reasoning about a more easily visualized situation—two heavily armed, suspicious blind men in a room. We can analyze the model of the blind men to draw conclusions about the superpowers.

Some analogies derive their strength from the number of similarities to which they appeal:

(26) Gwen and Isis are cats who like milk, hate dogs, like steak, and don't like fruit.

Gwen likes fish.

∴ Isis likes fish.

Other analogies, in contrast, draw their strength mainly from the number of instances to which the subject of the conclusion is compared. For example:

(27) Gwen, Snowball, Tina, Leela, and Isis are cats.

Gwen, Snowball, Tina, and Leela like fish.

∴ Isis likes fish.

The premises of this argument mention only one property that Isis and the others have in common, that of being a cat. Nevertheless, as we saw earlier in connection with arguments by enumeration, these premises provide at least some support for the generalization that all cats like fish. Thus they certainly provide support for one instance of that generalization (namely, that Isis, in particular, likes fish).[6]

Analogy (27) is actually not as different from (26) as it might seem. The likeness is evident when we rewrite (27) as (28):

(28) Being a cat and liking fish are attributes of Gwen, attributes of Snowball, attributes of Tina, and attributes of Leela.

Being a cat is an attribute of Isis.

∴ Liking fish is an attribute of Isis.

Example (28) draws an analogy between the properties of being a cat and liking fish on the basis of four characteristics these properties share. The argument concludes that since one of these properties (being a cat) has the characteristic of being an attribute of Isis, so does the other property (liking fish). But (28) is just a way of rephrasing (27); thus the same relationship between properties that is at work in the case of (28) underlies (27), as well. Together, (26) and (27) show that an argument by analogy draws upon similarities in two dimensions (to focus on just the first form of argument by analogy presented): similarities between the things $a_1, \ldots, a_i$, b having common properties $F_1, \ldots, F_j$; and also similarities among the properties $F_1, \ldots, F_j$, G, which have in common that they are each possessed by the things $a_1, \ldots, a_i$.

EVALUATING ANALOGICAL ARGUMENTS

Arguments by analogy are inductive: The existence of some similarities may make probable, but does not guarantee, the existence of further similarities. The strength of an analogical argument depends on several factors. They include the number and relevance of the similarities; the number and relevance of dissimilarities; the variety and number of analogous objects or circumstances; and the scope of the conclusion.

1. THE NUMBER AND RELEVANCE OF SIMILARITIES   The more relevant similarities the premises offer, the stronger the analogical inference. Arguments mentioning only a single similarity of dubious relevance to the desired conclusion are extremely weak:

---

[6]In fact, there is a sense in which the analogy (27) is more robust than the corresponding generalization, which, on the basis of the same premises, concludes that all cats like fish. If we add the sentence "Penelope is a cat, and Penelope doesn't like fish" to the premises in (27), the argument that Isis likes fish is weakened but not entirely undermined. And if we then introduce many more premises citing instances of cats who like fish, the argument would begin to revive. In the case of the generalization, on the other hand, once we add "Penelope is a cat, and Penelope doesn't like fish," the argument is clearly dead.

(29) Darryl is a man, and so is Jerryl.

Darryl is an engineer.

∴ Jerryl is an engineer.

This argument offers very little support for its conclusion. Adding similarities that have no connection to the conclusion does not help:

(30) Darryl and Jerryl are both men over 40 who like beets.

Darryl is an engineer.

∴ Jerryl is an engineer.

This argument is still terrible. Adding relevant similarities, however, strengthens the argument:

(31) Darryl and Jerryl dress similarly for work.

Darryl is an engineer.

∴ Jerryl is an engineer.

This argument is hardly overwhelming, but it's much better than (29) and (30); how people dress for work is connected to their occupation. If we add further similarities, the argument becomes stronger:

(32) Darryl and Jerryl dress similarly for work.

Both Darryl and Jerryl work for the local power company and have offices on the second floor of the same building.

Both Darryl and Jerryl took calculus from Professor Erdworm at Texas A&M.

Darryl is an engineer.

∴ Jerryl is an engineer.

2. THE NUMBER AND RELEVANCE OF DISSIMILARITIES Any two objects or circumstances differ in indefinitely many ways. The mere existence of differences does not undermine the strength of an analogical argument. When the differences are relevant to the issue at hand, however, they do count against the argument's strength. In general, the more relevant dissimilarities there are, the weaker the argument.

Consider the case of Darryl and Jerryl. By the time we reach (32), we have a fairly strong argument for the conclusion that Jerryl is an engineer. Adding information about irrelevant dissimilarities to the premises takes nothing away from the argument's strength:

(33) Darryl and Jerryl dress similarly for work.

Both Darryl and Jerryl work for the local power company and have offices on the second floor of the same building.

Both Darryl and Jerryl took calculus from Professor Erdworm at Texas A&M.

Darryl's favorite color is green, favorite food is meat loaf, and favorite drink is milk; Jerryl's favorites are blue, turkey, and beer, respectively.

Darryl is an engineer.

∴ Jerryl is an engineer.

Adding relevant dissimilarities, however, weakens the argument:

(34) Darryl and Jerryl dress similarly for work.

Both Darryl and Jerryl work for the local power company and have offices on the second floor of the same building.

Both Darryl and Jerryl took calculus from Professor Erdworm at Texas A&M.

Darryl has an M.S. in electrical engineering and works extensively with computers; Jerryl has a B.B.A. in accounting and works with spreadsheets, an adding machine, tax forms, and a green eyeshade.

Darryl is an engineer.

∴ Jerryl is an engineer.

This argument is no longer very strong. Indeed, we would be tempted to draw the conclusion that Jerryl is an accountant.

3. THE VARIETY AND NUMBER OF ANALOGOUS OBJECTS OR CIRCUMSTANCES

That Darryl and Jerryl dress similarly for work lends some support to the conclusion that they have the same occupation. We would strengthen this support by noting other engineers who dress in the same way. Darryl's habits of dress may be due to his personal preferences, his alma mater, or some limited local practice and be unusual for people in his profession. The more engineers we observe with similar habits, the stronger the argument. Also, the more varied a group we observe dressing the same way, the more reliable the argument. If we point to Darryl and his friends who are engineers, the argument is stronger than if we point to Darryl alone. But the argument becomes still stronger if we point to engineers in different places, companies, and social settings who dress similarly.

4. THE SCOPE OF THE CONCLUSION

The greater the scope of the conclusion we try to derive—the more informative and detailed we wish it to be—the weaker the argument becomes. If given the similarities between Darryl and Jerryl mentioned in (32) we conclude that Jerryl is an engineer, the argument seems reasonable. If mentioning that Darryl is an engineer working in transmission and distribution we conclude that Jerryl also works in transmission and distribution, the argument becomes much weaker. A more detailed conclusion requires more extensive similarities, while it opens the door more widely for relevant dissimilarities.

P R O B L E M S

Analyze the following analogies, explaining *(a)* what is being compared to what and *(b)* in terms of what similarities. Also, *(c)* mention some dissimilarities.

1. As my hand hath found the kingdom of the idols, and whose graven images did excel them of Jerusalem and of Samaria; Shall I not, as I have done unto Samaria and her idols, so do to Jerusalem and her idols? (*Isaiah* 10: 10–11)

2. Beware of false prophets, which come to you in sheep's clothing, but inwardly they are ravening wolves. (*Matthew* 7:15)

3. Would that you knew what the Disaster is! On that day men shall become like scattered moths and the mountains like tufts of carded wool. (*Koran* 101:1)

4. Unholy Mars bends all to his mad will:
   The world is like a chariot run wild
   That rounds the course unchecked and, gaining speed,
   Sweeps the helpless driver on to his doom. (Virgil)

5. We are such stuff
   as dreams are made on; and our little life
   Is rounded with a sleep. (William Shakespeare)

6. All the world's a stage,
   And all the men and women merely players;
   They have their exits and their entrances;
   And one man in his time plays many parts. . . .
   (William Shakespeare)

7. Like to the Pontic sea,
   Whose icy current and compulsive course
   Ne'er feels retiring ebb, but keeps due on
   To the Propontic and the Hellespont;
   Even so my bloody thoughts, with violent pace,
   Shall ne'er look back, ne'er ebb to humble love,
   Till that a capable and wide revenge
   Swallow them up. (William Shakespeare)

8. This city now doth, like a garment, wear
   the beauty of the morning. . . . (William Wordsworth)

9. It is beauteous evening, calm and free,
   The holy time is quiet as a Nun
   Breathless with adoration. . . . (William Wordsworth)

10. Then felt I like some watcher of the skies
    When a new planet swims into his ken;
    Or like stout Cortez when with eagle eyes
    He stared at the Pacific—and all his men
    Looked at each other with a wild surmise—
    Silent, upon a peak in Darien. (John Keats)

11. I love thee freely, as men strive for Right;
    I love thee purely, as they turn from Praise.
    (Elizabeth Barrett Browning)

12. Ah, love, let us be true
    To one another! for the world, which seems
    To lie before us like a land of dreams,
    So various, so beautiful, so new,
    Hath really neither joy, nor love, nor light,
    Nor certitude, nor peace, nor help for pain;
    And we are here as on a darkling plain
    Swept with confused alarms of struggle and flight,
    Where ignorant armies clash by night. (Matthew Arnold)

13. My heart is like a singing bird
    Whose nest is in a watered shoot;
    My heart is like an apple-tree
    Whose boughs are bent with thickset fruit;
    My heart is like a rainbow shell
    That paddles in a halcyon sea;
    My heart is gladder than all these
    Because my love is come to me.
    (Christina Rossetti)

14. Language is like a cracked kettle on which we beat out tunes for
    bears to dance to, while all the time we long to move the stars to
    pity. (Gustave Flaubert)

Explain the following analogies: *(a)* What is related to what? *(b)* In terms
of what similarities? *(c)* What are some dissimilarities between the
items? If the analogy is part of an analogical argument, *(d)* evaluate
the argument's strength.

15. Seeing Yankee fans for the first time is like waking up in a
    Brazilian jail. (Art Hill)

16. College is like an imprimatur. It's like having an American Express card. (Clare Booth Luce)

17. It struck me that this task force [Democratic Task Force on high interest rates, Senate Banking Committee] was a little like mosquitoes investigating the cause of malaria. (James Balog)

18. Television is to news what bumper stickers are to philosophy. (Richard Nixon)

19. Asking a working writer what he thinks about critics is like asking a lamppost what it feels about dogs. (John Osborne)

20. Going to church doesn't make you a Christian any more than going to the garage makes you a car. (Laurence J. Peter)

21. 'Tis with our judgments as our watches: none go just alike, yet each believes his own. (Alexander Pope)

22. My country, right or wrong, is a thing no patriot would think of saying, except in a desperate case. It is like saying, "my mother, drunk or sober." (G. K. Chesterton)

23. The universe of ideas is just as little independent of the nature of our experiences as clothes are of the form of the human body. (Albert Einstein)

24. Light and other forms of radiation are analogous to water ripples or waves, in that they distribute energy from a central source. (James Jeans)

25. Q. Some of the major carriers in the New York area want restrictions placed on private and corporate aviation. What about that? Let me draw a parallel. If, on the New Jersey Turnpike, it became crowded, would you like it if all automobiles were kept off the New Jersey Turnpike and only buses could run? (Rear Admiral Donald Engen)

26. High interest rates and slow growth reduce inflation in the way that chemotherapy works on cancer. It kills the good along with the bad cells and makes the patient dreadfully sick. Physicians, to their credit, are trying to find therapies that will kill cancer cells without killing the patient. But America's economic policymakers have not made the same intellectual leap. (Paul A. London)

27. . . . "You can't put the clock back." The simple and obvious answer is "You can." A clock, being a piece of human construction, can be restored by the human finger to any figure or hour.

In the same way society, being a piece of human construction, can be reconstructed upon any plan that has ever existed. (G. K. Chesterton)

28. For seven months now, ever since my arrest and trial, I have been living like a queen here in prison: doors are flung open before me wherever I go—into cells, interrogation rooms, the courtroom. . . . Other hands close these doors behind me, too. . . . Mostly I am driven from place to place. An impressive number of people are employed in "serving" me: just to get a pencil sharpened I summon a junior officer of the guard. (Irina Ratushinskaya)

29. My aim is to show that the celestial machine is to be likened not to a divine organism but rather to a clockwork. (Johannes Kepler)

30. The people who bind themselves to systems are those who are unable to encompass the whole truth and try to catch it by the tail; a system is like the tail of truth, but truth is like a lizard; it leaves its tail in your fingers and runs away knowing full well that it will grow a new one in a twinkling. (Ivan Turgenev)

31. The earth is an element placed in the middle of the world, as the yoke in the middle of an egg; around it is the water, like the white surrounding the yoke; outside that is the air, like the membrane of the egg; and around all is the fire, which closes it in as the shell does. (The Venerable Bede)

32. The easiest way to think of the electrons is as small "planets" circling round a central nucleus which is made up of neutrons and protons. . . . We can imagine electrons as planets whirling around the central sun in a solar system. . . . (A. M. Low)

33. As the mistletoe is disseminated by birds, its existence depends on them; and it may metaphorically be said to struggle with other fruit-bearing plants, in tempting the birds to devour and thus disseminate its seeds. (Charles Darwin)

34. Just as in astronomy the difficulty of admitting the motion of the earth lay in the immediate sensation of the earth's stationariness and of the planets' motion, so in history the difficulty of recognizing the subjection of the personality to the laws of space and time and causation lies in the difficulty of surmounting the direct sensation of the independence of one's personality. (Leo Tolstoy)

35. Supposing the light of any given colour to consist of undulations of a given breadth, or of a given frequency, it follows that these undulations must be liable to those effects which we have already examined in the case of the waves of water, and the pulses of sound. (Thomas Young)

36. Malcolm Forbes Jr. . . . attacked the Federal Reserve for clamping down too tightly on interest rates—slowing down and almost killing the economic recovery. "It would be as if a doctor had a patient who made a quick comeback from the operation, and the doctor thought the patient was getting too healthy and induced a relapse. Well, in medicine you'd get sued for malpractice. In economics they make you the head of the Federal Reserve." (WKRN-TV, Nashville)

37. The relation which the several parts or members, of the natural body have to each other and to the whole body, is here compared to the relation which each particular person in society has to the whole society: and the latter is intended to be illustrated by the former. And if there be a likeness between these two relations, the consequence is obvious: that the latter shows us we were intended to do good to others, as the former shows us that the several members of the natural body were intended to be instruments of good to each other and to the whole body. (Joseph Butler)

38. Electric displacement, according to our theory, is a kind of elastic yielding to the action of the force, similar to that which takes place in structures and machines owing to the want of perfect rigidity of the connexions. . . . Energy may be stored in the field . . . by the action of electromotive force in producing electric displacement. . . . [I]t resides in the space surrounding the electrified and magnetic bodies . . . as the motion and strain of one and the same medium. (James Maxwell)

39. For instance, there can hardly be a doubt that the animals which fight with their teeth, have acquired the habit of drawing back their ears closely to their heads, when feeling savage, from their progenitors having voluntarily acted in this manner in order to protect their ears from being torn by antagonists. . . . We may infer as highly probable that we ourselves have acquired the habit of contracting the muscles round the eyes, whilst crying gently, that is, without the utterance of any loud sound, from our progenitors, especially during infancy, having experienced during the act of screaming, an uncomfortable sensation in the eyeballs. (Charles Darwin)

40. And why are we not willing to acknowledge that the appearance of a daily revolution belongs to the heavens, its actuality to the earth? The relation is similar to that of which Virgil's Aeneas says: "We sail out of the harbor and the countries and cities recede." For when a ship is sailing along quietly, everything which is outside of it will appear to those on board to have a motion corresponding to the movement of the ship, and the voyagers are of the erroneous opinion that they with all that they have with them are at rest. This can without doubt apply to the motion of the earth, and it may appear as if the whole universe were revolving. . . . (Nicholas Copernicus)

41. Then I began to suspect whether the rays, after their trajection through the prism, did not move in curved lines, and according to their more or less curvity tend to divers parts of the wall. And it increased my suspicion, when I remembered that I had often seen a tennis ball struck with an oblique racket describe such a curved line. . . . For the same reason, if the rays of light should possibly be globular bodies, and by their oblique passage out of one medium into another, acquire a circulating motion, they ought to feel the greater resistance from the ambient ether on that side where the motions conspire, and hence be continually bowed to the other. (Isaac Newton)

For each of the following analogical arguments, consider whether adding each listed premise would make the argument stronger or weaker.

42. Twice before you asked Joyce for a favor and she agreed to help. You conclude that she will probably agree to help again.
    (a) Before, you asked her for rides; this time, you ask for a loan.
    (b) Before, Joyce was single; now she is married.
    (c) Before, you were single; now you are married.
    (d) Joyce is now extremely busy; she just had a baby.
    (e) Bart also asked Joyce for favors and she granted them.

43. Norbert has taken one philosophy course and hated it. He refuses to sign up for another, reasoning that he would hate it, too.
    (a) This course and that were taught by the same professor.
    (b) This course is on logic; that was on ancient skepticism.
    (c) Norbert has hated every humanities course he has taken.
    (d) This course and that both meet on Tuesdays and Thursdays.
    (e) Norbert has just spent a year off from school working on an oil rig.

44. Fanny has made As on both exams during the semester and infers that she will probably get an A on the final.
    (a) The three exams cover different, nonoverlapping material.
    (b) The same instructor made up all three exams.
    (c) Fanny studied hard for the first two exams but is not bothering to study for the final.
    (d) Fanny's grades on the two exams were the highest in the class.
    (e) Fanny's professor has gotten sick, and someone else is making up the final.

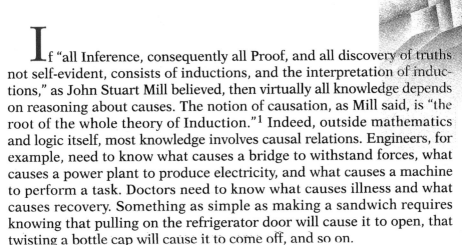

CHAPTER 11

CAUSES

If "all Inference, consequently all Proof, and all discovery of truths not self-evident, consists of inductions, and the interpretation of inductions," as John Stuart Mill believed, then virtually all knowledge depends on reasoning about causes. The notion of causation, as Mill said, is "the root of the whole theory of Induction."[1] Indeed, outside mathematics and logic itself, most knowledge involves causal relations. Engineers, for example, need to know what causes a bridge to withstand forces, what causes a power plant to produce electricity, and what causes a machine to perform a task. Doctors need to know what causes illness and what causes recovery. Something as simple as making a sandwich requires knowing that pulling on the refrigerator door will cause it to open, that twisting a bottle cap will cause it to come off, and so on.

Reasoning about causes and effects, therefore, is fundamental to logic. This chapter discusses the idea of causation and presents methods, first developed by Mill, for establishing or disproving a causal relation. These methods offer some general principles for evaluating causal arguments.

■ 11.1 KINDS OF CAUSES

To cause, in essence, is to bring about or produce. In English, we can speak of actions, events, objects, and perhaps even facts and states of affairs as causes:

(1) a. The Fed's announcement caused the dollar to drop.
  b. The dollar's slide caused investors to sell.
  c. Jill caused the accident.

---

[1]John Stuart Mill, *A System of Logic* (New York: Harper and Brothers, 1874), pp. 207, 236. Further page references to Mill will be to this work.

   d. The fact that she ignored him caused Karl to fly into a jealous rage.
   e. Lou's persistent slovenliness caused his wife to leave him.

As these examples suggest, however, what is caused is almost always an action or event. Sometimes we speak of something causing a state—as in "A genetic abnormality caused Frank to be obese"—but these seem to depend on causation of events ("A genetic abnormality caused Frank to *become* obese"). Throughout this chapter we will think of actions as events of a special kind and treat effects as events or collections thereof.

Similarly, we speak of objects, facts, and states as causes only when some event or collection of events acts as a cause. Jill caused the accident, for example, only if something Jill did caused it. Lou's persistent slovenliness caused his wife to leave him, similarly, only if a series of Lou's actions caused her departure. We may think of causation, then, as relating events or collections of events.

The notion of causation has provoked a lively debate among philosophers. Often, they analyze causation in terms of necessary and sufficient conditions.

> A condition is **necessary** for an event if and only if the event cannot occur when the condition does not hold.
> A condition is **sufficient** for an event if and only if the event must occur when the condition holds.

Consider a classic case: the lighting of a match. The match cannot light unless oxygen is present. The presence of oxygen, in other words, is a necessary condition of the lighting. If the match is dry and struck with enough force in the presence of oxygen, it must light. So, that joint condition is a sufficient condition of the lighting.

There may be more than one necessary condition of an event. The match's lighting requires, for example, not only the presence of oxygen but also the dryness of the match. There may be more than one sufficient condition as well. The match will light if we toss it, dry and in the presence of oxygen, into a blazing fire. By definition, all the necessary conditions must be included in each sufficient condition. If one were not included in some sufficient condition, the event could take place without it, in which case it would not be necessary after all.

Consider another example: A Texas university admits all Texas high school graduates who apply and who score at least 1200 on the SAT or finish in the top quarter of their high school class. Applying and graduating from a Texas high school in the top quarter of the class is a sufficient condition for admission. So is applying, graduating from a Texas high school, and scoring at least 1200 on the SAT. Neither, however, is necessary. The university admits some who scored below 1200, some who finished below the top quarter in their high school classes, and some who did not graduate from Texas high schools. Applying, however, is a necessary condition.

Some philosophers have defined causes as sufficient conditions. While this has appeal, it does not quite match our intuitive concept of causation. This exchange sounds odd:

(2) A: What caused the match to light?
    B: Striking it, its dryness, and the presence of oxygen.

Sufficient conditions usually include more than an event or even a set of events. They include some features of the background against which an event or set of events occurs. But we do not ordinarily count features of the background as causes or even parts of causes. Asked what caused the match to light, we would mention only the striking.

We might try to improve this proposed definition by requiring that causes be events that are part of sufficient conditions. This would allow us to mention the striking alone as the cause of the match's lighting. But it would also allow us to mention only the dryness. In general, many events, not all of which are causes, are parts of sufficient conditions. If we ask, "What caused it to snow today?" any full account of sufficient conditions probably must mention many events: a collision between cold and warm air masses, the rising of the warm air, the condensation of moisture, the movement of the front through our area, and so on. But not all these parts count as causes: "A warm air mass rose" and "Gravitational forces acted on water droplets in a cloud" are not very good answers.

Some philosophers have argued that causes are necessary conditions. But not all necessary conditions count as causes. The university admits only those whose applications have been delivered by the post office, but we do not consider the delivery a cause of a student's being admitted. Similarly, the university admits only humans, but it would be bizarre to answer the question "Why did you get in?" with "I'm human!"

Defining causation remains a controversial and difficult philosophical problem. Here we will outline a **pragmatic** account of causation. According to such an account, causation is a context-relative notion. If we ask, to use Mill's example, why a stone thrown into a pond sinks to the bottom, a huge and indefinite amount of information is relevant to the question. A complete explanation would have to include the gravitational attraction between the earth and the stone, the chemical composition of the stone, a comparison of the specific gravities of the stone and the water, and explanations for all of these. Yet, if asked what caused the stone to fall to the bottom of the pond, we would ordinarily respond with a simple answer: "Johnny threw it in," "It's heavy," or something similar, depending on context. We assume a certain amount of information as background and supply only whatever else the listener needs to understand why the event happened.

The participants in a conversation share some information. They have in common certain items of knowledge, beliefs, and assumptions. When we allege a causal relation, we rely on these to provide background information. We specify only the aspect of the relationship that changes prior circumstances and offers our listeners new information.

Moreover, we rely on the context of the conversation to indicate what range of alternatives is being considered. Ordinarily, if we ask why it snowed, we are asking, "Why did it snow, rather than remain fair?" But we might, in a different context, mean to ask why it snowed rather than rained or why it snowed here rather than there. Those require different responses ("Because temperatures remained below freezing at every altitude below the cloud" or "Because the jet stream altered course," for example).

So, we do not discuss all the necessary or sufficient conditions of an event in indicating its cause. Instead, we specify only what must be added to the context to yield a necessary and sufficient condition for the event.

> An event A *causes* an event B *in* a context C if and only if A, together with C, is a necessary and sufficient condition for B.

Thus, we say that striking the match causes it to light. Striking, by itself, is neither a sufficient nor a necessary condition of the lighting. But, relative to typical circumstances—in which the match is dry, oxygen is present, and there are no unusual, obscure, or bizarre processes taking place—the striking causes the match to light. Striking, in addition to the conditions that are part of the context, is necessary and sufficient for the event of the match's lighting.

Sometimes events occur in causal chains: A causes B, B causes C, C causes D, and so on. Striking the match, say, causes it to light; lighting the match causes the gas grill to start; starting the gas grill causes the temperature inside it to rise; the rising temperature causes the steak to cook. These chains play an extremely important role in explanations. We can often think of each member of a chain as an effect of the events preceding it. We can often think of D, that is, as an effect of A, of B, and of C. C is the **proximate** or **immediate** cause of D; A and B are **remote** causes. Whether we think of earlier members of a chain as causes of later events depends on how the context shifts as we move from considering one causal relation to considering the next. We might generally count starting the gas grill as a cause of the steak's cooking, for example, but we might not count striking the match as a cause.

PROBLEMS

Below are some typical sentences about causal relations. In each case, *(a)* is the cause a necessary condition for the effect, independently of the context? *(b)* a sufficient condition, independently of the context? *(c)* What context must we assume for the cause to be necessary and sufficient? *(d)* Is the cause proximate or remote?

1. Taking the drug caused the patient's recovery.

2. Imbibing poison was the cause of death.

3. A computer crash caused the delay in our sending you the information.

4. John's tardiness caused the meeting to start an hour late.

5. Turning the key caused the car to start.

6. Scattering salt on the ice caused it to melt.

7. Putting ice in the drink caused it to cool.

8. The heating of the oven caused the bread to rise.

9. The activity of the yeast caused the bread to rise.

10. A letter from the IRS caused Alex to leave the country.

11. Zeke's punch caused Charley's nose to bleed.

12. The flood caused widespread destruction.

13. Flipping the switch caused to light to come on.

14. Nora's testimony caused the jury to change its mind.

15. Hank's departure caused the team's fall into the second division.

The following are causal chains. *(a)* Say whether, in each link, the cause is a necessary or sufficient condition independently of the context. *(b)* Say whether, in your opinion, each event prior to the proximate cause of the last event in the chain should count as a cause of that event, and explain your judgment.

16. (a) The leak in the fuel line caused the engine failure.
    (b) The engine failure caused the plane to crash.
    (c) The plane crash caused the deaths of more than 100 people.

17. (a) Pulling the trigger caused the rifle to fire.
    (b) Firing the rifle caused its barrel to heat.
    (c) The heating of the barrel caused the rifle to emit an infrared glow.

18. (a) The scandal caused the party's defeat in the elections.
    (b) The party's electoral defeat caused a change in leadership.

    (c) The change in leadership caused a sweeping revision of government policy.

    (d) The sweeping revision of government policy caused widespread turmoil.

19. (a) The shooting of the archduke caused his government to issue an ultimatum.

    (b) The issuance of the ultimatum caused the Russians to mobilize.

    (c) The Russian mobilization caused the rejection of the ultimatum.

    (d) The ultimatum's rejection caused Austria-Hungary to declare war.

    (e) Austria-Hungary's declaration of war caused Serbia to declare war.

    (f) Serbia's declaration of war caused Russia's declaration of war.

    (g) Russia's declaration of war caused Germany to declare war.

20. Three desert travelers, A, B, and C, stopped for the night before setting out on separate journeys. Both A and B hated C. A poisoned C's only water supply, his canteen; B, not knowing what A did, drilled a small hole in C's canteen so that the water would leak out. Several says later, C died of thirst. Which action—A's or B's—caused C's death? Explain your answer.[2]

 # 11.2   AGREEMENT AND DIFFERENCE

The British philosopher Francis Bacon (1561–1626) recognized that there was more to induction than generalization. He saw the importance of causal reasoning and developed methods for devising and evaluating causal claims. John Stuart Mill developed these methods further, formulating them so clearly that they are now called "Mill's Methods." The methods pertain particularly to causal generalizations: assertions that events of one kind cause events of another kind. The first two of Mill's canons for inductive procedure are the methods of **agreement** and **difference.**

AGREEMENT

To find the cause of some phenomenon, consider various instances of it, and see where they agree—see what they have in common. As Mill puts his first canon:

> If two or more instances of the phenomenon under investigation have only one circumstance in common, the circumstance in which

---

[2]This puzzle is from Arthur Smullyan, *What Is the Name of This Book?* (Englewood Cliffs, NJ: Prentice-Hall, 1978), p. 9.

alone all the instances agree, is the cause (or effect) of the given phenomenon (*A System of Logic*, 280).

Schematically, suppose that we are seeking the cause of a kind of event E. In different circumstances, we find E following events of various other kinds, as detailed in this table:

CASE	ANTECEDENT EVENTS				FOLLOWED BY
1	A	B			E
2		B		D	E
3	A	B	C		E
4		B	C	D	E
5	A	B		D	E

Mill's method of agreement says that B is a likely candidate for being the cause of E. Events of kind E may or may not follow events of kinds A, C, or D, so, according to Mill's principle, those are unlikely to be causally connected to E.

In short, then, the method of agreement is this:

To find the cause of a kind of event, see where the antecedent circumstances agree. To find the effect of a given event, see where the subsequent circumstances agree.

In diagram form:

AGREEMENT

WHAT CAUSES E?

CASE	ANTECEDENT EVENTS		FOLLOWED BY
1	A	B	E
2	A		E

Conclusion: A causes E.

WHAT ARE THE EFFECTS OF C?

CASE	ANTECEDENT EVENTS	FOLLOWED BY	
1	C	E	F
2	C	E	

Conclusion: C causes E.

If, in every case in which E occurs, we find A has occurred, we have evidence that A is at least a necessary condition of E. Similarly, if we find that every occurrence of C is followed by an occurrence of E, we have evidence that C is sufficient for E.

An excellent example of applying the method of agreement occurred in a puzzling medical case in 1944, detailed by Berton Roueche in *Eleven Blue Men*.[3] Throughout the day on Monday, September 25, 1944, Beekman-Downtown Hospital in New York City received eleven ragged old men with similar and unusual symptoms. Each patient had collapsed. At first the extremities, and gradually the remainder of the body, of each patient became sky blue. The physician who examined the first patient at 8:30 A.M. attributed his blue color to cyanosis resulting from insufficient levels of oxygen in the blood and diagnosed carbon-monoxide poisoning. The tenth patient arrived at noon; the eleventh and last was found in a flophouse that evening. The Health Department had begun to investigate by afternoon. One patient died; the other ten, given heart stimulants, oxygen, and bed rest, began to recover.

Questioning the patients yielded the following information. All collapsed in the Bowery area. Five were stricken in the Globe Hotel, two in the Star Hotel, one in the Lion Hotel, and three on the street. Already Mill's method of agreement applied. It indicated that the cause was unlikely to involve any of the flophouses specifically but probably did involve the Bowery area. If indeed the cause were gas poisoning, the method would signal that the source must have been around the Bowery.

More interesting, all the patients had eaten breakfast at a single Bowery area eating place, the Eclipse Cafeteria. Agreement on this feature of the antecedent circumstances strongly suggested, by Mill's method of agreement, that the cafeteria was the source of the poisoning. All the men, moreover, had gotten sick within half an hour after eating; all but one had eaten oatmeal, rolls, and coffee, and the remaining one had eaten oatmeal. Again Mill's method applied: The cause probably involved the oatmeal.

The men had not, furthermore, been in the cafeteria at the same time. They had eaten at various times in a busy three-hour period, during which many other customers ate breakfast without incident. This led the doctors to abandon the hypothesis of gas poisoning. If carbon monoxide had filled the cafeteria, the gas would probably have been present for only a short time and would have affected everyone there. Again, the reasoning relied on the method of agreement. Suppose the Eclipse Cafeteria had contained toxic amounts of carbon monoxide. Then, to find the effects, we should look at all the relevant cases—all the people who were there while the gas was—and see what they have in common later. But most of the morning customers were fine; only eleven were ill.

---

[3](Boston: Little, Brown and Company, 1947, 1953).

The Health Department, therefore, conjectured that the oatmeal had been poisoned. Tests on blood samples from the affected men confirmed the conjecture, and chemical analysis of the oatmeal ingredients indicated that what should ordinarily have been sodium chloride—salt—was in fact sodium nitrite, a compound sometimes substituted for sodium nitrate in curing meats during World War II. The investigators thus concluded that the eleven old men were victims of sodium nitrite poisoning, an extremely rare type of poisoning. Before 1944, only ten cases had ever been reported in the medical literature.

This account of causal reasoning in the case of the eleven blue men is simplified, though accurate to the actual reasoning of the Health Department officials. A careful reader may have noticed, however, that the afflicted men had other things in common. All were old and all were heavy drinkers. These initially overlooked features became important at a later stage of the investigation. They also illustrate an important philosophical point. The method of agreement, like Mill's other methods, relies on an assumption that all and only potentially relevant features of the circumstances are being taken into account. Two cases, after all, never have just one feature or circumstance in common. Any two people, for example, have a great many things in common, simply because they are *homo sapiens*. Mill's description of the method makes sense, therefore, only if we understand it to concern potentially relevant features or circumstances. But this interpretation means that success of the method depends on judging what factors might be relevant to explaining the phenomenon in question. If we exclude as irrelevant items that could play a causal role, the method may lead us to incomplete or even incorrect conclusions.

Mill's methods also rely on another assumption concerning the phenomena to be discussed. We must assume that we have analyzed the circumstances correctly. That is, the success of the method of agreement, as well as Mill's other methods to follow, depends on a proper division of the circumstance into distinct causal factors. If we misidentify the factors that might play a causal role from the beginning, the methods are unlikely to yield any helpful results. For example, applying the method of agreement to chemical reactions is helpful only if we have properly distinguished and perhaps even identified the various compounds involved. If we have not—if, say, some of our samples marked X are sodium nitrate, while others are sodium nitrite—then the method may find no interesting causal relations at all. To work, the method requires that we sort objects and circumstances into appropriate kinds before searching for agreement and drawing causal conclusions.

The method of agreement is an excellent preliminary step in causal reasoning. It suggests hypotheses for further investigation. But, taken alone, it suffers from limitations. It helps identify necessary conditions. But identifying necessary and sufficient conditions requires additional investigation. An old joke points out the problem with the method of agreement taken in isolation. Dr. Bob is a traveling medicine man who claims to have a cure for the common cold. Take his elixir, he says, and your symptoms will disappear within ten days. He can support his claim by the method of agreement: "Here is a list of people who have taken the elixir! Not one still suffered from a cold ten days later!" But clearly this does not show that his elixir caused the recovery; almost everyone recovers from a cold within ten days, elixir or no elixir.

DIFFERENCE

Mill's second canon, the method of difference, seeks to discover causes and effects, not by looking at similar cases and seeing what they have in common, but by looking at cases that differ and seeing what the differences are. If one case shows a certain effect and another similar case does not, the method says to look for differences to account for the presence of the effect in only one case. As Mill himself puts it,

> If an instance in which the phenomenon under investigation occurs, and an instance in which it does not occur, have every circumstance in common save one, that one occurring only in the former; the circumstance in which alone the two instances differ, is the effect, or the cause, or an indispensable part of the cause, of the phenomenon (*A System of Logic*, 280).

In diagram form:

DIFFERENCE

WHAT CAUSES E?		
CASE	ANTECEDENT EVENTS	FOLLOWED BY
1	A B	E
2	A	E
Conclusion: B causes E.		

WHAT ARE THE EFFECTS OF C?		
CASE	ANTECEDENT EVENTS	FOLLOWED BY
1	C	D E
2		D
Conclusion: C causes E.		

Once again, we should understand Mill as referring to circumstances that might be relevant to the phenomenon. So, suppose that we can find two very similar cases: one in which the effect we are investigating occurs and another very similar case in which it does not. We look to see what other differences are present. The method of difference says that if we can find just one other difference, it is the cause, or "an indispensable part of the cause," of the effect. If we are seeking the cause of the effect E, then we seek two cases very similar except that only one has the effect. Suppose we find:

CASE	ANTECEDENT CIRCUMSTANCES			EFFECT	
1	A	B	C	E	
2	A	B			F

The method of difference indicates that C is the cause, or part of the cause, of E. We must say "part of the cause," for C might not produce the effect E except in the presence of B or under other special conditions. The same reasoning holds for finding effects. Suppose we find two similar cases and apply the cause we want to investigate only to one. We can observe differences in effects to determine the effect of that additional cause.

CASE	ANTECEDENT CIRCUMSTANCES				EFFECTS	
1	A	B	C	D	E	F
2	A	B	C		E	

The results described in the table indicate that D causes, or is part of the cause of, F.

The method of difference also played an important role in the case of the 11 blue men, but at a later stage of the investigation. Health Department officials traced the illness afflicting the 11 old Manhattan residents to oatmeal served at the Eclipse Cafeteria. This, however, introduced a new puzzle. The Eclipse made up enough oatmeal to serve around 125 breakfast customers, and, on September 25, 1944, all of it was consumed. So, the oatmeal could not be the entire explanation. Out of roughly 125 people who had eaten it, only 11 got sick. Why?

The method of difference says that to find the cause, we must consider differences between the 11 sick men and the approximately 114 oatmeal eaters who showed no ill effects. Health Department officials, making up a batch of oatmeal following the recipe of the Eclipse cook and chemically analyzing it, were able to explain why most customers did not get sick: The oatmeal contained only about 80 percent of the minimum toxic dose of sodium nitrite. The 11 sick men must have somehow received a greater dose. How? One doctor conjectured that if the salt container in the kitchen contained sodium nitrite, salt shakers on the tables might also. Testing them showed that all contained table salt but one, which contained a mixture of salt and sodium nitrite. So, the investigation uncovered two hypotheses about significant differences between the sick men and the healthy customers. The sick men differed in that they sat at one particular table and salted their oatmeal. (Unfortunately, this hypothesis could not be checked: By the time investigators had formulated it, the surviving patients had been released.) The method of difference indicates that their adding more sodium nitrite to the already poisoned oatmeal at that one table was the cause of the illness.

This hypothesis raised yet another question. Some people salt oatmeal. To obtain a toxic dose, however, the men had to add at least a teaspoon of salt to their oatmeal. Why would they add so much? Here, the method of agreement helped. The men had a number of things in common. They were found in the Bowery area; they had eaten oatmeal at the Eclipse, apparently at a single table; they were old; and they were heavy drinkers. Alcoholics tend to have lower sodium chloride levels in the blood than other people; they need to consume more salt to provide what their bodies require. So, the men probably used so much salt because they were alcoholics.

The method of difference is a powerful tool for isolating causes. But it, too, suffers some limitations. No two cases are alike in all but one respect; there are always other differences that might affect the outcome. The method of difference thus also depends on our recognizing potentially relevant factors. Moreover, a difference between very similar cases may not occur in other pairs of cases. The difference between the similar cases may be due to something else altogether. Consider again Dr. Bob and his "cure" for the common cold. He might use the method of difference: "Consider Meg and Peg McGillicutty, identical twins who came down with a cold on the same day. Meg took the elixir; Peg didn't. Now, two weeks later, Meg's cold is gone, but Peg is still sniffling and sneezing!" This is a legitimate application of the method, but it is not a very strong argument. Poor Peg probably caught another cold or developed a sinus infection.

Joint Method of Agreement and Difference

The methods of agreement and difference may be used together. As in the case of the eleven blue men, first one, then another may be used to isolate the cause of some phenomenon. But they may also be used jointly. Doing so produces an improvement of the method of agreement, the joint method of agreement and difference, which, in Mill's words, has as its canon:

If two or more instances in which the phenomenon occurs have only one circumstance in common, while two or more instances in which it does not occur have nothing in common save the absence of that circumstance; the circumstance in which alone the two sets of instances differ, is the effect, or the cause, or an indispensable part of the cause, of the phenomenon (*A System of Logic*, 284).

In diagram form:

AGREEMENT AND DIFFERENCE

WHAT CAUSES E?

CASE	ANTECEDENT EVENTS		FOLLOWED BY
1	A	B	E
2	A		E
3		B	
4			

Conclusion: A causes E.

WHAT ARE THE EFFECTS OF C?

CASE	ANTECEDENT EVENTS	FOLLOWED BY	
1	C	D	E
2	C		E
3		D	
4			

Conclusion: C causes E.

The joint method is in effect two applications of the method of agreement—one for the presence of E (cases 1 and 2), one for its absence (cases 3 and 4)—combined, in ideal circumstances, with two applications of the method of difference (cases 1 and 3 and cases 2 and 4). Its power comes not only from this combination but also from its ability to determine necessary and sufficient conditions. If cases in which E occurs have A in common, that is evidence that A is a necessary condition for E. If cases in which E does not occur have the absence of A in common, that is evidence that A's absence is a necessary condition of E's absence, that is, E would not be absent unless A were. And this means that A's presence would bring about E's presence—that, in other words, A is a sufficient condition for E.

This method corresponds to the experimental technique of using a **control group.** In the table, cases 1 and 2 are used to test the effects of A; cases 3 and 4 are controls used to see what happens without A. Using a control group of this kind greatly increases the reliability of a causal inference, for it indicates whether the effect—here, E— would have occurred even without A.

Here is a schematic example of a controlled experiment based on the joint method of agreement and difference:

HYPOTHESIS: A CAUSES E

CASE	ANTECEDENT CIRCUMSTANCES				EFFECTS		
EXPERIMENTAL GROUP							
1	A	B		E		G	
2	A		C	E			H
CONTROL GROUP							
3		B	C		F	G	
4			D			G	H

Here, the antecedent circumstances of cases 1 and 2, the experimental group, have only A in common; the antecedent circumstances of cases 3 and 4, the control group, have nothing in common but the absence of A. The experimental group exhibits the effect E, but the control group does not. The joint method of agreement and difference says that A is probably the cause (or, as Mill says, an indispensable part of the cause) of E.

To see why control groups are important—and, thus, why the joint method is an advance over the methods of agreement and difference taken separately—again consider Dr. Bob, who claims to have a cure for the common cold. If we were to test Dr. Bob's hypothesis by the joint method, we would set up an experimental group of people who take the elixir upon catching a cold and another, control group of people who do not. We then compare the results. Presumably, we find that, in both groups, almost everyone has no cold symptoms ten days later. The few who do are divided equally between the groups. This is evidence that Dr. Bob's elixir does not work. *If* we were to find that the experimental group showed no symptoms while the control group still suffered, we would have a strong argument in favor of Dr. Bob's claim that his elixir cures the common cold.

A dramatic instance of the joint method of agreement and difference is Louis Pasteur's experiment before the agricultural society of Melun to test his anthrax vaccine. He collected several dozen healthy sheep, goats, and cattle and vaccinated half of them. Some weeks later he injected all with deadly doses of anthrax bacilli. All the vaccinated animals remained healthy, but, within two days, all the unvaccinated animals were dead.

P R O B L E M S

Find and assess applications of the methods of agreement, difference, and agreement and difference in these passages.

1. Texas is a key to Republican presidential victories. No Democrat has ever won the White House without carrying Texas.

2. In the past few years, dozens of countries have cut marginal tax rates. Almost everywhere, the cuts have been followed by increased economic growth and, in the presence of monetary policy keeping the currency stable, vastly reduced inflation.

3. Other societies differ in so many respects that it is impossible to say whether a given situation is a result of socialism or capitalism or something else. In the two Germanys, however, until 1945, peoples, cultures, and histories were much alike. And the two are still strikingly similar in such basic aspects as industrial structure, trade and demographic patterns, educational levels and social services. . . .

   In 1962, West Germany's per capita output was 1.3 times higher than East Germany's, and by 1982 it had grown to 1.9 times more. . . . In 1980, the average East German worker had only 52 percent of the disposable income that the average West German had. . . .

   . . . it takes 1.39 more hours of work in [East Germany] to produce 53 percent of West German output. Thus, the command economy's overall labor productivity—the key to how much a society produces—is only 38 percent of the free-market economy. (Richard J. Willey, 1984)

4. A drunk reasons: "Monday, I drank Scotch and water and got drunk. Tuesday, I drank rye and water and got drunk. Wednesday, I drank bourbon and water and got drunk. Clearly, I've been drinking too much water."

5. A glance was all he needed. It was enough to convince him that he was indeed up against a series of related cases of tetanus. It was also enough to give him an excellent idea of how they must have originated. Like Williams, Bab Miller, Ida Metcalf, Juanita Jackson, Josephine Dozier, and Ruby Bowers had all been firmly addicted to heroin. (Berton Roueche)

6. "I don't know what she cut it with, but you know how those addicts operate. They'll use anything that's handy. My guess is she mixed in a pinch of dust."

    "I suppose that's as good a guess as any," Dr. Greenberg said. "Except for one thing. It doesn't explain why Lulu is still alive."

    "I was coming to that," Dr. Clarke said. "As a matter of fact, it does. It's about the only explanation that seems to stand up. The last few times Lulu and her friends met, Lulu didn't get a regular shot. She didn't have any money for drugs. All she got, she says, was what she could cadge from the others. And they weren't overly generous. They only gave her just enough to keep her going." (Berton Roueche)

7. Both were delivered at the same instant; so that both were constrained to allow the same constellations, even to the minutest points, the one for his son, the other for his new-born slave. . . . and yet Firminus, born in a high estate in his parents' house, ran his course through the gilded paths of life, was increased in riches, raised to honours; whereas that slave continued to serve his masters, without any relaxation of his yoke. . . . (Saint Augustine)

8. [Battista Grassi] cleared a dozen or twenty different mosquitoes of the suspicion of the crime of malaria—he was always finding these beasts where there was no malaria. (Paul de Kruif)

9. . . . everywhere where there was malaria, there *were* mosquitoes. And *such* mosquitoes! . . . Always, where the "zan-za-ro-ne" buzzed, there Grassi found deep flushed faces on rumpled beds, or faces with chattering teeth going toward those beds. . . . Then Grassi went back to Rome to his lectures, and on September 28 of 1898, before he had ever done a single serious experiment, he read his paper before the famous and ancient Academy of the Lincei: "It is the anopheles mosquito that carries malaria. . . ." (Paul de Kruif)

10. In 1947, there was an outbreak of smallpox in New York City. Health Department officials studied the cases of two smallpox patients at Willard Parker Hospital, Acosta and Patricia. Acosta was married, from the East Bronx, and worked as a porter at another hospital. He had entered Willard Parker on March 27th. Patricia had been admitted to Willard Parker on March 21st. Both patients had been in the hospital previously. Acosta had been in Willard Parker from February 27th to March 11th, suffering from mumps; Patricia, suffering from croup, had been there from February 28th to March 13th. Soon, a two and half year old boy, John, contracted smallpox; he had been in the hospital, suffering from whooping cough, since March 6th. The

Health Department officials concluded that Acosta, Patricia, and John had contracted smallpox at Willard Parker, from somebody who had been there between March 6th and March 13th.[4]

11. Pasteur was urged by Napoleon III to investigate the diseases of wines. . . . Pasteur turned his microscope onto the diseased wines. He saw that, in addition to the wild yeasts originally present on the skins of grapes and responsible for the healthy fermentation of the grape sugar, there were other microorganisms, and he was convinced that these gave rise to the products causing the bad flavor. (A. E. E. McKenzie)

12. [Pasteur] first tried to kill the contaminating microorganisms by antiseptics, but eventually found that heating to 55°C would do so without destroying the flavor and bouquet of the wine. This heating process, now called pasteurization, was tested by a commission. Five hundred liters of wine, half of it heated and half unheated, were put aboard a ship at Brest and taken on a ten month's cruise. At the end of the cruise the heated wine had an excellent flavor, while the unheated wine was acid and sour. (A. E. E. McKenzie)

13. In 1945, there was an outbreak of psittacosis, parrot fever, on Long Island. Initially, everyone who contracted the disease worked on a duck farm. Additional cases began to appear, however, among people who had no link to duck farms. One patient worked on the Long Island Railroad; one ran a farm in Riverhead. Both kept chickens, and a neighbor of the first owned pigeons. Ten other patients raised chickens, and a few patients enjoyed feeding wild pigeons.[5]

14. For instance, there can hardly be a doubt that the animals which fight with their teeth, have acquired the habit of drawing back their ears closely to their heads, when feeling savage, from their progenitors having voluntarily acted in this manner in order to protect their ears from being torn by antagonists; for those animals which do not fight with their teeth do not thus express a savage state of mind. (Charles Darwin)

15. John T. Molloy describes an experiment concerning the importance of ties:

> . . . I panhandled money around the Port Authority Bus Terminal and Grand Central Station in New York. My approach was to stop people, say I was terribly embarassed, but had

---

[4]See Berton Roueche, "A Man from Mexico," *Eleven Blue Men* (Boston: Little, Brown, 1953).

[5]Roueche, "Birds of a Feather," *Eleven Blue Men*.

left my wallet home, and needed 75 cents to get home. I did this for two hours at rush hour. During the first hour, I wore a suit, but no tie, for the second hour, I added my tie. In the first hour, I made $7.23, but in the second, with my tie on, I made $26.00, and one man even gave me extra money for a newspaper.

(a) Which of Mill's methods is Molloy using? What can we conclude? *(b)* Imagine repeating this experiment with other articles of clothing or aspects of grooming: with and without glasses, watch, jacket, shirt, belt, shoes, socks, pants; hair combed and uncombed, shaved and unshaved, and so on. What results would you predict? What does this allow us to conclude about these items?

16. The starting roster of the Cincinnati Reds, 1960:

PLAYER	POS.	HR	RBI	AVG.	SB
Ed Bailey	c	13	67	.261	1*
Frank Robinson	1b	31	83	.297	13
Billy Martin	2b	3	16	.246	0*
Eddie Kasko	3b	6	51	.292	2
Roy McMillan	ss	10	42	.236	9*
Wally Post	lf	19	50	.282	4
Vada Pinson	cf	20	61	.287	32
Gus Bell	rf	12	62	.262	0*

		W	L		ERA
Bob Purkey	p	17	11		3.59
Jim O'Toole	p	12	12		3.81
Jay Hook	p	11	18		4.50*
Cal McLish	p	4	14		4.17*

Players marked with an asterisk were traded before the 1961 season. A sixth-place team thereby rose to win the National League pennant. Using Mill's methods, explain why the players who were kept were kept and why those who were traded were traded.

17. An early use of the method of difference is Elijah's contest with the priests of Ba'al at Mount Carmel, described in I *Kings* 18:20–40. How is this an application of that method? What is

Elijah's conclusion? In some ways, Elijah deviates from the method. How? Do these deviations weaken or strengthen his inference?

11.3 RESIDUES AND CONCOMITANT VARIATION

The methods we have considered so far draw causal conclusions from the presence or absence of possible causes or effects. Often, however, things are a matter of degree. An economist trying to understand the causes of inflation, for example, never finds unemployment, the money supply, interest rates, government spending, investment, and the like *absent;* they are always present, but to varying degrees. Even when a factor may be absent, we may be unable to understand its causal power without considering how it varies. There could be bodies having no forces acting upon them, for example; the methods we have would allow us to infer that forces cause acceleration. But we cannot infer that acceleration is proportional to force (as expressed in Newton's law $F = ma$—force equals mass times acceleration) without considering degrees of force and acceleration. Mill's last two methods allow us to consider variations by degree in addition to the presence or absence of causal factors.

RESIDUES

Mill's fourth method is the method of **residues.** Aptly named, it rests on the idea that if we "subtract out" all effects of known causes, the remainder or residue must be the effect of some other cause or causes. If we can explain some aspects of the phenomenon, in other words, we can look at the other, so far unexplained aspects as effects of causes yet to be determined. As Mill puts this canon of inductive reasoning:

> Subduct from any phenomenon such part as is known by previous inductions to be the effect of certain antecedents, and the residue of the phenomenon is the effect of the remaining antecedents (*A System of Logic,* 285).

"Of all the methods of investigating laws of nature," Mill declared, "this is the most fertile in unexpected results" (*A System of Logic,* 285).

The method of residues is so familiar that it counts as common sense. Suppose that Meg wants to weigh a fairly light object—a baby, say—but has a bathroom scale accurate only for fairly heavy objects. If Meg knows how much she weighs, she can step on the scale with the baby to determine their combined weight. Subtracting her weight from the combined weight yields the baby's weight. Here, Meg's weight is the portion of the phenomenon known to be the effect of a certain cause—the gravitational attraction between the earth and her own body—while the difference between the combined weight and her weight is the residue to be explained by the additional factor of the baby's weight.

Schematically, we can see the pattern behind the method of residues:

RESIDUES

CASE	ANTECEDENT CIRCUMSTANCES		EFFECTS	
1	A	B	E	F

Known: A causes E

Conclusion: B causes F

This method is unlike the others in at least two respects. First, it applies to even a single case. Second, it requires reference to prior knowledge of causal relationships. The other methods allow us to infer causal relations from examining several cases; the method of residues allows us to draw causal inferences from examining one or more cases in conjunction with already understood causal laws.

CONCOMITANT VARIATION

Mill's fifth and final method is that of **concomitant variation.** Mill expresses the canon of this method as follows:

> Whatever phenomenon varies in any manner whenever another phenomenon varies in some particular manner, is either a cause or an effect of that phenomenon, or is connected with it through some fact of causation (*A System of Logic,* 287).

In more modern terms, we might call this the method of correlations. When variations in the antecedent circumstances correlate with variations in effects, we can infer some causal connection between these variations.

Schematically, we can see the pattern behind applications of the method of concomitant variation. Here, "A+" represents an increased quantity of A, while "A−" represents a decreased quantity.

CONCOMITANT VARIATION

CASE	ANTECEDENT CIRCUMSTANCES		EFFECTS
1	A	B	E
2	A+	B	E+
3	A−	B	E−

Conclusion: A and E are causally linked.

We conclude that A and E are causally linked—that is, that A causes E, E causes A, or some C causes both A and E.

The harder a driver presses on a car's gas pedal, for example, the faster the car moves. Movements of the accelerator correlate with increases in the car's velocity. We can infer that there is a causal connection between pressing on the gas pedal and increasing the speed of the car. The former, in particular, causes the latter. Another simple example concerns summer temperatures and electricity use. The hotter the day, the more electricity people use. We can infer, by the method of concomitant variation, that there is a causal link between temperatures and electricity use. In this case, the link is slightly more complicated. The hotter the day, the more electricity air conditioners require to maintain comfortable indoor temperatures.

As this example shows, the method of concomitant variation allows us to infer the existence of causal connections but not their specific form. We cannot assume that one varying factor must be the cause while the other must be the effect. Suppose that we find a correlation in various economies between rates of inflation and interest rates. This allows us to infer that inflation and interest rates are causally connected. We cannot, however, infer that increases in inflation cause increases in interest rates. Nor can we infer that increases in interest rates cause increases in inflation. Both kinds of variation might result from some third factor, or some collection of factors, that causes increases in both. So, from the kind of information in our previous table, we can infer that A and E are causally linked, but not that A causes E or that E causes A. Both A and E may be effects of some common cause.

The kind of variation occurring in that table is **direct:** Increases in one quantity occur in conjunction with increases in another. But the variation may also be **inverse:** Increases in one quantity may occur in conjunction with decreases in another. That is, we might find this pattern:

CASE	ANTECEDENT CIRCUMSTANCES		EFFECTS
1	A	B	E
2	A+	B	E−
3	A−	B	E+

The volume of a quantity of gas, for example, varies inversely with the pressure applied to it. The more pressure, the smaller the gas's volume; the less pressure, the larger the volume. We can infer a causal connection between the pressure applied and the gas's volume. To take another example: The harder a driver presses on the brake pedal, the slower the car moves. The movement of the brake pedal correlates with the car's deceleration. Again, we can infer a causal link. Finally, demand for electricity varies inversely with its price. The more electricity costs, the less people tend to use. The method of concomitant variation allows us to conclude that some causal relation lies behind the correlation.

Mill's method of concomitant variation is immensely powerful, but it has some limitations. First, the method allows us to infer some causal connection between correlated events but cannot determine whether one causes the other or both are effects of some common cause. Determining that requires the use of other methods. There is a substantial correlation between wearing boots and wearing scarves, for example, but wearing boots does not cause anyone to wear a scarf or vice versa; they are effects of cold weather. Similarly, there is a strong correlation between the amount of water on streets and the amount of water on sidewalks, but neither causes the other; they are both effects of rain. To determine the nature of the causal connection, we need to see whether people who are given only boots or only a scarf will wear one without the other or whether wetting the streets will wet the sidewalks. In short, we need to apply the method of difference or the joint method of agreement and difference to establish a definite cause-effect relationship.

Second, the method of concomitant variation requires that we vary two quantities, holding everything else steady. Sometimes this is possible. We can apply forces of varying strength, for example, to the same object to measure acceleration. But sometimes it is impossible. An economist studying inflation cannot hold all other economic factors constant to study the link between inflation and unemployment. To study more complicated cases such as this, twentieth-century mathematicians and economists have extended Mill's method to a technique called **multivariate analysis** or **linear regression,** which allows us to draw conclusions from the variations of several different quantities at the same time.

P R O B L E M S

Discuss the applications of Mill's methods in these bits of reasoning.

1. Since going from 65-mph and 70-mph state speed limits to a national 55-mph rule, traffic fatality rates have dropped from 3.5 per 100 million vehicle miles in 1975 to 2.9 in 1982, with or without safety devices, the National Safety Council says. *(Forbes)*

2. Well, true enough—as far as it goes. But the statement leaves us with a serious question: Did that editor not know what else the Council says, that fatality rates dropped from 18.2 in 1925 to 3.6 in 1974, with no national speed laws at all? (Kevin Smith)

3. Ancient Rome declined because it had a Senate; now what's going to happen to us with both a Senate and a House? (Will Rogers)

4. From 1981 to 1983, the federal deficit skyrocketed—from 2.6 percent of national income in 1981 to 6 percent in 1982 and 6.9 percent in 1983. At the same time, nominal interest rates fell sharply—the three-month T-bill rate, for example, averaged 14 percent in 1981, 10.7 percent in 1982 and 8.6 percent in 1983. Real interest rates—the excess of nominal interest rates over the rate of inflation—remained roughly constant, averaging 4.7 percent in 1981 and 1982 and 4.4 percent in 1983, if inflation is measured by the GNP deflator. Nonetheless, neither this dramatic counterexample nor the more extensive empirical studies by respected economists that have found no historical relation between deficits and interest rates has shaken the confidence of Wall Street and Washington that deficits raise interest rates. Unexamined repetition works wonders. (Milton Friedman)

5. Every time we have reduced taxes, we have increased the total revenue paid in taxes by the people to the government, because there is an incentive for people to earn more, and to go out and experiment and so forth. And so, no, the deficit has not been caused by the cut in taxes. The deficit would increase if we yielded to those who want us to increase taxes. (Ronald Reagan)

6. Mama believed that garlic was the cure-all for any disease. Every morning she'd line us all up and she'd rub garlic on a little hankie and tie it around our necks. We'd say, "Mama, don't do that." She'd say, "Shut up." (She was a very loving woman.) She'd send us off to school with this garlic around our necks and we stunk to high heaven. But I want to tell you a secret; I was never sick a day. My theory about it is that no one ever got close enough to me to pass the germs. (Leo Buscaglia)

7. Liberals often assert that punishment does not deter crime, but James Q. Wilson and Richard J. Herrnstein, in *Crime and Human Nature*, cite evidence to the contrary from several countries. What does fail, it seems clear, is *not* punishing—putting offenders on probation, for example. Failing to punish sends pernicious messages—encouraging offenders, perplexing nonoffenders, subverting parents and impairing the prospects of civilized life. (William Bowen)

8. Advertisement appearing in magazines on San Diego's newsstands:

> "Last year, Handguns killed
>
> 48 people in Japan
>
> 8 in Great Britain

34 in Switzerland

52 in Canada

58 in Israel

21 in Sweden

42 in West Germany

10, 728 in the United States" (James McClure)

9. There is no connection between New Hampshire's reputation as an outstanding ski state and the fact that we make 75 percent of all wooden crutches. (State Planning and Development Commission)

10. Q: What if I'm still not convinced helmets work? A: Try this experiment. Put on a helmet and have a friend whack you on the head with a baseball bat. Now try the same experiment without the helmet. If you're still not convinced, you're probably too hard-headed to need a helmet. (Jerry Smith)

11. In 1980, Dr. Richard B. Shekelle of Rush-Presbyterian-St. Luke's Medical Center in Chicago analyzed deaths among 1,900 middle-aged American men whose diets were first examined more than 20 years earlier. The analysis showed that those who consumed large amounts of cholesterol had a much greater chance of dying prematurely of a heart attack. And a national study completed in 1983 among 3,806 men by the National Heart, Lung, and Blood Institute showed that reducing cholesterol in the blood can indeed be life-saving. When the men, all of whom faced a high risk of developing heart disease, were treated with a cholesterol-reducing drug and dietary advice, each 1-percent fall in cholesterol produced a 2-percent drop in their rate of coronary heart disease. (Jane Brody)

12. . . . researchers studying the relationship between married couples' work commitments and their love lives found that people with M.B.A. degrees have "significantly better" sex lives than other advanced-degree professionals. Even more surprising is the group found at the bottom of the sexual-satisfaction scale: Doctors of Philosophy, or Ph.D.s. Is this yet another sign of the times—that lucre is a greater turn-on than learning? *(Forbes)*

13. [Jacob and Esau were] two twins born so near together that the second held the first by the heel; yet in their lives, manners, and actions, was such a disparity, that that very difference made them enemies to one another. (Saint Augustine)

14. Beethoven played a more decisive role in the evolution of music than any other single figure, not excepting Bach. It is only necessary to compare his earliest works with his last ones to recognize what progress the art of music made in his time—and largely owing to him. (David Ewen)

15. The generally accepted theory that both the Greeks and the Jews owe their systems of writing to the Phoenicians is strongly supported by the similarity in the names of the symbols: compare the Greek *alpha, beta, gamma,* with the Hebrew *aleph, beth, ghimel.* (Tobias Dantzig)

16. This tide increases with the declination of the moon till the 7th or 8th day; then for the 7 or 8 days following it decreases at the same rate as it had increased before, and ceases when the moon changes its declination, crossing over the equator to the south. (Isaac Newton)

17. Take away oxygen from the blood. The mind loses its reasoning ability. (Waldemar Kaempffert)

18. The ultimate fate of an ounce of uranium may be expressed by the equation:

$$1 \text{ ounce uranium} = \begin{cases} 0.8653 \text{ ounce lead,} \\ 0.1345 \text{ ounce helium,} \\ 0.0002 \text{ ounce radiation.} \end{cases}$$

The lead and helium together contain just as many electrons and just as many protons as did the original ounce of uranium, but their combined weight is short of the original uranium by about one part in 4000. Where 4000 ounces of matter originally existed, only 3999 now remain; the missing ounce has gone off in the form of radiation. (James Jeans)

19. During the [1988] campaign, the "racism" issue coalesced around Bush's attacks on the Massachusetts prison-furlough program and in particular around the case of Willie Horton, a murderer who escaped while on furlough and committed a brutal assault and rape upon a Maryland couple. . . . [T]he implication that voters hate blacks *per se* more than they fear criminals, or that weekend passes for murderers and other forms of coddling felons would stir little protest if the convicts

were white, is absurd. The state of Maryland is currently being convulsed by a scandal: a rehabilitation-oriented penal institution is in danger of being closed down by legislators irate over the discovery that weekend furloughs were granted to a triple murderer, and work release to a rapist suspected of having used the occasion to commit another rape. Both convicts were white, and, to boot, the murderer was upper-class while the victim of his most heinous killing was black. But the people of Maryland—surprise?—are outraged nevertheless. (Joshua Muravchik)

20. Throughout the history of the game, almost every significant increase in offense has been accompanied by an increase in attendance, and almost every decrease in offense has been accompanied by a decrease in attendance. With the sole exception of the 1930s, every hitter's era in baseball history has been a period of growth, and every pitcher's era has been a period of stagnation. When runs per game dropped from 4.5 in 1911 to 3.7 in 1914, attendance dwindled from 6.6 million to 4.5 million. . . . When runs per game jumped from 3.8 in 1919 to 4.4 in 1920, attendance exploded. . . . When the ball was deadened in 1931, runs per game dropped from 5.5 to 4.8 and attendance dropped from 10.1 million to 8.5. By 1933, runs were down to 4.4 per game, and attendance was down to 6.1 million. . . . In the late 1940s, runs per game were back up in the 4.7 range, and attendance shot up to over 20 million; in 1952, runs were down to a little below 4.2, and attendance was down under 15 million. (Bill James)

How might the following causal claims be defended, using Mill's methods?

21. The United States has become great not because of things but because of ideas. (James Michener)

22. Kindness in words creates confidence, kindness in thinking creates profoundness, kindness in giving creates love. (Lao-tzu)

23. Things are interesting because we care about them, and important because we need them. (George Santayana)

24. As a humanist, I am bound to reply that almost all important questions are important precisely because they are not susceptible to quantitative answers. (Arthur Schlesinger, Jr.)

25. Beggars get handouts before philosophers because people have some idea what it's like to be blind and lame. (Diogenes of Sinope)

26. Neighbors praise unselfishness because they profit by it. (Friedrich Nietzsche)

27. The ultimate result of shielding men from the effects of folly is to fill the world with fools. (Herbert Spencer)

28. You are free and that is why you are lost. (Franz Kafka)

29. It sounds too simple to say that lack of self-insight causes our griefs, but that is the plain fact. (Vernon Howard)

30. In the following well-known passage, William James reverses the usual conception of the causal order of emotions and physiological reactions:

> Common-sense says, we lose our fortune, are sorry and weep; we meet a bear, are frightened and run; we are insulted by a rival, are angry and strike. The hypothesis here to be defended says that the order of this sequence is incorrect, that one mental state is not immediately induced by the other, that the more rational statement is that we feel sorry because we cry, angry because we strike, afraid because we tremble, not that we cry, strike, or tremble because we are sorry, angry, or fearful as the case may be.

How, using Mill's methods, might we test James's hypothesis?

Devise studies or experiments, using Mill's methods, to answer the following questions; explain their use of Mill's methods, and explain difficulties that might be encountered.

31. Does the death penalty deter crime and, specifically, murder?

32. What causes people to commit violent acts against others?

33. How does racial or ethnic background affect economic success?

34. Do welfare payments increase or decrease poverty?

35. Why do girls get better grades in high school than boys?

36. Why do minority students drop out of college at higher rates than nonminorities?

37. To what extent is intelligence hereditary?

CHAPTER 12

EXPLANATIONS

The goal of science is to achieve reliable knowledge and understanding. Almost all science involves careful observation—of the natural world, of people and their behavior, of social organizations, or of events in a laboratory. Scientists generalize on the basis of these observations, constructing laws and theories to explain what they observe. Because science seeks reliable knowledge, logic plays an important role in this process. Scientists seek to gain knowledge by reliable means.

Science goes beyond generalizations from experience—"What goes up must come down," for example—to ask not only *what* happens, and under what circumstances, but also *why*. Answering the "Why?" questions is crucial to understanding, rather than merely cataloging, the world. Consequently, explanation is one of science's most important concepts.

Often, explanation requires constructing a theory about entities, such as atoms, electrons, electromagnetic fields, and the like, that we do not experience directly. Often, too, explanation is causal; it proceeds by finding causes for the phenomena to be explained. This chapter, therefore, relates closely to the preceding chapter. But not all explanations are causal. Indeed, scientists disagree about whether, in the final analysis, physical laws will make any reference to causation at all. In this and other ways, explanation, like causation, has provoked lively debate among philosophers. This chapter addresses basic and relatively noncontroversial aspects of explanation.

■ 12.1 GENERALIZATIONS AND LAWS

On November 22, 1963, President John F. Kennedy rode in a motorcade through downtown Dallas. Shots rang out; the presidential limousine raced to Parkland Memorial Hospital. Half an hour later, the president was dead. Why? asked a shocked and mourning nation.

This question divides into many others. Why did Kennedy die? The immediate answer was discovered by attending physicians almost immediately. Kennedy suffered a massive gunshot wound to the head. Why and how was he wounded? Here, the answer is far less clear. The Warren Commission held that a single assassin hit the president with two bullets, one in the neck or upper back and one in the head. Critics have argued that as many as three assassins were involved. Whether the fatal shot was fired from the Texas School Book Depository Building or the grassy knoll remains controversial. Why did the assassin or assassins kill the president? Here, too, controversy reigns. If Lee Harvey Oswald was the lone gunman, as the Warren Report argued, his motives are still obscure. If other conspirators were involved, it remains unknown who they were, much less why they shot Kennedy.

Consider the first question: Why did Kennedy die? The doctors explained his death by saying that he suffered a massive gunshot wound to the head. This explanation is causal, for the gunshot caused the wound that caused his death. It also exhibits a common feature of explanations: It explains a particular occurrence by pointing to a generalization under which it falls. We might explain why leaves, once detached from trees, fall to the ground by pointing out that all objects dropped near the earth fall toward it. Similarly, we might explain why a particular crow is black by pointing out that all crows are black. Such explanations of particular events or states of the world thus have the logical form of simple arguments:

(1) Generalization

Particular circumstances

∴ Particular event or state

(Explanations and arguments both have structures that can be revealed by logic, though they function differently, in fact in opposite ways, in most contexts; see section 12.4.) In the case of the crow, for example, the explanation has the form

(2) Crows are black.

This is a crow.

∴ This is black.

In the case of President Kennedy, it has the form

(3) People who suffer massive gunshot wounds to the head die.

Kennedy suffered a massive gunshot wound to the head.

∴ Kennedy died.

Similar explanations can account for general as well as particular truths. We might explain why crows on campus are black by saying that crows are black. Similarly, we might explain why batted baseballs travel in parabolic arcs by pointing out that all projectiles travel in parabolic arcs. The form of these explanations is syllogistic. For example, in the case of the campus crows:

(4) Crows are black.

All crows on campus are crows.

∴ Crows on campus are black.

This example suggests that covering law explanations of general truths have the form

(5) Broader generalization

Universal affirmative relating narrower to broader generalization

∴ Narrower generalization

Often, this sort of explanation is said to **subsume** one generalization under another.

Explanations in terms of generalizations are either deductive or inductive. Sometimes, the broader generalization that explains the phenomena under discussion is universal, applying to everything of a certain kind. Then, the explanation is deductive; what is to be explained follows deductively from the universal generalization and the particular circumstances. This explanation, for example, is deductively valid:

(6) All opossums are marsupials.

This is an opossum.

∴ This is a marsupial.

In other cases, the generalization is generic or statistical, applying in general to things of a certain kind or to a certain percentage of things of that kind. Then, the explanation is inductive and must be judged strong or weak to some degree rather than valid or invalid. These arguments, for example, are inductively strong but not deductively valid.

(7) What goes up must come down.

The bullet you fired went up.

∴ The bullet you fired will come down.

(8) People who smoke have an increased chance of contracting emphysema.

You smoke.

∴ You have contracted emphysema.

The principle that what goes up must come down is generally true, but there are exceptions (space probes sent to Jupiter, for example). And it is not even generally true that people who smoke contract emphysema. But smokers have enough of an increased risk that smoking is a reasonable explanation for a smoker's getting the disease.

Deductive explanations are called **covering law** explanations when the universal generalization they use is a law of nature. Philosophers debate what laws of nature are. But most agree that laws are universal generalizations that are invariant and necessary. They are invariant in that they apply to all places and times and necessary in that they apply not only to the actual world, the way things actually are, but also to any way the world could be. Here, for instances, are some laws of nature:

(9) $s = \frac{1}{2} g t^2$

(Galileo's law: distance of free fall = ½ the gravitational constant times time squared)

$F = ma$

(Newton's second law: force = mass times acceleration)

$f = gm_1 m_2 / r^2$

(Newton's law of gravitation: the gravitational force between objects 1 and 2 = the gravitational constant times the mass of 1 times the mass of 2, divided by the square of the distance between 1 and 2)

$e = mc^2$

(Einstein's law: energy = mass times the velocity of light squared)

All copper objects are electrical conductors.

All opossums are marsupials.

These hold anywhere, anytime, no matter what happens. They allow us to reason **counterfactually,** that is, to reason about what would happen if something-or-other were to take place. Galileo's law, for example, allows us to deduce that *if* we were to drop something from a second-story balcony, sixteen feet in the air, it *would* (ignoring friction and air resistance) take one second to fall. The law that copper conducts electricity allows us to deduce that *if* we were to place a penny in the fuse box, the circuit *would* be completed.

So far we have talked about simple explanations involving one generalization or law. But scientific explanations often involve more than one law. The medical explanation of Kennedy's death portrayed in (3) summarizes a much more complete medical explanation that would cite damage to parts of the brain, the importance of their functioning, and so on. Explanations, like arguments, can be complex.

P R O B L E M S

Analyze the structure of each of the following explanations, identifying the generalizations and covering laws used.

1. They said today we should stock up on canned goods. So I went out and bought a case of beer. (John Gretchen III, Galveston carpenter, preparing for a hurricane)

2. People were forced to be with their loved ones. (Police officer Dave Gaouette, explaining an increase in domestic disturbances during two days of heavy snow)

3. I never want to be the best at anything. Anybody who wants to be the best at anything in the world must spend 80% of their waking time on it. And you lose the pursuit of other things. (Warren Avis, founder of Avis Rent-A-Car)

4. If a man is wise, he gets rich an' if he gets rich, he gets foolish, or his wife does. That's what keeps the money movin' around. (Finley Peter Dunne)

5. Woman to umbrella vendor, in the pouring rain: "I don't want that one. It's wet." *(New York)*

6. People in capitalist countries do not earn enough money to buy products and therefore they remain on the shelves. The income of the Soviet peoples has been rising steadily so that now they can buy everything they desire. It is the buying power of the Soviet people that keeps the store shelves empty. (Soviet schoolteacher, 1986)

7. The moon's orbit is elliptical, and departs by an angle of about five degrees from the earth's orbit around the sun. This explains why eclipses of the sun do not occur every month. (Daniel J. Boorstin)

8. The "housing crisis" agitating Mayor Koch and Governor Cuomo is actually a product of local attempts to suspend economic law. Everybody who isn't a tenant admits how destructive the city's World War II rent controls have been, but tenants have the most votes. An extralegal free market in abandoned factory-loft conversions compensated somewhat, but City Hall's first instinct was to prohibit most conversions. Anyone fool enough to want to build new apartments has to pay double the standard price for concrete because of a Mafia stranglehold on the commodity. Vast areas of New York City are a wasteland of abandoned housing and rubble. *(Wall Street Journal)*

9. Furthermore the sphericity of the earth is proved by the evidence of our senses, for otherwise lunar eclipses would not take such forms; for whereas in the monthly phases of the moon the segments are of all sorts—straight, gibbous and crescent—in eclipses the dividing line is always rounded. Consequently, if the eclipse is due to the interposition of the earth, the rounded line results from its spherical shape. (Aristotle)

10. Now it is scarcely possible to conceive how the aggregates of dissimilar particles should be so uniformly the same. If some of the particles of water were heavier than others; if a parcel of the liquid on any occasion were constituted principally of these heavier particles, it must be supposed to affect the specific gravity of the mass, a circumstance not known. Similar observations may be made on other substances. Therefore, we may conclude that the ultimate particles of all homogeneous bodies are perfectly alike in weight, figure, etc. In other words, every particle of water is like every other particle of water; every particle of hydrogen is like every other particle of hydrogen; etc. (John Dalton)

11. Their [the Mayans'] world, once so certain, stable, dependable, and definite is gone. And why? Here of course is a first-rate mystery for modern skill and knowledge to unravel. The people were not exterminated, nor their cities taken over by an enemy. Plagues may cause temporary migrations, but not the permanent abandonment of established and prosperous centers. The present population to the north has its share of debilitating infections, but its ancestors were not too weak or wasted to establish the Second Empire after they left the First. Did the climate in the abandoned cities become so much more humid that the invasion of dense tropical vegetation could not be arrested, while fungous pests, insects, and diseases took increasing toll? This is hard to prove. Were the inhabitants starved out because they had no steel tools or draft animals to break the heavy sod which formed over their resting fields? Many experts think so. (Paul B. Sears)

12. Are not the rays of light very small bodies emitted from shining substances? For such bodies will pass through uniform mediums in right lines without bending into the shadow, which is the nature of the rays of light. They will also be capable of several properties and be able to conserve their properties unchanged in passing through several mediums, which is another condition of the rays of light. (Isaac Newton)

13. . . . the country now called Hellas had in ancient times no settled population; on the contrary, migrations were of frequent occurrence, the several tribes readily abandoning their homes

under the pressure of superior numbers. Without commerce, without freedom of communication either by land or sea, cultivating no more of their territory than the exigencies of life required, destitute of capital, never planting their land (for they could not tell when an invader might not come and take it all away, and when he did come they had no walls to stop him), thinking that the necessities of daily sustenance could be supplied at one place as well as another, they cared little for shifting their habitation, and consequently neither built large cities nor attained to any other form of greatness. (Thucydides)

14. This morning the windowsills were wet. Overnight, the outside temperature fell, cooling the window glass. The air inside contained water vapor, which condenses on any surface that's cold enough. So, the water vapor condensed when it came in contact with the cooler glass, and the resulting drops of water fell to the sill.

15. Wood floats on water, because its density is lower than that of water. Archimedes discovered that a liquid supports a body immersed in it with a force equal to the weight of the liquid the body displaces. So, any body floats if its weight is less than the weight of an equal volume of water.

16. Children of blue-eyed parents invariably have blue eyes. Why? Eye color is determined by a pair of genes, and having blue eyes is a recessive trait. So, if the parents have blue eyes, both genes of each must be for blue eyes. The child gets one eye color gene from each parent. The child, therefore, must end up with two blue-eye genes, making his or her eyes blue.

17. Why did Hitler wait until June 22, 1941, to attack Russia, when his generals pleaded for an early spring invasion? There were at least two factors. He had had to send Rommel to Africa to bail out Mussolini in February and wanted to wait until that campaign was over. And Hitler underestimated Soviet strength; he thought that Russia would fall within weeks, calling the Red Army "no more than a joke." So, he hardly felt constrained to maintain a rigid timetable.

18. In 1955, my father maintained our family of five on a wage of $82 a week as a bookbinder. In 1963 in Las Vegas, three years out of school, I bought an upscale, brand-new three-bedroom home, with a two-car garage on a quarter acre. It cost $20,000.
    Where did this good economy go? It was inflated away. The price of gold was $35 in those years. It is roughly ten times that level today. To appreciate the numbers above, multiply by 10.

There is a secondary answer, which is that inflation [moved] the entire work force into higher tax brackets, with consequent reductions in after-tax purchasing power. My father's bindery job paid $80 a week after deductions; today, at $820 a week the net would be $662. The $20,000 house now costs at least $200,000, but with interest, property taxes and wages being what they are, it's way out of the reach of young people just out of college, not to mention high school. (Jude Wanniski)

19. By 1978 John Lennon had ceased to resemble himself. Wasted by dieting, fasting, and self-induced vomiting, he weighed only 130 pounds. Totally enervated by lack of purpose and exercise, he rarely left his bed. Drugged all day on Thai stick, magic mushrooms, or heroin, he slept much of the time and spent his waking hours in a kind of trance.

    What had happened to John? The same thing that happened to Howard Hughes and many another wealthy, self-indulgent recluse. Lennon had simply refused to pay the price for staying alive, the toll levied in terms of involvement, responsibility and effort. (Albert Goldman)

20. "Why do all these bright students keep flocking to law school?"

    "Simple. It's greed. They want BMWs and gold credit cards. If you go to a good law school, finish in the top 10% and get a job with a big firm, you'll be earning six figures in a few short years, and it only goes up. It's guaranteed. At the age of 35, you'll be a partner raking in at least $200,000 a year. Some earn much more."

    "What about the other 90%?"

    "It's not such a good deal for them. They get the leftovers."

    "Most lawyers I know hate it. They'd rather be doing something else."

    "But they can't leave it because of the money." (John Grisham)

21. The popularity of video cameras arises from a simple but potent misunderstanding. Somehow people have gotten the idea that they won't mind being old so much if they can turn on the TV and see what they were like when they were young. Not true. The best memories are ones that have been allowed to evolve unhindered by documentary proof. When I feel weary and infirm, I often cheer myself up by thinking back on my days as the star of my junior high school football team. These thrilling recollections would be less compelling if they were accompanied by a videotape showing that I weighed 80 pounds and spent most of my time on the bench. Memory is better than a video camera, because, in addition to being free, it doesn't work very well. (David Owen)

22. The English sponsors of the Pilgrims had made an agreement with them in 1620, as reported by William Bradford, governor of the Plymouth Colony, that "all profits," that is, all crops, fish, and trade goods, would "remain still in the common stock," from which all colonists were to take their food and goods. The economic system was pure communism, in accord with their religious convictions.

The colonists starved. Sharing everything in common "was found to breed much confusion and discontent." Finally, they decided to make a drastic change in their organization. Bradford wrote that from then on, "they should set corn every man for his own particular." Each family was assigned a parcel of land. Though the land still remained common property, each family could keep whatever it could grow on it.

Bradford wrote: "It made all hands very industrious; so much corn was planted than otherwise would have been." To celebrate the abundant harvest, the Pilgrims "set apart a day of thanksgiving." (Fred Foldvary)

Bas van Fraassen has summarized some traditional puzzles for theories of explanation.[1] Can you explain how the following two might be analyzed? Which pose problems for the sketch of an account in this section (and the sketch of an account of causation in section 11.1)?

23. Suppose that a flagpole, 100 feet high, casts a shadow 75 feet long. Given the angle of elevation of the sun, we can compute one quantity when given the other. Yet we explain the length of the shadow by saying that the pole is 100 feet high; we do not explain the height of the pole by saying that the shadow is 75 feet long.

24. No one contracts paresis, a disease leading to paralysis, without having latent, untreated syphilis. One could explain why someone contracted paresis by saying that he had latent, untreated syphilis. But only a small percentage of people with latent, untreated syphilis contract paresis.

25. In van Fraassen's view, and in the view of causation presented in section 11.1, causation (and thus explanation) are context dependent. There should be a context in which one can explain the height of a pole by appeal to the length of its shadow. Can you elaborate such a context?[2]

---

[1]Bas van Fraassen, *The Scientific Image* (Oxford: Clarendon Press, 1980), pp. 104–106.

[2]For van Fraassen's answer, see his story, "The Tower and the Shadow," in *The Scientific Image*, pp. 132–134.

■ 12.2   THE HYPOTHETICO-DEDUCTIVE METHOD

Scientific explanations arise by a process called the **hypothetico-deductive method,** or, more simply, **hypothetical reasoning.** Scientists accept or reject statements depending on whether there is evidence to support them. Before they assess the evidence, proposed statements are **hypotheses:** statements not yet believed but to be adopted tentatively for purposes of testing. Hypothetical reasoning is the process of proposing hypotheses; testing them; and accepting, rejecting, or modifying them in light of evidence. With the goal of reaching an explanation for certain phenomena, such reasoning consists of four stages, which are often repeated:

HYPOTHETICAL REASONING

1. Formulate a *hypothesis* to explain the phenomena in question.

2. *Deduce* from the hypothesis statements that can be tested.

3. *Test* these implications or predictions.

4. Accept, reject, or modify the hypothesis.

We can express this in a diagram:

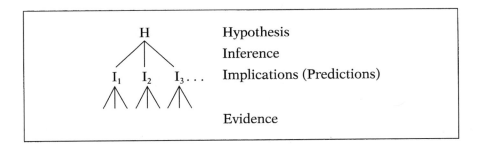

Consider again the Kennedy assassination. The Dallas police and later the Warren Commission hypothesized that Lee Harvey Oswald, acting alone, killed the president. To test this hypothesis, they explored its implications and tested them against the available evidence. For example, films of the assassination show that the shots were all fired within six seconds. A single gunman probably could have fired only three shots during that time. So, the hypothesis entails that there were at most three shots. Is this conclusion compatible with the evidence? Unfortunately, the evidence is quite unclear. Witnesses in Dealey Plaza disagreed about how many shots were fired. A Dallas police radio was left on inadvertently during the shooting, but acoustic tests have not established conclusively whether three or four shots were fired. A single gunman would have taken about three seconds to reload and fire a rifle. The hypothesis also entails, therefore, that about three seconds transpired between shots. Does this square with the evidence? Again, the answer is unclear. The Zapruder film of the assassination seems to show Kennedy react to being hit about a second before Governor John Connally reacted—too long for a single bullet to have hit both but too short for a second bullet to have been fired by the same gunman. Critics of the Commission have argued on this basis for revising the single-gunman hypothesis. Defenders of the Commission have maintained that Connally's reaction was delayed.

The first stage—formulating a hypothesis—requires imagination. There is no rule for devising hypotheses; thinking of them demands intelligence, insight, ingenuity, creativity, and hard work. Some philosophers have thought that we need only to collect the facts, generalize on them to obtain a hypothesis, and test that. But where do we begin collecting facts? There are infinitely many facts we could collect; we need to know where to begin. We might say, "Collect the *relevant* facts"—but relevant to what? Even if we begin with a well-defined problem to be solved— who killed the president? for example—we will not know what is relevant to it without having some proposal for solving it. Hundreds of people were in Dealey Plaza when the president was shot; thousands of people might have had some motive for killing him. Where do we begin?

Taking a different example, suppose we want to find the cause of a newly discovered disease. What facts are relevant? Until we have a hypothesis about whether it is contagious, how it may be transmitted, and so on, we cannot begin to know how to distinguish relevant from irrelevant facts. Hypotheses, therefore, are crucial to scientific activity. They are not dictated by the data, but are proposals for organizing and explaining the data.

The second stage—determine what the hypothesis implies or predicts—is largely deductive, though it may also involve inductive reasoning. Once we have formulated a hypothesis, we need to determine how to test it. So, we must see what testable implications it has. These implications may concern the future, in which case they are predictions. But they may also concern the present or the past.

Often, the hypothesis alone will not imply anything we can test. To derive something testable, we need to bring in other assumptions, items of knowledge, hypotheses, principles, and facts. These are called **auxiliary assumptions.** They may be well established, or they may be conjecture. But we assume them to be true for the purposes of testing the hypothesis. For example, in inferring from the single-gunman hypothesis that at most three shots were fired, we used the auxiliary assumptions that all the shots were fired within six seconds and that about three seconds were needed between shots.

The third stage, testing the hypothesis, is fundamental to scientific method. Having derived some testable implications of our hypothesis, we proceed to test them. This may be easy; if we hypothesize that limestone is heavier than water, we can put some limestone in water and see that it sinks. It may also require great cleverness and technical skill. Testing the single-gunman hypothesis, for example, has required sophisticated work in acoustics, photographic analysis, ballistics, and forensics. Testing hypotheses in modern particle physics tends to require very complex, sophisticated, and expensive equipment.

If we discover, through testing, that the implications are true, our test **confirms** the hypothesis (together with any auxiliary assumptions we used); it gives us reason to believe that the hypothesis is true. If we discover that at least one implication is false, our test **disconfirms** or **infirms** our hypothesis (assuming that all auxiliary assumptions hold); it gives us reason to believe that the hypothesis is false.

In general, confirmation and disconfirmation are inductive matters. Hypotheses are not proved or disproved; tests give us reasons to believe or disbelieve—often *good* reasons, but not **conclusive** reasons.

Nevertheless, confirmation and disconfirmation of a hypothesis having the form of a law of nature are quite different from a logical point of view. Tests that confirm a hypothesis constitute evidence in its favor; they support it inductively. Tests that disconfirm a hypothesis of this form, however, show conclusively that it, or some auxiliary assumption accompanying it, is false. Confirming tests, that is, have inductive force. Disconfirming tests have deductive force. Disconfirming tests, therefore, require revision or outright rejection of the hypothesis or some auxiliary assumption.

To see why, recall the definition of implication. The truth of a hypothesis, together with auxiliary assumptions, guarantees the truth of implications or predictions drawn from it. So, suppose H (hypothesis) and A (auxiliary assumptions) imply P (prediction). (We might write this, in short, as $H \cdot A \Rightarrow P$.) Then the truth of H and A guarantees the truth of P. If P is false, then H or A must be false. ($\sim P \Rightarrow \sim H \vee \sim A$.) The falsehood of some implications or predictions thus guarantees the falsehood of the hypothesis or an auxiliary assumption.

All this assumes that H and A imply P deductively. If they merely support P inductively, then the falsehood of P constitutes nothing more than inductive evidence against them. Whether the explanation is deductive or inductive in form, moreover, confirming instances have inductive force. If P is true, we can deduce nothing at all about H or A. The truth of P may constitute evidence for H and A, but it does not conclusively establish their truth. Hence, the truth of implications or predictions does not guarantee the truth of the hypothesis, though it does give a reason to believe it.

Consider once again the single-gunman hypothesis. If we were able to show that there were only three shots, then we would have evidence in favor of the hypothesis. We would not, however, have proved it to be true. Several shooters might have fired a total of three shots between them. If we were to show that there were more than three shots, however, we would have established that either the hypothesis (that there was a single gunman) or the auxiliary assumption (that it takes about three seconds to reload, aim, and fire a rifle) is false.

To analyze another case of scientific reasoning, recall, from Chapter 11, the case of the 11 blue men. The story involves repeated applications of the hypothetico-deductive method. It begins with a policeman, early on the morning of September 25, 1944, discovering an old man on the street having vomited and collapsed. The policeman, accustomed to Bowery residents who drank too much, conjectured that the old man was suffering from a bout with alcohol. Here is a first hypothesis about the cause of illness. It has implications for the old man's condition. (Auxiliary assumption: Consuming large amounts of alcohol can cause vomiting and collapse.) A closer examination, however, did not bear out those implications: The old man's fingers and nose were blue. This disconfirmed the policeman's hypothesis.

The doctor who treated the old man upon his arrival at Beekman-Downtown Hospital tentatively diagnosed carbon-monoxide poisoning. That explained the sky-blue color of the man's extremities, as well as his collapse. (Auxiliary assumptions: Carbon-monoxide poisoning can cause sky-blue skin coloration, dopiness, headache, and collapse.) But it also had implications that were not borne out. The old man did not exhibit the dopiness and headache typical of gas poisoning but did experience abdominal cramps and vomiting. As these aspects of the patient's condition became clear, the doctor was forced to abandon the hypothesis of carbon-monoxide poisoning.

Once Health Department officials learned that all 10 surviving patients suffering from the same symptoms had eaten breakfast at the Eclipse Cafeteria, they conjectured that food poisoning caused the old men's afflictions. This explained the cramps and vomiting. But food poisoning takes from 5 to 24 hours for the onset of its symptoms; the old men collapsed within 30 minutes after eating. So, that hypothesis, too, had to be rejected.

Tracing the problem to the oatmeal eaten by all the men and, in particular, to the accidental substitution of sodium nitrite for sodium chloride in it, the Health Department officials hypothesized that the men suffered from sodium nitrite poisoning, which was induced by eating chemically contaminated oatmeal. The sodium nitrite poisoning hypothesis correctly accounted for all their symptoms. It was confirmed. But the hypothesis that the poisoning was induced by eating contaminated oatmeal had some false consequences. The men could have been poisoned by ingesting a toxic amount of sodium nitrite, but the oatmeal, mixed according to the cook's recipe, contained less than a toxic dose. Moreover, if the oatmeal had been toxic, all 125 people who ate it should have become ill. Yet only 11 were sick.

These facts required revising the hypothesis. The officials hypothesized that the men suffered from sodium nitrite poisoning, the source of which was the Eclipse Cafeteria oatmeal and "salt," actually sodium nitrite, from a table shaker. This no longer implied that the other patrons should have become sick. It continued to explain the men's symptoms. And it predicted that at least one table at the Eclipse would have a shaker containing a significant amount of sodium nitrite. Testing the shakers confirmed that prediction. So, the hypothesis was confirmed.

The process of investigation in the case of the 11 blue men is fairly typical of scientific inquiry. Actually, it is simpler than many cases. In essence, it is a case of puzzle solving: The case of the 11 blue men occurs against a background of well-developed medical knowledge. Sometimes, there is no theoretical framework in the background. Then, the problem is not just to apply a framework to a particular case, but to devise a framework and then apply it.

P R O B L E M S

1. Review the explanations in the problems at the end of section 12.1. Identify uses of hypothetical reasoning. In each case where such reasoning is explicit, identify the problem, the hypothesis proposed, and the test procedure used. In each case where no such reasoning is explicit, explain a procedure for testing the hypothesis used in the explanation.

2. Write a short (3–5 page) paper on the use of hypothetical reasoning by one of these scientists on the topic listed. Discuss the problem, the hypotheses formulated, implications drawn, auxiliary assumptions used, and test procedures followed.

   (a) Nicolaus Copernicus—the solar system
   (b) William Harvey—circulation of blood
   (c) Charles Darwin—natural selection
   (d) J. J. Thomson—the electron
   (e) Alexander Fleming—penicillin

3. Write a short (3–5 page) paper on one of Sir Arthur Conan Doyle's Sherlock Holmes stories, outlining Holmes's use of hypothetical reasoning in that story.

■ 12.3   CONFIRMATION AND AUXILIARY ASSUMPTIONS

Earlier, we saw that tests confirming a hypothesis support it but do not conclusively establish it. The same is true of tests disconfirming generic or statistical hypotheses. Tests disconfirming a universal hypothesis, in contrast, show that the hypothesis *or an auxiliary assumption* must be false. This raises the question: Can evidence ever conclusively prove or disprove a scientific hypothesis?

Clearly, evidence cannot conclusively prove a hypothesis. No matter how many implications of a hypothesis we observe to hold, no matter how great the inductive evidence in favor of the hypothesis, its truth is not guaranteed. John Hunter (1728–1793), an English biologist, hypothesized that the same biological agent was responsible for both syphilis and gonorrhea. He tested the hypothesis on himself, infecting himself with gonorrhea. He thought he could cure both diseases; he was tragically wrong. But he did come down with syphilis, confirming his hypothesis. Despite this development, his hypothesis was false: Different agents cause the two diseases. Classical physics, developed by Isaac Newton (1642–1727), was confirmed by observation after observation, experiment after experiment, for more than a century; we now believe it to be literally false, although it approximates the truth very closely. Several dramatic tests during this century have confirmed Albert Einstein's (1879–1955) theory of relativity, but it could still turn out to be false.

Some tests are nevertheless very compelling. Very good confirming experiments can tempt us to think that the hypothesis is the only explanation for the data. But this is never literally true. Scientific laws are always asserted on the basis of incomplete evidence. At any point, only finitely many tests of a law could have been performed. But any law has infinitely many consequences. So, the body of empirical evidence at any stage of investigation is incomplete; many different hypotheses would explain the same data.

To take a simple example, suppose that we have conducted experiments relating the mass of an object to its acceleration in free fall. Say that we have recorded our observations in a graph:

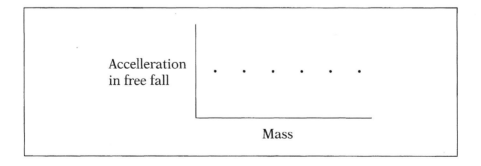

No matter how many points we have recorded, there are infinitely many curves that would pass through those points. Here, for example, are several.

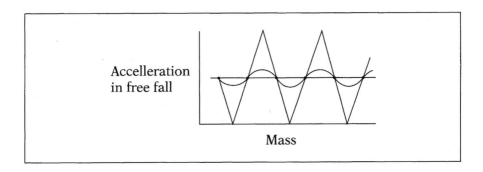

Accelleration in free fall

Mass

Even a million points do not determine what relation holds between mass and free-fall acceleration.

This example has historical significance. Aristotle had held that free-fall acceleration depended on the mass of the falling object. Galileo (1564–1642) formulated a law holding that acceleration in free fall toward the earth is independent of mass and, in fact, constant. (It corresponds to the straight line in the graph above.) Newton's law of gravitation explained this. According to Newtonian physics, gravitational attraction depends on the mass of the falling object and the distance between the object and the center of the earth. ($F = gmm_e/r^2$, where F is gravitational force, m the mass of the object, m_e the mass of the earth, and r the distance between the object and the earth's center.) So, the earth will attract a heavier object with greater force than it will attract a lighter object. But, according to Newton's second law, force—in this case, gravitational attraction—is equal to mass times acceleration. ($F = ma.$) The effects of mass thus cancel each other out ($ma = gmm_e/r^2$, so a $= gm_e/r^2$, which does not depend on m); the greater gravitational force accelerates the greater mass at the same rate as lighter objects.

The Newtonian theory does differ from Galileo's in an important way. Newton's law of gravitation predicts that gravitational force is stronger near the earth's surface than it is farther away. (As r, the distance between the object and the center of the earth, increases, the acceleration decreases.) For medium-sized objects relatively near the earth's surface, these effects are tiny: The predictions of Newton's law differ only minutely from those of Galileo's law. In the absence of extremely sophisticated and sensitive equipment, tests using such objects would seem to confirm both laws. Two stones dropped from different heights will accelerate at what appears to be the same rate. But the laws differ noticeably in their predictions when one of the objects involved is very far from the earth.

Experimental findings, then, cannot prove laws conclusively. Laws say something about the future and about unexamined circumstances in the past; in this sense they go beyond the evidence. No scientific laws, therefore, can be proved from observations alone. But this does not mean that one law is as good as another or that there can be no scientific knowledge. It simply means that there are degrees of confirmation. The evidence may support—indeed, overwhelmingly support—a particular theory or hypothesis.

Negative results, as we have seen, are more powerful than positive results with respect to proposed universal, natural laws in that they prove that something is wrong somewhere. The argument form

(10) $(H \cdot A) \supset P$

   P

   $\therefore H \cdot A$

is not valid, but

(11) $(H \cdot A) \supset P$

   $\sim P$

   $\therefore \sim(H \cdot A)$

is. If evidence can prove anything conclusively, it seems, it can prove proposed laws false.

Nevertheless, negative results almost never conclusively refute a hypothesis. This is true for three reasons. First, between a hypothesis and falsified predictions are auxiliary assumptions. The negative results show that the hypothesis and auxiliary assumptions are not all true. But they do not locate the source of the falsehood. So, disconfirming tests do not prove that the hypothesis is false. They prove only that it or an auxiliary assumption is false. Recall the reaction of the defenders of the Warren Commission to the Zapruder film, claiming that Governor Connally showed a delayed reaction to being hit by a bullet. Faced with recalcitrant evidence, they reject, not the hypothesis of the lone gunman, but the auxiliary assumption that people react to bullets as soon as those bullets strike.

Second, the negative results might themselves be due to error. The observations may be mistaken. So, in the face of a disconfirming test, we can always deny the accuracy of the test. A well-known acoustic study of the Kennedy assassination, for example, found evidence of four shots, leading a congressional committee to conclude that there was a conspiracy. But others claimed that the acoustic findings were mistaken, the fourth "shot" being due to cross talk between channels rather than an actual gunshot.

Third, the deduction of the predictions from the hypothesis and assumptions may contain a mistake. The hypothesis may not really imply what the test shows to be false. Initial discussions of the Kennedy assassination often assumed that since the shots were fired within six seconds and a single gunman would take roughly three seconds between shots, a single gunman could have fired only two shots. (6 ÷ 3 = 2.) But this ignores the fact that the six-second interval begins with the first shot. So, mistakes happen, in evidence and in reasoning.

In short, faced with the argument

(11) (H · A) ⊃ P

~P

∴ ~(H · A)

there are three ways to save the hypothesis H: deny an auxiliary assumption A, reject the evidence ~P, or deny the conditional (H · A) ⊃ P.

These possibilities are not idle. Mistakes have played an important role in the history of science. Consider first the problem posed by auxiliary assumptions. Tycho Brahe (1546–1601), contemplating the controversy between theories of the solar system devised by Ptolemy (second century A.D.) and Copernicus (1473–1543), devised an interesting test to determine which theory was accurate. If the earth revolved around the sun, as Copernicus claimed, then fixed stars should not remain in the same position in the sky, as judged by a sextant or other navigational device; they should appear at different angles, depending on the position of the earth in its orbit. This phenomenon is called the parallax of the stars. Brahe recorded the stars' positions and found, over the course of six months, that they did not change. He therefore rejected the Copernican hypothesis and tried to combine the Copernican and Ptolemaic approaches.

But the parallax of the stars does not follow from Copernicus's theory alone. Auxiliary assumptions are needed. Brahe had no telescope; the parallax, as he well knew, would be observable only if the stars were fairly close to the earth. He thought they must be close to earth, since they appear as small disks rather than mere points of light. In fact, the stars are very distant. If we were to model the solar system with an orange representing the sun and a pea, about 20 feet from the orange, representing the earth, the nearest star would be about 500 miles away! Stars appear as disks, not points, only because the atmosphere diffuses their light. Brahe's negative result, then, showed that the Copernican theory, together with some auxiliary assumptions, could not be entirely correct. But the problem was in the auxiliary assumptions, not the Copernican theory. The parallax was observed, with a telescope, in 1838.

The possibility of observational error is also real. Reports of observation may err because of an observer's fatigue, distraction, inebriation, incompetence, wishful thinking, or falsification of data. High school students routinely obtain results in chemistry and physics laboratory experiments that contradict current scientific theory. No one takes this to refute modern chemistry or physics; the errors are chalked up to incompetence. Even great scientists, however, have made serious mistakes in observation. Gregor Mendel (1822–1884), who founded modern genetics, counted flowers having various traits to confirm his genetic hypotheses. His counts so precisely conform to his theory's predictions that his observations are unlikely to have been accurate. Some modern astronomers have claimed that Ptolemy's data contain substantial errors. In these cases, the source of the error is not clear.

A researcher may also make a mathematical or logical mistake about what a hypothesis predicts or implies. Newton's theory failed to explain some aspects of the perturbations in the moon's orbit. But this was due to a mathematical error in Newton's work; if the mistake is corrected, the theory's predictions are accurate. (Newton himself found this error, but did not bother to publicize it.)

For all these reasons, then, negative results do not unequivocally refute a hypothesis. The problem may lie in auxiliary assumptions, observations, or deductions. Disconfirming tests definitively indicate the presence of falsehood, but not its location.

P R O B L E M S

What is some confirming evidence for the following widely accepted hypotheses?

1. The earth is round.

2. The earth has a gravitational field.

3. Electricity is a form of energy.

4. The moon revolves around the earth.

5. The earth rotates on its axis.

What is some disconfirming evidence for each of the following false claims?

6. Tomatoes are poisonous.

7. Witches are the main cause of crop failures.

8. Everyone arrested by the police is guilty.

9. High-fat diets are good for you.

10. Misfortunes are punishments for evil deeds.

Discuss the appeals to confirming or disconfirming evidence in the following. What hypothesis do these appeals address? What do they imply about that hypothesis? What auxiliary assumptions might be involved?

11. Ptolemy's astronomy predicted the motions of the stars and planets in the sky fairly successfully but had at least one unfortunate consequence. Ptolemy's theory required that the moon's orbit brought it, sometimes, twice as close to the earth as at other times. It follows that the moon should sometimes appear twice as big as at other times. But, of course, it does not.

12. Ptolemy's theory also held that everything in the heavens revolved around the earth. Galileo, however, discovered the moons of Jupiter, which revolve around Jupiter, not the earth.

13. Newtonian physics held that there were infinitely many "fixed" stars that seemed not to change position, relative to each other, in the sky. But since they exert gravitational attraction for each other, why do they not move closer and eventually collapse?

14. If there were infinitely many "fixed" stars, roughly evenly distributed throughout the universe, every line of sight would end at a star. So, the heavens as a whole would be as bright, from our perspective, as the sun. Obviously, that's not the case.

15. Einstein proposed three tests for the general theory of relativity. First, the theory predicts that light should bend, or be deflected, in the vicinity of massive objects like the sun to a greater extent than Newtonian physics allows. On May 29, 1919, a day that one historian has called the beginning of the modern world, a British expedition in Principe confirmed this experimentally by photographing a solar eclipse. Second, general relativity also implies that Mercury's orbit should turn, its perihelion advancing very slightly with each trip around the sun. Observations have indicated that this prediction is correct. Third, the theory predicts a reddening of light from strong gravitational fields. Observations of "white dwarfs," small, very dense stars, have borne out the theory's prediction.

16. The wave theory of light had been attacked on various grounds. One was that the theory had bizarre consequences. Waves, for example, intersect to form characteristic patterns of interference; if light were passed through two small slits in a screen and light consisted of waves, the result would be a pattern of alternating bands of light and shadow:

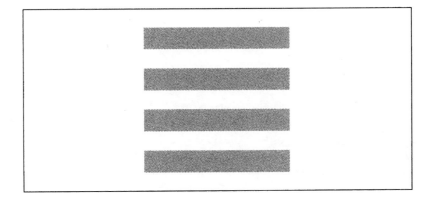

This was thought to be absurd. But Thomas Young did the experiment and found that the pattern was exactly what the wave theory predicted.

17. He hath brought many captives home to Rome,
Whose ransoms did the general coffers fill:
Did this in Caesar seem ambitious?
When that the poor have cried, Caesar hath wept:
Ambition should be made of sterner stuff:
Yet Brutus says he was ambitious;
And Brutus is an honorable man.
You all did see that on the Lupercal
I thrice presented him a kingly crown,
Which he did thrice refuse: was this ambitious? (Shakespeare)

18. The great opportunity for the Jesuits came when an eclipse was expected on the morning of June 21, 1629. The Imperial astronomers predicted that the eclipse would occur at 10:30 and would last for two hours. The Jesuits forecast that the eclipse would not come until 11:30 and would last for only two minutes. On the crucial day, as 10:30 came and went the sun shone in full brilliance. The Imperial astronomers were wrong, but were the Jesuits right? Then, just at 11:30, the eclipse began and lasted for a brief two minutes, as the Jesuits had predicted. (Daniel J. Boorstin)

19. Sir: For the past six months, miniskirts have been coming back into style and look what has happened to the market. Just how high do skirts have to go for the market to act in the traditional fashion? (Alan R. Greenwald)

20. The value of the instrument [the sextant] has been proved in a splendid way by the fact that the distances found by its aid in Kassel agree within a minute, indeed within one half minute with those found by us in Denmark with our sextants. (Tycho Brahe)

 # 12.4   Evaluating Explanations

Explanations are not arguments, but they have a similar form. Explanations and arguments play different roles in communication. Arguments begin with premises believed to be true, from which we seek to infer a conclusion. Explanations begin, in effect, with a conclusion believed to be true—the thing to be explained—and seek premises from which we might infer it. In the order of knowledge, as Aristotle would say, arguments and explanations are opposites. But inference is fundamental to both. We can therefore evaluate explanations using many of the same criteria we use to evaluate arguments.

Recall that arguments should satisfy several conditions. The premises should offer evidence for the conclusion. The argument should be relevant to the issue at hand and assume only what both speaker and audience are willing to accept. The premises should be true. And, finally, the forms of inference used in the argument should be reliable; they should be deductively valid or inductively strong.

Explanations should satisfy similar conditions. We want our explanations to use reliable forms of inference. An explanation should take for granted only what is accepted by the scientific community in general. That is not to say that new assumptions can never arise but that if an assumption is not already common currency, it must be identified and argued for independently. Most important, we want our laws, descriptions of circumstances, and auxiliary assumptions to be true. In scientific practice, because of limitations of measurement and interfering factors such as friction, results are rarely exact; the demand for truth becomes a demand for accuracy.

EVIDENCE

In Chapter 3, we stipulated that the premises of an argument should be more evident than the conclusion. In an explanation, the reverse is true. The thing to be explained is evident; the explanation remains to be postulated. The fact to be explained is often obvious—the president is dead, or the crow is black, for example—while the hypothesis introduced to explain it may be theoretical and complex. Consider, for instance, the falling of an apple. The fact is quite plain, but the laws explaining it— a universal law of gravitation, in Newton's theory, or the curvature of space, in Einstein's—are quite abstract.

This is true of all explanations. Arguments start with premises and try to justify conclusions; explanations start with things to be explained and try to find general principles that explain them. In the order of knowledge, therefore, explanations are the reverse of arguments. The fact to be explained should provide evidence for the generalizations used in the explanation. To do that, the conclusion must be more evident than the generalization. So, we want the conclusion to be more evident than at least one premise. Otherwise, the explanation would be little more than a list of facts.

This criterion rules out a kind of explanation often called "dormitive virtue" explanation. It is so called because of a well-known example in a Molière (1622–1673) play:

> I, a learned scholar, am asked the cause and reason why opium puts me to sleep. To which I reply: Because it possesses a dormitive virtue whose nature it is to make the senses drowsy.

This, plainly, is no explanation. The doctor Molière ridicules tries to explain opium's power to cause sleep in terms of its power to cause sleep. Expressing this explanation as an argument, we obtain

(12) Opium has dormitive virtue.

∴ Opium has the power to cause sleep.

This argument begs the question; the premise possesses the same degree of evidence as the conclusion. Another example of a dormitive virtue explanation would be this:

(13) Cities tend to have high rates of violent crime, because large population centers experience high levels of violence.

The explanation does little more than repeat the fact to be explained.

RELEVANCE

Explanations must be relevant to the issue at hand. This demands more than that the fact to be explained follow from laws, auxiliary assumptions, statements describing the circumstances, and so on. As with arguments, it demands that the explanation be directed at the proper question. What the proper question is depends on context. Just as with arguments, a conversation or context of inquiry contains information about the problem that the argument or explanation addresses. To be relevant, an explanation must address that problem.

Seeing how to violate it may make this requirement clearer. A famous bank robber, Willie Sutton, was asked why he robbed banks. "That's where the money is," he replied. To most of us, that is a peculiar explanation. We want to know why he *robbed* banks, not why he robbed *banks*. So, the explanation addresses the wrong issue. It may nevertheless be a perfectly good explanation in another context. If we imagine that a fellow thief asked Sutton the question, his reply makes good sense. So, an explanation must be relevant; what is relevant depends on the context and, in particular, on the interests of the participants in the conversation or inquiry.[3]

Another example: A professor is found stark naked in the girl's dormitory at midnight. To explain his presence, we say, "He was there, stark naked, one billionth of a second before midnight, and he couldn't put his clothes on or leave without going faster than the speed of light. But nothing travels faster than light." The covering law here—"nothing travels faster than light"—is true, as is the fact that he could not move much in a billionth of a second without going faster than light. And, it is less evi-

---

[3]This and the following example are from Hilary Putnam, *Meaning and the Moral Sciences* (Boston: Routledge and Kegan Paul, 1978), p. 42. He credits the use of the Willie Sutton example in this context to Alan Garfinkel.

dent than the fact to be explained. Together with the information that he was there at a billionth of a second before midnight, the premises entail that he was there, stark naked, at midnight. Nevertheless, the explanation is unhelpful because we want an explanation, not of why he was *still* in the girl's dormitory at midnight, but of why he was there at all. The explanation addresses the wrong problem.

INDUCTIVE CRITERIA

There are other, peculiarly inductive criteria for evaluating explanations. Unlike the first two, these are matters of degree. An explanation does not simply satisfy, or fail to satisfy, these conditions but satisfies them to a greater or lesser degree. The most important are consistency, confirmation, power, and simplicity.

Consistency. An explanation should be consistent, internally and with established results.

Confirmation. An explanation should be confirmed from above and below.

Power. An explanation should lead to conclusions or predictions about more than the case at hand.

Simplicity. An explanation should be uniform and economical.

Confirmation, power, and simplicity, because they are matters of degree, are not absolute requirements of inductive explanations. But, all other things being equal, the more highly confirmed, powerful, and simple an explanation is, the better it is.

CONSISTENCY

Explanations must be consistent, internally and with established results. That means that they must not contain or imply contradictions, and they must not contradict well-confirmed facts or theories. The reason is obvious. We want explanations to be true. An inconsistent theory cannot possibly be true. And since we have good reason to believe well-confirmed statements, we have reason to reject anything that contradicts them.

A famous and ancient example of internal inconsistency is pointed out by Zeno's paradoxes of motion. To get from one place to another, Zeno observed, you must first get halfway there. But before you get to the halfway point you must get halfway to it, and so on. To move at all, therefore, you must complete infinitely many motions. But each motion would take time. So, going anywhere would require an infinite amount of time. No one has *that* much time; so, Zeno concluded, no one can move. The result, of course, is absurd. Zeno showed that the ancient Greeks' theory of motion is inconsistent.

Another ancient paradox, Epimenides' *liar paradox*, shows that our intuitive notion of truth is internally inconsistent. Consider the statement

(14) This statement is false.

It is true if and only if it is false—a contradiction.

More recently, internal inconsistency has struck the foundations of mathematics. German mathematician Georg Cantor (1845–1918) developed set theory in the late nineteenth century. He advanced a principle called the axiom of abstraction, which said that for any property, there is a set of things having the property—that, to put it in different terms, the extension of every term is a set. Bertrand Russell showed that the axiom is inconsistent. Most sets, surely, do not belong to themselves. The set of elephants, for example, is not itself an elephant. The set of red things is not itself red. But one might think that some sets are elements of themselves—the set of all sets, for example (if there were such a thing). So, Russell said, consider the property of not belonging to oneself. According to the axiom of abstraction, there must be a set of all, and only those, sets that do not belong to themselves. Does it belong to itself? It is the set of things that do not belong to themselves, so it can belong to itself only by not belonging to itself—a contradiction. This contradiction, called Russell's paradox, crumbled the foundations of mathematics and spurred a vast amount of work in twentieth-century logic and mathematics.[4]

External consistency—consistency with established results—is also important. Aristotle theorized that the heavenly bodies were unchangeable; this was consistent with established results until the invention of the telescope, which enabled Galileo to discover sunspots. At that point, the theory became untenable. Madame Curie's discovery of radium was consistent with Dmitry Mendeleyev's periodic table of the elements. In fact, the periodic table led scientists to expect a variety of new elements, which were subsequently isolated in the laboratory. Scientists have recently resisted the possibility of cold fusion because it contradicts established theories and results. Consistency with established results is important, but that is not to say that one can never overthrow scientific orthodoxy. The history of science contains many instances of such change, even of scientific revolutions. The point is rather that any hypothesis that contradicts well-established results must have weighty evidence in its favor, being well confirmed by evidence that is reproducible by other scientists.

---

[4]Russell devised a simple, intuitively appealing version of the paradox: Suppose that there is a village with a barber who shaves all, and only those, who do not shave themselves. Who shaves the barber? He shaves himself if and only if he does not shave himself—a contradiction.

CONFIRMATION

Explanations typically employ generalizations and auxiliary assumptions. The more evidence supports these principles, the stronger the explanation. The quantity of evidence is important, but so is its quality and variety. Reliably observing more objects and kinds of objects, in various circumstances, and obtaining positive results add to a principle's confirmation.

The confirmation criterion says that an explanation should be confirmed *from above and below*. An explanation is confirmed to the degree its laws, assumptions, and other principles are confirmed. They may be confirmed as hypotheses are confirmed by confirming tests: Their implications may be confirmed through observation. This is confirmation from below. Principles may also be confirmed from above, by being inferred from other principles with inductive support. A typical law, then, may be confirmed by tests of its implications as well as by being inferred from, and thus explained by, more general laws.

More general laws

$\downarrow$ from above

Law

$\uparrow$ from below

Tests

Consider, for example, the earth's rotation. Initially considered an absurd supposition by Ptolemaic astronomers, the rotation of the earth is now highly confirmed. Much confirmation comes from below: the rising and setting of the sun, stars, and planets; the Coriolis effect (i.e., the deflection of airplanes and projectiles from their otherwise-expected routes); the rotation of free-swinging pendulums; and the trajectories of objects falling from high places. But confirmation from above—from Newtonian physics and Kepler's (1571–1630) laws—is also very important.

The importance of confirmation from below is obvious. But confirmation from above also matters. Without it, principles tend to look weak and *ad hoc*. Consider, for example, the traditional generalization that hemlines correlate with the stock market. We can check this by checking skirt lengths and market averages, but, no matter how good a correlation we find, we are unlikely to count the generalization as a law until we have an explanation of it. The lack of support from above makes us distrust the principle. The moral, then, is that explained principles are, all other things being equal, more trustworthy and better confirmed than unexplained principles.

POWER

The more powerful or fruitful an explanation is, the better it is. That is, all other things being equal, we prefer explanations that have implications for more than the case at hand. The more to which the explanation could apply—the more general it is, in other words—the better.

Explaining the earth's motion by appeal to Newton's theory, for example, is preferable to explaining the same thing by appeal to Kepler's laws. Newton's theory is more general; indeed, it implies a very close approximation of Kepler's laws. It also explains much more: Galileo's law of falling bodies, the orbit of the moon, the paths of comets, the shape of the earth, the wobble of its axis, and the action of its tides. For this reason, a Newtonian explanation is deeper than one in terms of Kepler's laws.

The goal of explanation is understanding. And a more powerful explanation provides a deeper understanding. Explaining the earth's motion by appeal to Newton's theory explains not only why the earth moves as it does but also why other planets obey Kepler's laws and why many other phenomena occur as they do.

To put this another way, Newton's theory is richer and more fruitful than Kepler's. Kepler's laws explain the orbits of the planets: in particular, why those orbits are roughly elliptical. But they have no implications about other phenomena. Newton's theory bears more fruit. It explains much more and has many more observational consequences.

Another factor makes power preferable. Theories that explain other theories, as Newton's explains Kepler's and Galileo's, often show that those theories are true only approximately and in a limited realm. As we have seen, Newton's law of gravitation implies that the gravitational attraction between two objects depends on the masses of the objects and the distance between them. So, it implies that the attraction between the earth and an object depends on its distance from the earth. Galileo's law ignores this dependency. Within the realm of medium-sized objects near the earth's surface, it makes little difference. Outside that realm, the difference can be significant. Thus, Newton's theory not only explains everything Galileo's law does but also shows that the law is a simplification. The same is true of Kepler's laws. Newton's theory shows that they hold only approximately. Each planet is affected by the gravitational pull of the others as well as that of the sun. These other forces distort each planet's orbit slightly.

Einstein's theory of relativity, in turn, shows that Newtonian physics is only an approximation. When velocities are small relative to the velocity of light, relativistic and classical mechanics yield very similar results. At high velocities, however, they differ considerably. More powerful explanations, then, are often more accurate as well as deeper.

SIMPLICITY

The criterion of simplicity says that explanations should be uniform and economical. Given two hypotheses explaining the same class of phenomena, and otherwise equally acceptable, we should prefer the simpler.

The classic formulation of the demand for economy is *Ockham's razor*, so named for the English philosopher William of Ockham (1300–1349): "Entities should not be multiplied beyond necessity." That is, omit unnecessary entities. Explanations should posit entities, kinds, and processes only when necessary. Newton formulated a similar principle: "We are to admit no more causes of natural things than such as are both true and sufficient to explain their appearances." The demand for economy, therefore, is a demand that an explanation appeal only to what it must. In general, the fewer entities, processes, and events an explanation postulates, the better.

Newton also formulated a principle of uniformity: "To the same natural effects we must, as far as possible, assign the same causes." We should prefer explanations that explain the phenomena more uniformly, that attribute more uniformity to nature. To stress the link with Ockham's razor, we might express the demand for uniformity this way: Omit unnecessary *kinds* of entity. More generally still: Omit unnecessary complications.

The Copernican revolution succeeded largely because of the greater simplicity of Copernicus's theory. Ptolemy's astronomy held that the earth was the center of the universe, with everything else revolving around it in a circular orbit. This did not explain the movements of the planets observed in the heavens, so the theory over time became more and more complicated. Ptolemaic astronomers hypothesized that the planets moved in epicycles, that is, circles around points revolving on a circle, at different tilts and with different amounts and directions of eccentricities. The resulting theory was ingenious and quite accurate but immensely complex. Copernicus advanced the radical but much simpler theory that the earth and other planets revolve around the sun in elliptical orbits. Copernicus's theory had various advantages over Ptolemy's, but its simplicity was perhaps the most important.

Simplicity is a way of minimizing the risk of error. Introducing unnecessary things, kinds, and complications into a theory or explanation adds to the possibility of falsehood, with no balancing gain in explanatory power. So, simplicity is a form of safety.

Simple explanations also tend to be more powerful than more complicated explanations. Explanations attributing a higher degree of uniformity explain a range of phenomena with fewer independent hypotheses. Whenever a single explanation can be found for what had been two independent hypotheses, we gain power and insight. Indeed, this sort of increase in uniformity has marked most of the great scientific advances of the past few centuries. Newton's theory, for example, reduces the laws of motion to three. The quest for uniformity is one of science's most important driving principles. Scientists often proceed by considering distinct principles or classes of phenomena and trying to find a single account for all. The search for field theories and theories of quantum gravity in twentieth-century physics has been motivated, in large part, by a desire for uniformity.

Nevertheless, making the criterion of simplicity precise is difficult. What, for example, is the simplest function linking two variables? What is the simplest two-dimensional curve? How can we count the hypotheses a theory or explanation invokes? (Any statement is equivalent to a conjunction of two other statements, and any two statements are equivalent to a single statement, their conjunction.) There are various kinds of simplicity, which further complicates matters.

Nevertheless, simplicity is an important goal of scientific inquiry. Indeed, it is an important goal of thinking in general. Consider a simple numerical progression:

1   4   9

What is the next number? Most will say 16, thinking that the underlying rule is that the nth number is n^2. This is a simple hypothesis. But there are many other possibilities. Perhaps it's all 9s from here on. Perhaps it's n^2 until the 100th place, when the answer becomes 9999 from then on. Perhaps the next numbers are 12, 17, 20, and 25; we add 3 and 5 in alternation to each previous entry. Ordinarily, we do not even consider these possibilities; we assume that there is a simple rule to the progression that does not change.

The assumption of simplicity is not only an important goal but also is fundamental to scientific practice. Suppose we test two variables to determine their relation to each other. We find the following points:

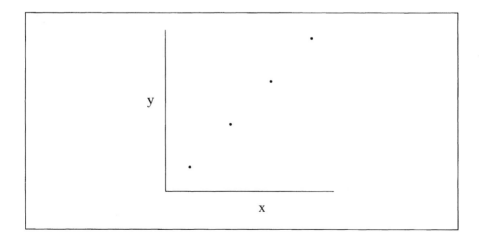

There are many lines that could be drawn between these points:

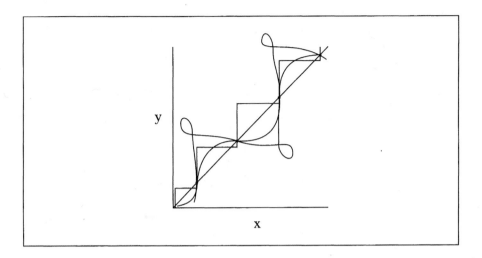

But nobody, in the absence of additional evidence, would draw anything but a straight line.

A LOGICAL EPILOGUE

This book has presented various hypotheses and theories that attempt to explain logical relations between statements. Some of the theories—of definitions, for example, or fallacies or emotive language—have been informal, while others—Aristotelian, propositional, and predicate logic—have been formal. The last few chapters have presented theories of inductive relationships. These hypotheses and theories may be evaluated using the criteria outlined in this section. If they are good theories, they should pass the tests we have laid out for explanations in general.

We think they do—up to a point. Many arguments rely on logical relations that we have not investigated in this book. Some point to further developments in predicate logic:

(15) All horses are animals.

∴ All heads of horses are heads of animals.

Such arguments, known since Aristotle's time, require a theory of relational predicates finally developed in the late nineteenth century that we have not presented here. Other arguments use connectives that are not truth-functional:

(16) If you were to drop out, you'd become unemployable.

You should avoid becoming unemployable.

∴ You shouldn't drop out.

Still others use more sophisticated techniques of inductive reasoning.

Even within the realm we have investigated, the theories we have presented in this book are idealizations. Just as physicists ignore friction in order to present an elegant theory, we have idealized away from some important features of language. We have assumed, for example, that all statements are true or false. But many terms in natural languages are vague (*red,* for example). Statements containing vague terms may seem to be somewhere in between true and false. (Is the setting sun red? Well, *sort of.*) Statements may lack a truth value altogether if they have presuppositions that are not met. In various other ways, what a statement means, and what inferences it licenses, depends on context. We have discussed little of this.

The limitations of our book, however, are not limitations of logic. Logicians have developed theories of the phenomena we have mentioned, and more. Some of those theories are by now well established; others are still controversial, and some are in early stages of development. Beyond the simple logic of our introduction lies a world of logic that extends the basic approaches we have outlined to the full realm of human reasoning.

P R O B L E M S

1. Evaluate, by the criteria of this section, the explanations appearing on pages 452–456.

2. Using the criteria of this section, discuss the three formal logical theories presented in this book: Aristotelian logic, propositional logic, and predicate logic. How do they compare in evidence, relevance, consistency, confirmation, power, and simplicity?

ANSWERS TO
SELECTED PROBLEMS

1. Reasoning,

    1.1 Premises and Conclusions

    3. Argument. Conclusion: The chance of being in a serious accident is nearly one in two.
    6. Argument. Conclusion: Science is necessary for true happiness.
    9. Not an argument, since the sentence "Do not love your neighbor as yourself," is not a statement. (If interpreted as a statement, for example, "You should not love your neighbor as yourself," then it is the conclusion of an argument.)
    12. Not an argument, as stated, since there is no conclusion or relation of intended support, but implies an unstated conclusion, "Everything must be paid for."
    15. Not an argument, as stated, since "Is it not awful to admit that the master cannot live without the slave?" is not a statement. But the question seems rhetorical, that is, tantamount to the statement, "It is awful to admit that the master cannot live without the slave." So interpreted, there is an argument with the conclusion "it is absurd to bring back a runaway slave."
    18. Argument. Conclusion: Culture dominates science.
    21. Not an argument. If the questions are interpreted as rhetorical, there is an argument in the neighborhood with an unstated conclusion: "If I were alive, I wouldn't be in a coffin. If I were dead, I wouldn't have to go to the bathroom. I'm in a coffin, and I have to go to the bathroom. Therefore, I'm neither alive nor dead."
    24. Argument. Conclusion: The pure sciences form a body of propositions with whose genesis experience has nothing to do.

    1.2 Recognizing Arguments

Because discourse relations are often implicit, these are not the only possible correct answers.

    3. Explanation.
    6. Argument. Conclusion: A bad person can't write a good book.
    9. Argument. Conclusion: The statist society can't work.
    12. Argument. Conclusion: Deficits are bad.
    15. Description.
    18. Explanation.

21. Narrative.
24. Explanation and elaboration.
27. Argument. Conclusion: If one sees someone drowning, mauled by beasts, or attacked by robbers, one is obligated to save him.
30. Narrative.
33. Narrative.
36. Narrative.
39. Description.
42. Explanation.
45. Argument. Conclusion: Let us be true to one another!
48. Explanation, illsutration.

1.3 Extended Arguments

Extended arguments are difficult to analyze, for relations of support are often left implicit. These answers are thus not the only justifiable answers.

3. Simple argument, with the conclusion: The "housing crisis" agitating Mayor Koch and Governor Cuomo is actually a product of local attempts to suspend economic law.
6. Simple argument, with the unstated conclusion: "Individual manufacturers must agree on the uniform standards under which computers will 'talk' to one another."
9. Extended argument, consisting of three simple arguments: (1) In 1983 government agents in south Florida seized some six tons of cocaine and 850 tons of marijuana (which tends to come by the boatload). In 1985 the figures were twenty-five and 750 tons respectively. Therefore, seizures of cocaine, a potentially lethal drug, have quadrupled, while seizures of marijuana, a substance that looks benign in comparison, have fallen off. (2) Seizures of cocaine, a potentially lethal drug, have quadrupled. The amounts of drugs seized reflects the amounts coming in. Therefore more cocaine is being imported now than ever before. (3) Seizures of cocaine, a potentially lethal drug, have quadrupled, while seizures of marijuana, a substance that looks benign in comparison, have fallen off. The heightened risk of interdiction has prompted smugglers to favor drugs that are compact and expensive, like cocaine, over drugs that are bulky and relatively cheap, like marijuana. Therefore, the government's strategy in the war on drugs may be partly to blame.
12. Extended argument. (1) The recent changes are little more than a public relations campaign aimed at getting naive Americans to make unilateral concessions that will allow Castro to weather Cuba's current crisis. Castro is no longer seen abroad as a charismatic revolutionary hero, but rather as a ruthless dictator and an abuser of human rights who has ruined Cuba. . . . Therefore, U.S. policy toward Cuba had worked and continues to work.

    (2) If Cuba becomes more open economically and politically, that is good for the United States. And if the Soviets have to keep bankrolling Cuba, that is better than what they believe the more liberal policy would lead to—the American subsidy of a Cuba that remains under Castro's control and militarily allied with the Soviet Union. Therefore, the United States cannot lose.

15. Extended argument. (1) The lady had been quite willing to undergo the wedding ceremony. She had repented of it within a few minutes of returning home. Therefore something had occurred during the morning to cause her to change her mind. (2) Something had occurred during the morning to cause her to change her mind. She could not have spoken to anyone when she was out, for she had been in the company of the bridegroom. Therefore, she might have seen someone. (3) She had spent so short a time in this country that she could hardly have allowed anyone to acquire so deep an influence over her that the mere sight of him would induce her to change her plans so completely. (Earlier, she had been in America.) Therefore, if she had seen someone, it must be someone from America. (4) She might have seen someone. If she had seen someone, it must be someone from America. Therefore, she might have seen an American. (5) Seeing the American caused her to change her mind about the wedding. Therefore, this American possessed much influence over her. (6) This American possessed much influence over her. Therefore, he was a lover or a husband. (7) Her young womanhood had been spent in rough scenes and under strange conditions. There was a man in a pew, a change in the bride's manner, so transparent a device for obtaining a note as the dropping of a bouquet, her resort to her confidential maid, and her very significant allusion to claim-jumping, which in miners' parlance means taking possession of that which another person has a prior claim to. Therefore, she had gone off with a man. (8) That man was the one she had seen. She had seen a lover or a previous husband. Therefore, the man was either a lover or was a previous husband, the chances being in favor of the latter.

18. Simple argument. Conclusion: There is no idea or conception of anything we call infinite.

21. Simple argument. Conclusion: The choice of profession, therefore, is a thing in which great caution is required.

24. Extended argument. (1) By Matter we are to understand an inert, senseless substance, in which extension, figure, and motion do actually subsist. But it is evident from what we have already shown, that extension, figure, and motion are only ideas existing in the mind, and that an idea can be like nothing but another idea. So, neither they nor their archetypes can exist in an unperceiving substance. (2) Neither they nor their archetypes can exist in an unperceiving substance. So, it is plain that the very notion of what is called matter or corporeal substance involves a contradiction in it.

27. Extended argument. (1) If *a* is identical with *b*, whatever is true of the one is true of the other, and either may be substituted for the other in any proposition without altering the truth or falsehood of that proposition. Now George IV wished to know whether Scott was the author of *Waverley;* and, in fact, Scott *was* the author of *Waverley.* So, we may substitute *Scott* for *the* author of *Waverley.* (2) Now George IV wished to know whether Scott was the author of *Waverley;* and, in fact, Scott *was* the author of *Waverley.* We may substitute *Scott* for *the* author of *Waverley.* So, George IV wished to know whether Scott was Scott. (3) Yet an interest in the law of identity can hardly be attributed to the first gentleman of Europe. [So, we have a puzzle; something is wrong here.]

30. Extended argument. (1) It is by the excellence of the eye that we see well. So, the excellence of the eye makes both the eye and its work good. (2) The excellence of the eye makes both the eye and its work good. Similarly, the excellence of the horse makes a horse both good in itself and good at running and at carrying its rider and at awaiting the attack of the enemy. So, if this is true in every case, every virtue or excellence both brings into good condition the thing of which it is the excellence and makes the work of that thing be done well. (3) Every virtue or excellence both brings into good condition the thing of which it is the excellence and makes the work of that thing be done well . So, the virtue of man also will be the state of character which makes a man good and which makes him do his own work well.

33. Extended argument. (1) Moral virtue is a mean. It is a mean between two vices, the one involving excess, the other deficiency. It is such because its character is to aim at what is intermediate in passions and in actions. In everything it is no easy task to find the middle, for example, to find the middle of a circle is not for every one but for him who knows; so, too, any one can get angry—that is, easy—or give or spend money; but to do this to the right person, to the right extent, at the right time, with the right motive, and in the right way, that is not for every one, nor is it easy. So, it is no easy task to be good. (2) It is no easy task to be good. So, goodness is both rare and laudable and noble.

36. Extended argument. (1) The conclusions which reason draws from considering one circle are the same which it would form upon surveying all the circles in the universe. But no man, having seen only one body move after being impelled by another, could infer that every other body will move after a like impulse. So, reason is incapable of any such variation. (2) Reason is incapable of any such variation. So, the hypothesis seems even the only one which explains the difficulty, why we draw, from a thousand instances, an inference that we are not able to draw from one instance, that is, in no respect, different from them. (3) The hypothesis seems even the only one which explains the difficulty, why we draw, from a thousand instances, an inference which we are not able to draw from one instance, that is, in no respect, different from them. So, after the constant conjunction of two objects—heat and flame, for instance, weight and solidity—we are determined by custom alone to expect the one from the appearance of the other. (4) After the constant conjunction of two objects—heat and flame, for instance, weight and solidity—we are determined by custom alone to expect the one from the appearance of the other. So, all inferences from experience, therefore, are effects of custom, not of reasoning.

39. Simple argument. Conclusion: A first and supreme being exists.

42. Extended argument. (1) If a mind could think of something better than You, the creature would rise above its creator and judge its creator, and that is completely absurd. [So, a mind can think of nothing better than You.] (2) In fact, everything else, except You alone, can be thought not to exist. A mind can think of nothing better than You. [It is better not to be able to be thought not to exist than to be able to be thought not to exist.] So, You exist so truly, Lord my God, that You cannot even be thought not to exist. (3) You exist so truly, Lord my God, that You cannot even be thought not to exist.

So, You alone of all things most truly exist. (4) You alone of all things most truly exist. Anything else does not exist as truly, and possesses existence to a lesser degree. Therefore, of all things You possess existence to the highest degree.

45. Extended argument. (1) We observe things springing up and dying away. So, we observe them being and not being. (2) We observe things being and not being. So, we observe in things something that can be, and cannot be. (3) Whatever cannot be, once was not. So, if all things could not be, at one time there was nothing. (4) If all things could not be, at one time there was nothing. Something that does not exist can be brought into being only by something that already exists. So, if all things could not be, there would be nothing even now. (5) If there had been nothing, it would have been impossible for anything to come into being, and there would be nothing now, which is patently false. So, not all things, therefore, are possible but not necessary; something is necessary. (6) Something is necessary. What is necessary may or may not have its necessity caused by something else. It is impossible to go on to infinity in a series of necessary things having a cause of their necessity, just as with any series of causes. It is therefore necessary to posit something that is itself necessary, having no other cause of its necessity, but causing necessity in everything else.

48. Extended argument. (1) Whatever has being must either have a reason for its being or have no reason for it. If it has a reason, then it is contingent. If on the other hand it has no reason for its being in any way whatsoever, then it is necessary in its being. So, any being is contingent or necessary. (2) If a being is contingent, it cannot enter upon being except for some reason which sways the scales in favor of its being and against its not-being. If the reason is also contingent, then there is a chain of contingents linked one to the other. This being which is the subject of our hypothesis cannot enter into being so long as it is not preceded by an infinite succession of beings. But an infinite succession of beings is absurd. So, [if everything is contingent], there is no being at all. (3) [There are beings.] So, something is not contingent. (4) Something is not contingent. But everything is contingent or necessary. So, something is necessary. (4) If a being has a reason for its being, then it is contingent. Something is not contingent. So, something has no reason for its being.

1.4   Validity and Strength

3. Valid.
6. Valid.
9. Valid.
12. Invalid. Suppose that everything is easy, and only logic problems give me headaches. In that circumstance, all logic problems are easy, since everything is; everything that isn't easy gives me a headache (trivially, since there are no such things). Nevertheless, everything that gives me a headache is a logic problem.
15. Valid.
18. Valid.
21. Valid.

1.5   Implication and Equivalence

3. Equivalent.
6. Neither implies the other.
9. (a) implies (b).
12. (b) implies (a).
15. Equivalent.
18. (b) implies (a).
21. Equivalent.
24. Equivalent.
27. Ralph will fail the course.
30. Ralph won't fail the final exam.
33. Nothing.
36. No.
39. Neither follows from nor implies.
42. Equivalent.
45. Equivalent.
48. Neither follows from nor implies.
51. Nothing.
54. Your gross income is less than $3,560.
57. Jeff won't see Mutt first.
60. Yes.
63. Yes. (The invitation implies that if you can come, you should not respond; it says nothing about what to do if you cannot come.)

1.6   Form and Invalidity

3. *Direct:* Things other than persons might have rights to life as well.
   Indirect: Form: If *A* then *B*. So, if not *A*, then not *B*. Example: If your head is chopped off, you will die. So, if your head isn't chopped off, you won't die.
6. *Direct:* These institutions may be analogous in some ways but not as means of social control.
   Indirect: Form: *A* is analogous to *B*. *B*s are *C*. So, *A*s are *C*. Example: Swimming pools are analogous to ponds. Ponds have fish. So, swimming pools have fish.
9. *Direct:* Things can retain their identity even when they change. You cooked the meat, didn't you?
   Indirect: Form: *A* was *B*. *C* is *A*. So, *C* is *B*. Example: Joan was a baby; Joan is the brightest student in class. So, the brightest student in class is a baby. (What *would* follow is that the brightest student in class *was* a baby. Similarly, we could conclude that the meat we ate tonight was raw—but raw *yesterday*, not raw when we ate it.)
12. *Direct:* The arts and inquiries may aim at different goods, so that there is no one thing at which all things aim.
   Indirect: Form: Every *A* relates to some *B*. So, there is a *B* to which all *A* relate. Example: Every person has a mother. So, someone is mother of all people.
15. *Direct:* That each of us wants to be happy ourselves doesn't mean that we collectively want all of us to be happy, or want anything at all, for that matter.

Indirect: Form: Each *A* relates to his *B*. So, the aggregate of *A*s relates to the aggregate of *B*s. Example: Each team wants to win. So, the aggregate of teams (the league?) wants everyone to win.

2. Language

    2.1 Reason and Emotion

Emotional impacts differ; these are not the only acceptable answers.

3. politician: statesman, vote–grubber
6. Emotionally loaded: powerful, action, coolest. (Notice also the italics and boldface type.) Argument: NHL Hockey is fast, powerful, with lots of action. It's cool. So, you ought to watch it.
9. Emotionally loaded: Power, out of it, #1, top, unparalleled, expertise, successful, outstanding, reliability, mission–critical, opportunity, challenging, finesse, innovation. Argument: We're the #1 choice of the best companies. We have unique abilities. Many companies have chosen us as their preferred Web hosting provider. We use Compaq ProLiant™ servers for their performance, reliability, and manageability. Internet commerce is projected to grow from $8 billion now to $327 billion by 2002, on an Internet that will multiply by 100 times. Our commitment is to help you. So, hire us as your Web provider.
12. Emotionally loaded: blood, treasure, matured, wisdom, enlightened, enjoyed, felicity, devoted, amicable, declare, dangerous, peace, safety.
15. Emotionally loaded: disdain, conceal, openly declare, forcible overthrow, tremble, revolution, chains, win.
18. Emotionally loaded: destroy, fair, save their lives, precautions, meagre, haphazard, distressing, cruel, unmanly, swept, ruthlessly, help, mercy, friendly, belligerents, relief, sorely bereaved and stricken, reckless, lack of compassion or of principle.

    2.2 Goals of Definition

3. description
6. description
9. persuasion
12. description
15. persuasion
18. persuasion
21. description
24. description
27. album: book for mounting photographs, stamps, etc.; long-playing record; holder for long-playing record; book consisting mostly of a collection of pictures. pest: insect, animal or person that is a nuisance or troublemaker.
30. wealthy: applied to those who exceed most others in discrimination, taste, and, most importantly, money; or, applied to those who have successfully exploited their fellows for their own aggrandizement. kind-hearted: generous of spirit; or seeing the best while looking at the worst in people.

2.3   Means of Definition

3. description, listing
6. description, synonymy
9. description, analysis
12. persuasion, analysis
15. stipulation or precision, analysis
18. description, analysis
21. persuasion, analysis
24. persuasion, analysis
27. persuasion, analysis
30. description, analysis
33. stipulation, listing
36. description, ostension: Triangle: △
39. description, analysis: Century: period of 100 years
42. stipulation, synonymy: Zick: neurotic
45. precision, listing: World power: United States, United Kingdom, Russia, China
48. persuasion, listing: Ruler: Hitler, Stalin, Lenin, Pol Pot, Ivan the Terrible, . . .
51. New England state: Maine, New Hampshire, Vermont, Massachusetts, Rhode Island, Connecticut.
54. livestock: cattle, sheep, goats, pigs, chickens. rock star: Paul McCartney, Mick Jagger, Madonna, Alanis Morrisset, . . .
57. lunatic: person who is mentally ill. psychologist: person who practices psychology. These definitions are analytic. (Your definitions, of course, may be synonymous.)
60. acorn: seed of an oak. sow: female pig. grandfather: father of a parent. professor: teacher at an institution of higher education. church: organized religious body.

2.4   Criteria for Definitions

3. description, analysis. Figurative.
6. description, analysis. Too narrow? Does virtue require demanding the fullest development from oneself?
9. description, analysis. Figurative.
12. persuasion, analysis. Too narrow; people may elect representatives, none of whom will individually be blamed, or may vote on propositions rather than candidates.
15. persuasion, analysis. Too broad and too narrow. Some journalists have jobs, and some out-of-work reporters do not count as journalists.
18. description, analysis.
21. (a) description, analysis. (b) changed. Too narrow: anecdotes may be published. And too broad: unpublished recipes, philosophical essays, and class notes are not anecdotes.
24. (a) description, analysis. (b) changed. Too narrow; bugs need not stink or be bred in household stuff.
27. (a) description, analysis. (b) unchanged.
30. (a) description, synonymy. (b) changed. Now refers to panic, depression, or earthiness.
33. (a) description, analysis. (b) changed, and now common.

36. (a) description, analysis. (b) unchanged.
39. (a) description, synonymy. (b) unchanged,
42. (a) description, synonymy. (b) unchanged.
45. (a) description, analysis. (b) unchanged, but the definition was always too narrow, since no insect bite is cured by music and nothing else.
48. (a) description, analysis. (b) unchanged.
51. Too broad (bullies may be unjust) and too narrow (justice may require supporting the underdog).
54. Too narrow (one may know things one does not see, hear, etc.) and too broad (illusions show that perception does not always yield knowledge).
57. Too narrow; there are other settings in which courage may be displayed. And too broad; staying and fighting may be based on ignorance or recklessness instead.
60. Too broad and too narrow. Knowing what would be courageous and doing it are two different things.
63. Too broad? What if *A* and *B* have met only once?

3.   Informal Fallacies

     3.1   Fallacies of Evidence

3. Begging the question
6. Begging the question

9. a. Reconstruction was shamefully harsh or surprisingly lenient.
   b. Johnson failed.
   c. Johnson was a miserable bungler or a heroic victim.
   d. The freedman had new responsibilities.
   e. Racial segregation hardened into an elaborate mold.

12. Napoleon III: Enlightened statesman or Proto-Fascist? or reasonably successful leader, fortunate politician, or bumbling reformer?
    The Causes of the War of 1812: National honor or national interest? or a mixture of both, or idealistic crusade, or unavoidable defense?
    The Abolitionists: Reformers or fanatics? or Idealists, or practical revolutionaries, or cynical manipulators?
    Plato: Totalitarian or Democrat? or Constitutional Monarchist, Oligarchist, or Republican?
    John D. Rockefeller—Robber baron or industrial statesman? or strongman, or wealthy innovator, or lucky gambler?
    The Robber barons—Pirates or poneers? or specialists, or captains of industry, or fortunate winners?
    The New Deal—Revolution or evolution? or idealistic reform, or practical adjustment, or experiment?
    Renaissance man—Medieval or modern? or a bit of both, or neither, or a mixture of these and other elements?
    What is history—Fact or fancy? or a bit of both, or theory, or rational guess?

     3.2   Fallacies of Relevance: Credibility

3. ad hominem circumstantial
6. ad hominem abusive
9. ad hominem abusive

12. ad hominem abusive
15. begging the question

3.3   Fallacies of Relevance: Confusion

3. Red herring.
6. Straw man.
9. Straw man.
12. Mr. Fast accuses pro-lifers of a tu quoque; Mr. Buckley accuses him of a red herring.

3.4   Fallacies of Relevance: Manipulation

3. Appeal to common practice.
6. Appeal to the people.
9. Appeal to force.
12. Appeal to the people.
15. Appeal to the people.
18. Complex question; incomplete enumeration; appeal to the people.

3.5   Inductive Fallacies

3. False cause (post hoc ergo propter hoc).
6. Accident.
9. False cause.
12. False cause.
15. Several fallacies hover around this example. The verdict's dependence on a coin flip exemplifies the fallacy of accident. The judge's response may exemplify false cause or ad hominem circumstantial.
18. Misapplication.
21. False cause.
24. Begging the question.
27. Appeal to common practice.
30. False cause (post hoc ergo propter hoc).

3.6   Fallacies of Clarity

Keep in mind throughout the following that there are many ways to eliminate ambiguities.

3. Division.
6. Amphiboly. (Or, equivocation.)
9. Composition.
12. Equivocation.
15. Leibniz assails the objection as relying on equivocation.
18. Accident.
21. Appeal to the people.
24. Hasty generalization or false cause.
27. Appeal to common pratice; tu quoque.
30. al-Ghazali alleges a fallacy of composition.
33. art: skill, craft, creative work, products of creative work
36. funny: humorous, peculiar
39. jog: run slowly, stimulate
42. mind: organ of thought, pay attention to

45. run: go by moving legs rapidly, flow, function, continue
48. well: healthy, source of water, adverbial form of "good"
51. Equivocation on "house-keeper."
54. Equivocation on "the blind"—the blind in general, or the blind umpire?
57. Amphiboly ("want in the worst way" vs. "dance in the worst way").
60. Equivocation ("anti-Communist" as a type of Communist rather than opponent of Communism).
63. Equivocation on "picture-taking."
66. Equivocation on "seats."
69. Equivocation on "violator."
72. Amphiboly: does "his high-school sweetheart" modify "his wife," or is it a separate item on the list?
75. Amphiboly: "Charitable" as modifying "Eye" and "Ear" or "Hospital."
78. Equivocation on "springing up." To eliminate, replace with unanmiguous phrase like "proliferating."
81. Amphiboly. (Also, equivocation on "case.") Is the jury getting a case or getting drunk? To disambiguate, add "a," or replace "drunk driving" with "DWI."
84. Equivocation on "growth": economic or population? Add one of these adjectives to disambiguate.
87. Equivocation on "antique stripper": old stripper or stripper of antiques?
90. Equivocation: Is the complete dope the weatherman or his information?
93. Amphiboly (or equivocation—ellipsis). Do they know no better fowl or know no better than to buy these?

96. a. Someone else generally bets on the Seahawks.
    a. Larry generally has some other relation to the Seahawks.
    b. Larry generally bets on some other team.
    c. Larry doesn't bet on the Seahawks generally.

99. This is confusing, because it literally says the opposite of what Bush apparently means. "Nobody said it would be easy, and nobody was right" literally asserts that nobody said it would be easy; evidently everyone who said anything about it said it would be difficult. But nobody was right, so it must have been easy after all. Yet Bush apparently means to say that it was difficult.

4. Categorical Propositions

3. Term.
6. Term.
9. Term.
12. Not a term.
15. Term.

4.1 Kinds of Categorical Proposition

3. Not categorical.
6. Not categorical.
9. Not categorical.
12. Converts to a particular affirmative categorical proposition: Some dean was a person who was late for the meeting.
15. Converts to a universal affirmative categorical proposition: All my friends are people who showed up.

4.2   Categorical Propositions in Natural Language

3. Some D are B. (D: dogs; B: barking at my cat)
6. All W are V. (W: people who could have written that letter; V: vindictive people)
9. All D are B. (D: people who deserve the fair; B: brave)
12. All N are C. (N: people identical to Nell; C: people who are considering what to do)
15. All T are K. (T: times; K: times when John thinks he knows the answer)
18. No P are F. (P: places; F: places where I can find it)

4.3   Diagramming Categorical Propositions

3. All S are O. (S: sciences; O: things that have been outcasts)

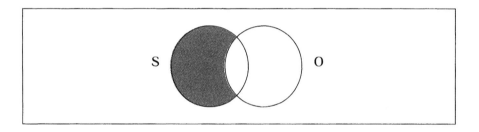

6. No B are M. (B: locations in Britain; M: places more than 65 miles from the sea)

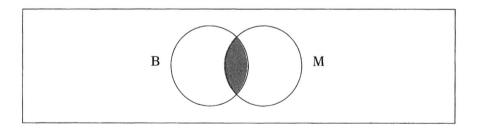

9. All E are N. (E: people with the fame to be an early riser; N: people who may sleep until noon)

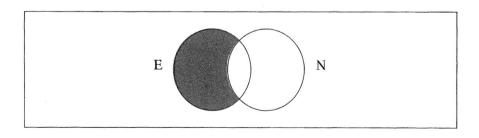

12. No K are T; No T are K. (K: those who know; T: those who tell)

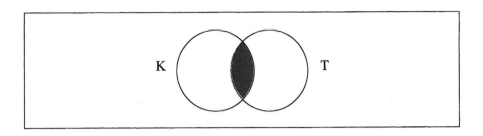

15. All G are W. (G: things that are good or evil; W: things in the will)

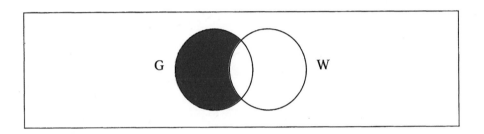

18. All N are S. (N: people who do not judge by appearances; S: shallow people)

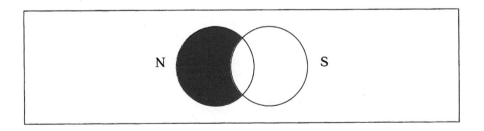

4.4 Immediate Inference

3. Some B are W. (B: books; W: things worth reading twice)
   a. Some W are B. Equivalent.
   b. Some B are not nonW. Equivalent.
   c. Some nonW are nonB.
   d. No B are W.

6. No G are E. (G: Greeks at Thermopylae; E: people who escaped)
   a. No E are G. Equivalent.
   b. All G are nonE. Equivalent.

   c. No nonE are nonG.
   d. Some G are E.

9. Some R are V. (R: Republicans; V: those who voted against the motion)
   a. Some V are R. Equivalent.
   b. Some R are not nonV. Equivalent.
   c. Some nonV are nonR.
   d. No R are V.

12. No M are W. (M: men; W: people who were ever wise by chance)
    ∴ It is not true that some M are W.
    Valid.

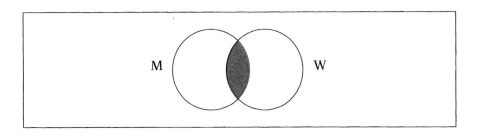

15. All R are D. (R: people who remember; D: people who doubt)
    ∴ All nonD are nonR.
    Valid.

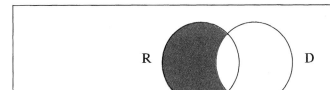

18. No I are R. (I: things they have imagined to do; R: things that will be re-
    strained from them)
    ∴ All I are nonR.
    Valid.

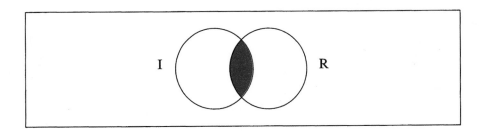

4.5   The Aristotelian Square of Opposition

There are many acceptable answers. The key is that the contrary is stronger than the contradictory, while the subcontrary is weaker. So, any contradiction (such as "It's raining and it's not raining") will work for a contrary; any tautology (such as "It's either raining or not raining") will work for a subcontrary. Below are less trivial answers.

3. (a) Verne likes Clarence. (b) Verne likes Clarence a lot; everyone likes Clarence; etc. (c) Someone doesn't like Clarence; Verne doesn't like someone; Verne doesn't like Clarence very much; etc.

6. (a) It is not the case that Albania and Bulgaria are both Balkan countries. (Or, either Albania or Bulgaria is not a Balkan country.) (b) Neither Albania nor Bulgaria are Balkan countries. (c) Albania is a Balkan country; Bulgaria is a Balkan country; etc.

9. (a) All that glitters is gold. (b) Everything is gold; all that glitters is gold and extremely valuable; etc. (c) Not everything is gold; some things glitter; etc.

12. a. ?
    b. ?
    c. T
    d. ?
    e. ?
    f. F
    g. F
    h. ?
    i. ?

15. a. F
    b. ?
    c. ?
    d. T
    e. ?
    f. ?
    g. F
    h. ?
    i. ?

18. a. T
    b. F
    c. T
    d. F
    e. T
    f. ?
    g. ?
    h. T
    i. F
    j. F

4.6   The Modern Square of Opposition

3. (a) Verne likes Clarence. (b) Verne likes Clarence a lot; everyone likes Clarence; etc. (c) Someone doesn't like Clarence; Verne doesn't like someone; Verne doesn't like Clarence very much; etc.

6. (a) It is not the case that Albania and Bulgaria are both Balkan countries. (Or, either Albania or Bulgaria is not a Balkan country.) (b) Neither Albania nor Bulgaria are Balkan countries. (c) Albania is a Balkan country; Bulgaria is a Balkan country; etc.

9. (a) All that glitters is gold. (b) Everything is gold; all that glitters is gold and extremely valuable; etc. (c) Not everything is gold; some things glitter; etc.

12. a. ?
    b. ?
    c. T
    d. ?
    e. ?
    f. F
    g. F
    h. ?
    i. ?

15. a. F
    b. ?
    c. ?
    d. T
    e. ?
    f. ?
    g. F
    h. ?
    i. ?

18. a) T
    b. ?
    c. ?
    d. F
    e. T
    f. ?
    g. ?
    h. T

i. ?
j. ?

5. Syllogisms

    5.1 Standard Form

   3. Major: H; minor: F; middle: G. 1st figure, EEO. Not valid.
   6. Major: H; minor: F; middle: G. 1st figure, OOO. Not valid.
   9. Major: H; minor: F; middle: G. 2nd figure, EAE. Valid.
  12. Major: H; minor: F; middle: G. 2nd figure, OAO. Not valid.
  15. Major: H; minor: F; middle: G. 3rd figure, AEE. Not valid.
  18. Major: H; minor: F; middle: G. 3rd figure, EOI. Not valid.
  21. Major: H; minor: F; middle: G. 4th figure, AAA. Not valid.
  24. Major: H; minor: F; middle: G. 4th figure, EEA. Not valid.
  27. All D are I; All D are C; ∴ Some C are I. (D: the infinite divisibility of matter; I: incomprehensible things; C: most certain things) 3rd figure, AAI; not valid.
  30. All M are O; All O are F; ∴ Some F are M. (M: miracles of nature; O: ordinary things; F: things that fail to catch our attention) 4th figure, AAI; valid under Aristotelian interpretation only.
  33. No U are H; Some H are P; ∴ Some P are not U. (U: unfortunate people; H: happy people; P: poor people) 4th figure, EIO; valid.

    5.2 Venn Diagrams

   3. No G are H
       No F are G
       ∴ Some F are not H
       Not valid.

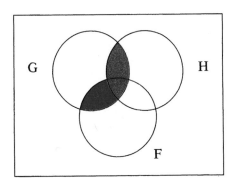

   6. Some G are not H
       Some F are not G
       ∴ Some F are not H
       Not valid.

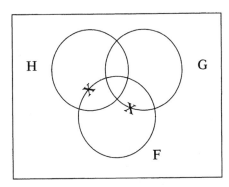

9. No H are G
All F are G
∴ No F are H
Valid.

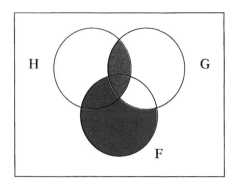

12. Some H are not G
All F are G
∴ Some F are not H
Not valid.

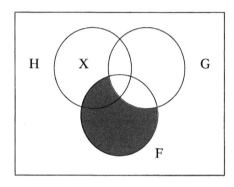

15. All G are H
No G are F
∴ No F are H
Not valid.

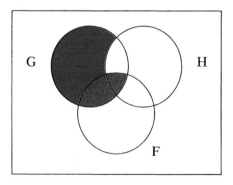

18. No G are H
    Some G are not F
    ∴ Some F are H
    Not valid.

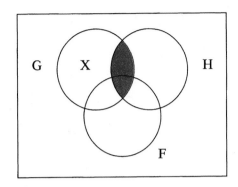

21. All H are G
    All G are F
    ∴ All F are H
    Not valid.

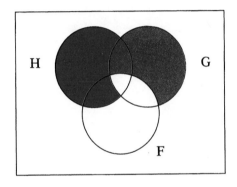

24. No H are G
    No G are F
    ∴ All F are H
    Not valid.

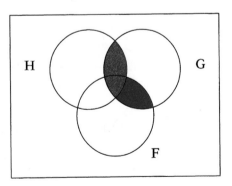

27. All D are I; All D are C; ∴ Some C are I. (D: the infinite divisibility of matter; I: incomprehensible things; C: most certain things.) Not valid.

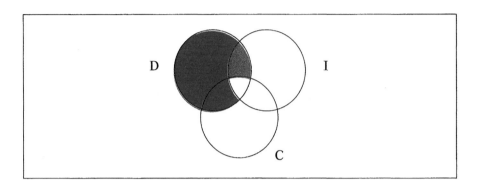

30. All M are O; All O are F; ∴ Some F are M. (M: miracles of nature; O: ordinary things; F: things that fail to catch our attention.) Not valid.

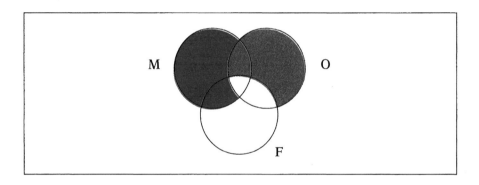

33. No U are H; Some H are P; ∴ Some P are not U. (U: unfortunate people; H: happy people; P: poor people.) Valid.

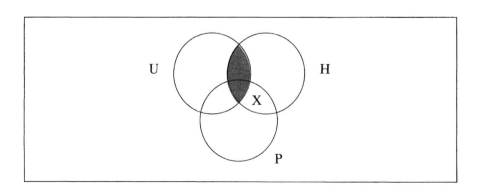

5.3 Distribution

3. Some F^U are G^U
6. Someone I know^U is looking for you^U.
9. Anybody^D could play this game better than Harry^U.
12. Only Republicans^U campaigned on the issue^D.
15. Few Congressmen^D are eager to pursue an investigation^D.
18. Joan^D usually calls before 10^U.

5.4 Rules for Validity

Terms are undistributed unless marked.

3. All G^D are H
   Some F are not G^D
   ∴ Some F are not H^D
   Not valid; violates rule 2.

6. Some G are H
   Some F are not G^D
   ∴ Some F are not H^D
   Not valid; violates rule 2.

9. All H^D are G
   Some F are not G^D
   ∴ Some F are not H^D
   Valid.

12. Some H are not G^D
    No F^D are G^D
    ∴ Some F are H
    Not valid; violates rules 3 and 4.

15. All G^D are H
    Some G are not F^D
    ∴ Some F are not H^D
    Not valid; violates rule 2.

18. Some G are H
    No G^D are F^D
    ∴ Some F are not H^D
    Not valid; violates rule 2.

21. Some G are not H^D
    Some G are F
    ∴ Some F are not H^D
    Not valid; violates rule 1.

24. No H^D are G^D
    Some G are not F^D
    ∴ Some F are not H^D
    Not valid; violates rule 3.

27. All S^D are M; Some B are not S^D; ∴ Some B are not MD. (S: soldiers; M: people who can march; B: babies) Not valid; violates rule 2.
30. All A^D are W; Some B are W; ∴ Some B are A. (A: people who are anxious to learn; W: people who work hard; B: these boys) Not valid; violates rule 1.
33. No M^D are S^D; All M^D are C; ∴ Some C are not S^D. Not valid; violates rule 5.
36. Every syllogism with mood EIO is valid. Proof: A syllogim with mood EIO has this pattern of distribution among its terms:

D	D
U	U
U	D

    No matter what figure the syllogism is in, all rules are satisfied. The middle term must be distributed exactly once, satisfying rule 1. The major is distributed in both occurrences, and the minor is undistributed in both occurrences, satisfying rule 2. There is an affirmative premise (the minor), satisfying rule 3; the major premise and the conclusion are negative, satisfying rule 4. Finally, there is a particular premise and a particular conclusion, satisfying rule 5.
39. The conclusion of every valid third figure syllogism is particular. Proof: Syllogisms in the third figure have the pattern of terms:

G	H
G	F
F	H

    Suppose that the conclusion were universal. Then its subject term, the minor term, would be distributed in both occurrences (by rule 2):

G	H
G	F^D
F^D	H

    But then the minor premise would be negative. By rule 4, the conclusion would have to be negative as well.

G	H
G	F^D
F^D	H^D

    Since H is distributed in the conclusion, it must also be distributed in the premise, by rule 2:

G	H^D
G	F^D
F^D	H^D

    But then both premises are negative, violating rule 3.
42. Every valid syllogism has at least one universal premise. Proof: Say there were a valid syllogism with two particular premises. The conclusion would have to be affirmative or negative. If it were affirmative, then the pattern of distribution of terms would have to be:

By rule 4, if the conclusion is affirmative, both premises must be affirmative. So, the predicate terms in the premises must be undistributed as well.

```
U     U
U     U
─────────
      U
```

But then the middle term cannot be distributed, violating rule 1.
So, suppose the conclusion is negative:

```
U
U
─────
   D
```

By rule 2, the major term cannot be distributed in the conclusion without also being distributed in the major premise:

```
U     D (major term)
U
─────
   D
```

By rule 3, at least one premise must be affirmative, so the predicate term of the minor premise must be undistributed:

```
U     D (major term)
U     U
─────
   D
```

But then the middle term must be undistributed in both occurrences, violating rule 1.

45. No two valid syllogisms with contradictory minor premises have the same major premise. Proof: Suppose there are two such syllogisms. There are two pairs of contradictories: A and O propositions and E and I propositions. So, there are two possibilities to consider.

    First, consider the possibility that the minor premises are A and O propositions. Then the distribution of the terms of the minor premises in the two syllogisms would be:

```
U     D                    D     U
```

The left syllogism has a negative premise, so it must have a negative conclusion, by rule 4, and the other premise must be affirmative, by rule 3:

```
      U
U     D                    D     U
─────
   D
```

But since the major term is distributed in the conclusion, it must be distributed in the major premise as well:

```
D     U (middle)
U     D                    D     U
─────
   D
```

By rule 1, the middle term must be distributed at least once, so the syllogism must be in second figure:

D   U (middle)
U ____ D (middle)   D ____ U
    D

But these two syllogisms have the same major premise, and contradictories have the same terms:

D   U (middle)   D   U (middle)
U ____ D (middle)   D ____ U (middle)
    D

But now the right syllogism cannot be valid, for it violates rule 1, committing the fallacy of undistributed middle.

    Second, then, consider the possibility that the contradictory minor premises are E and I propositions:

D ____ D   U ____ U

The left syllogism has a negative premise, so its conclusion must be negative, while the other premise is affirmative:

    U
D ____ D   U ____ U
    D

But then, as before, the major must be distributed in the premise, and the predicate of the major premise must be the middle term:

D   U (middle)
D ____ D   U ____ U
    D

Since the syllogisms have the same major premise, this holds of the right syllogism as well:

D   U (middle)   D   U (middle)
D ____ D   U ____ U
    D

The right syllogism thus has an undistributed middle term, violating rule 1.

    5.5   Reduction

3. 1. All M are S
   2. Some P are M   /∴ Some S are P
   3. Some P are S   D, 1, 2
   4. Some S are P   C, 3

6. 1. No M are S
   2. All P are M   /∴ No S are P

  3. All M are nonS   O, 1
  4. All P are nonS   B, 3, 2
  5. No P are S   O, 4
  6. No S are P   C, 5

9. 1. No P are M
   2. All S are M   /∴ Some S are not P
   3. No M are P   C, 1
   4. All M are nonP   O, 3
   5. All S are nonP   B, 4, 2
   6. Some S are nonP   S, 5
   7. Some S are not P   O, 6
   (Valid only on the Aristotelian interpretation.)

12. 1. All P are M
    2. No S are M   /∴ Some S are not P
    3. No M are S   C, 2
    4. All M are nonS   O, 3
    5. All P are nonS   B, 4, 1
    6. No P are S   O, 5
    7. No S are P   C, 6
    8. All S are nonP   O, 7
    9. Some S are nonP   S, 8
    10. Some S are not P   O, 9
    (Valid only on the Aristotelian interpretation.)

15. 1. No N are R
    2. All B are N   /∴ No R are B
    3. All N are nonR   O, 1
    4. All B are nonR   B, 3, 2
    5. No B are R   O, 4
    6. No R are B   C, 5

18. 1. All W are R
    2. Some E are not R   /∴ Some E are not R
    3. No W are nonR   O, 1
    4. No nonR are W   C, 3
    5. All nonR are nonW   O, 4
    6. Some E are nonR   O, 2
    7. Some E are nonW   D, 5, 6
    8. Some E are not W   O, 7

21. 1. All H are G
    2. All F are nonG   /∴ All F are nonH
    3. No H are nonG   O, 1
    4. No nonG are H   C, 3
    5. All nonG are nonH   O, 4
    6. All F are nonH   B, 5, 2

24. 1. No H are nonG
    2. Some F are not G   /∴ Some F are nonH

   3. No nonG are H   C, 1
   4. All nonG are nonH   O, 3
   5. Some F are nonG   O, 2
   6. Some F are nonH   D, 4, 5

27. 1. Some nonG are H
   2. All nonG are F   /∴ Some F are H
   3. Some H are nonG   C, 1
   4. Some H are F   D, 2, 3
   5. Some F are H   C, 4

30. 1. All nonH are nonG
   2. All F are G   /∴ All F are H
   3. No nonH are G   O, 1
   4. No G are nonH   C, 3
   5. All G are H   O, 4
   6. All F are H   B, 5, 2

33. 1. All nonI are nonG
   2. No G are nonW   /∴ Some W are I
   3. All G are W   O, 2
   4. No nonI are G   O, 1
   5. No G are nonI   C, 4
   6. All G are I   O, 5
   7. Some G are I   S, 6
   8. Some I are G   C, 7
   9. Some I are W   D, 3, 8
  10. Some W are I   C, 9
   (Valid only on the Aristotelian interpretation.)

36. 1. No nonC are nonF
   2. No L are C   /∴ All L are F
   3. All L are nonC   O, 2
   4. All nonC are F   O, 1
   5. All L are F   B, 4, 3

39. 1. All P are H
   2. No B are nonP   /∴ No B are nonH
   3. All B are P   O, 2
   4. All B are H   B, 1, 3
   5. No B are nonH   O, 4

42. 1. No B are C
   2. No nonB are G   /∴ No C are G
   3. No C are B   C, 1
   4. All C are nonB   O, 3
   5. All nonB are nonG   O, 2
   6. All C are nonG   B, 5, 4
   7. No C are G   O, 6

6. Propositional Logic

    6.1 Connectives

    3. truth–functional:

p	q	p, BUT q
T	T	T
T	F	F
F	T	F
F	F	F

    6. not truth–functional:

p	JOHN REALIZES THAT p
T	?
F	F

    If p is true, John may or may not realize it.

    9. not truth–functional:

p	ALLEGEDLY p
T	?
F	?

    Whether p is true is independent of whether p has been alleged to be true.

    12. not truth–functional:

p	IT IS PROBABLE THAT p
T	?
F	?

    Improbable things can happen.

    15. truth–functional:

p	q	p, NEVERTHELESS q
T	T	T
T	F	F
F	T	F
F	F	F

    6.2 Truth Functions

    3. $(\sim p \vee q)$
    6. $\sim(\sim p \cdot \sim q)$
    9. $((p \supset q) \cdot (q \supset p))$
    12. $p \downarrow p$
    15. $((p \cdot q) / (p \cdot q))$
    18. $((p / p) / (q / q))$

6.3   Symbolization

3. A · C
6. ~G ⊃ A
9. A · ~(M · D)
12. ~A ⊃ ~R
15. ~A · H
18. ~F · L (F: The trouble is that there is too many fools; L: The trouble is that the lightning ain't distributed right); alternatively, ~M · ~D (M: There is too many fools; D: The lightning is distributed right)
21. P · H (P: The idea that all wealth is acquired through stealing is popular in prisons; H: The idea that all wealth is acquired through stealing is popular at Harvard)
24. D · ~F (D: Ambition has its disappointments to sour us; F: Ambition has the good fortune to satisfy us)
27. O ⊃ ~C (O: Other people are going to talk; C: Conversation remains possible)
30. ~S · B (S: I am sure what the unpardonable sin is; B: I believe it is a disposition to evade the payment of small bills)

6.4   A Symbolic Language

3. a; ∨
6. d
9. d
12. d
15. b; ∨

6.5   Logical Properties of Statements

3. Tautologous
6. Contingent
9. Contingent
12. Tautologous
15. Tautologous
18. Tautologous
21. Tautologous
24. Tautologous
27. Contingent, though it could be considered contradictory if one takes it as tautologous that everyone knows him/herself.
30. This appears contradictory—being tan is not the kind of thing that can be heard—but could be seen as contingent, if one can be so tan that one's skin crackles.
33. This sounds contradictory, but it is better viewed as contingent, for two reasons: (a) it is a conditional—if the antecedent and consequent contradict each other, as in A ⊃ ~A, the whole is not contradictory but contingent (in fact, equivalent to ~A or, in this case, the contention that there is not one major cause for the spread of mass illiteracy); (b) In speaking of mass illiteracy, de Vries evidently means to imply not that many people cannot read or write, but that they cannot do so well.
36. (a) a tautology (b) contingent or a tautology (c) satisfiable (d) nothing.

39. True: If A were tautologous, and B followed from A, then B would have to be tautologous as well. If B were contingent, it would be possible for B to be false; it would thus be possible for B to be false while A was true, since A is always true.

42. True: Say that A is a tautology. Then A is always true. So, A is always true when B is, no matter what B is. Thus, A follows from any B.

6.6   Truth Tables for Statements

3. F
6. T
9. F
12. F
15. T

18.

P	$P \equiv {\sim}P$
T	T F FT
F	F F TF

Contradictory

21.

P	Q	$P \supset (Q \supset P)$
T	T	T T T T T
T	F	T T F T T
F	T	F T T F F
F	F	F T F T F

Tautologous

24.

P	Q	${\sim}P \supset (Q \supset P)$
T	T	FT T T T T
T	F	FT T F T T
F	T	TF F T F T
F	F	TF T F T F

Contingent

27.

P	Q	$(P \supset Q) \cdot {\sim}(Q \supset P)$
T	T	T T T F FT T T
T	F	T F T F FF T T
F	T	F T T T TT F F
F	F	F T F F FF T F

Contingent

30.

P	Q	$(P \vee Q) \equiv \sim(\sim P \cdot \sim Q)$
T	T	T T T T T FT F FT
T	F	T T F T T FT F TF
F	T	F T T T T TF F FT
F	F	F F F T F TF T TF

Tautologous

33.

P	Q	R	$(P \vee (Q \supset R)) \equiv ((P \vee Q) \supset (P \vee R))$
T	T	T	T T T T T T T T T T T T T
T	T	F	T T T F T T T T T T T T F
T	F	T	T T F T F T T T F T T T T
T	F	F	T T F T F T T T F T T T F
F	T	T	F T T T T T F T T T F T T
F	T	F	F F T F F T F T T F F F F
F	F	T	F T F T T T F F F T F T T
F	F	F	F T F T F T F F F T F F F

Tautologous

36.

P	Q	$((P \supset \sim Q) \cdot P) \supset \sim(P \supset Q)$
T	T	T F FT F T T FT T T
T	F	T T TF T T T TT F F
F	T	F T FT F F T FF T T
F	F	F T TF F F T FF T F

Tautologous

39.

P	Q	R	$((P \supset (Q \supset R)) \cdot (Q \supset \sim R)) \supset \sim P$
T	T	T	T T T T T F T F FT T FT
T	T	F	T F T F F F T T TF T FT
T	F	T	T T F T T F F T FT T FT
T	F	F	T T F T F T F T TF F FT
F	T	T	F T T T T F T F FT T TF
F	T	F	F T T F F T T T TF T TF
F	F	T	F T F T T T F T FT T TF
F	F	F	F T F T F T F T TF T TF

No; contingent

42. a) contingent; (b) satisfiable; (c) not tautologous; (d) satisfiable; (e) satisfiable; (f) nothing
45. a) contingent; (b) satisfiable; (c) not tautologous; (d) satisfiable; (e) satisfiable; (f) nothing

6.7 Truth Tables for Symbolic Arguments

3. M: I am mistaken; F: I am a fool. M ∨ F; F ⊃ M; ∴ M. Valid.

M	F	M ∨ F	F ⊃ M	M
T	T	T T T	T T T	T
T	F	T T F	F T T	T
F	T	F T T	T F F	F
F	F	F F F	F T F	F

6. A: The members of this department agree; S: The issue at hand is significant; E: The end of the earth is at hand. A ≡ ~S; S · A; ∴ E. Valid.

A	S	E	A ≡ ~S	S · A	E
T	T	T	T F FT	T T T	T
T	T	F	T F FT	T T T	F
T	F	T	T T TF	F F T	T
T	F	F	T T TF	F F T	F
F	T	T	F T FT	T F F	T
F	T	F	F T FT	T F F	F
F	F	T	F F TF	F F F	T
F	F	F	F F TF	F F F	F

9. C: Something is conceivable as greater than God; I: God exists in the imagination; R: God exists in reality. ~C; (I · ~R) ⊃ C; ∴ I ⊃ R. Valid.

C	I	R	~C	(I · ~R) ⊃ C	I ⊃ R
T	T	T	FT	T F FT T T	T T T
T	T	F	FT	T T TF T T	T F F
T	F	T	FT	F F FT T T	F T T
T	F	F	FT	F F TF T T	F T F
F	T	T	TF	T F FT T F	T T T
F	T	F	TF	T T TF F F	T F F
F	F	T	TF	F F FT T F	F T T
F	F	F	TF	F F TF T F	F T F

12.

p	q	p · q	Valid
T	T	T T T	
T	F	T F F	
F	T	F F T	
F	F	F F F	

15.

p	q	p ⊃ q	~q	~p	Valid
T	T	T T T	FT	FT	
T	F	T F F	TF	FT	
F	T	F T T	FT	TF	
F	F	F T F	TF	TF	

18.

p	q	r	p ⊃ q	q ⊃ r	p ⊃ r	Valid
T	T	T	T T T	T T T	T T T	
T	T	F	T T T	T F F	T F F	
T	F	T	T F F	F T T	T T T	
T	F	F	T F F	F T F	T F F	
F	T	T	F T T	T T T	F T T	
F	T	F	F T T	T F F	F T F	
F	F	T	F T F	F T T	F T T	
F	F	F	F T F	F T F	F T F	

21. We may symbolize the motion made by Representative McClory, "to post-pone for ten days further consideration of whether sufficient grounds exist for the House of Representatives to exercise constitutional powers of im-peachment unless by 12 noon, eastern daylight time, on Saturday, July 27, 1974, the president fails to give his unequivocal assurance to produce forth-with all taped conversations subpoenaed by the committee. . . ." as P ∨ ~A, where P: Further consideration is postponed or ten days; A: The president gives his unequivocal assurance to produce all subpoenaed tapes. Repre-sentative Latta summarizes this as ". . . you get ten days providing the pres-ident gives his unequivocal assurance to produce the tapes by tomorrow noon," which we may symbolize as A ⊃ P. These are equivalent:

P	A	P ∨ ~A	A ⊃ P
T	T	T T FT	T T T
T	F	T T TF	F T T
F	T	F F FT	T F F
F	F	F T TF	F T F

For the following, let P: A fetus is a person; R: A fetus has a right to life.

24.

P	R	P ⊃ R	P
T	T	T T T	T
T	F	T F F	T
F	T	F T T	F
F	F	F T F	F

Neither follows from nor implies.

27.

P	R	P ⊃ R	R ⊃ P
T	T	T T T	T T T
T	F	T F F	F T T
F	T	F T T	T F F
F	F	F T F	F T F

Neither follows from nor implies.

30.

P	R	P ⊃ R	~R ⊃ ~P
T	T	T T T	FT T FT
T	F	T F F	TF F FT
F	T	F T T	FT T TF
F	F	F T F	TF T TF

Equivalent; follows from and implies.

33.

P	R	P ⊃ R	~R ∨ P
T	T	T T T	FT T T
T	F	T F F	TF T T
F	T	F T T	FT F F
F	F	F T F	TF T F

Neither follows from nor implies.

36.

P	R	P ⊃ R	R ∨ ~P
T	T	T T T	T T FT
T	F	T F F	F F FT
F	T	F T T	T T TF
F	F	F T F	F T TF

Equivalent; follows from and implies.

39.

P	Q	~(P · Q)	P	~Q
T	T	FT T T	T	FT
T	F	TT F F	T	TF
F	T	TF F T	F	FT
F	F	TF F F	F	TF

Valid.

42.

P	Q	~(P · Q)	~P	Q
T	T	FT T T	FT	T
T	F	TT F F	FT	F
F	T	TF F T	TF	T
F	F	TF F F	TF	F

Not valid.

45.

P	Q	P ⊃ Q	P ⊃ ~Q	~P
T	T	T T T	T F FT	FT
T	F	T F F	T T TF	FT
F	T	F T T	F T FT	TF
F	F	F T F	F T TF	TF

Valid.

48.

Q	P	~(Q · P)	~Q · ~P
T	T	FT T T	FT F FT
T	F	TT F F	FT F TF
F	T	TF F T	TF F FT
F	F	TF F F	TF T TF

~Q · ~P implies ~(Q · P).

51.

Q	P	Q · P	(Q ∨ P) · (Q ⊃ P)
T	T	T T T	T T T T T T T
T	F	T F F	T T F F T F F
F	T	F F T	F T T T F T T
F	F	F F F	F F F F F T F

Q · P implies (Q ∨ P) · (Q ⊃ P).

54.

P	P ∨ P
T	T T T
F	F F F

57.

Q	P	~(Q ∨ P)	~Q · ~P
T	T	F T T T	FT F FT
T	F	F T T F	FT F TF
F	T	F F T T	TF F FT
F	F	T F F F	TF T TF

60.

P	Q	P ⊃ Q	P ⊃ (P · Q)
T	T	T T T	T T T T T
T	F	T F F	T F T F F
F	T	F T T	F T F F T
F	F	F T F	F T F F F

63.

Q	P	Q ≡ P	(Q ⊃ P) · (P ⊃ Q)
T	T	T T T	T T T T T T T
T	F	T F F	T F F F F T T
F	T	F F T	F T T F T F F
F	F	F T F	F T F T F T F

66. False: If a symbolic argument has a contradictory premise, the premises could never be all true while the conclusion is false. So, it must be valid.

69. True: A satisfiable formula can be true, and anything that follows from it will be true when it is.

72. True: A tautology is always true, and anything that follows from it is true when it is and, so, must also be always true.

75. False: Consider P and ~P.

78. True: Suppose P implies ~P. Then, if P were ever true, it would also be false, which is impossible. So, P can never be true.

7. Semantic Tableaux

7.1 Motivation

3. M: I'm mistaken; F: I'm a fool.

M	F	M ∨ F	F ⊃ M	M	
F	T	F T T	T F F	F	Valid

6. A: The members of this department agree; S: The issue at hand is significant; E: The end of the earth is at hand.

A	S	E	A ≡ ~S	S · A	E	
T	T	F	T F FT	TTT	F	Valid

9. C: Something can be conceived as greater than God; I: God exists in our imaginations; R: God exists in reality.

C	I	R	~C	(I · ~R) ⊃ C	I ⊃ R	
F	T	F	TF	TT TF F F	T F F	Valid

12.

p	q	p · q
T	T	TTT

15.

p	q	p ⊃ q	~q	~p
T	F	T FF	TF	FT

18.

p	q	r	p ⊃ q	q ⊃ r	p ⊃ r
T	T	F	T TT	T FF	T F F
T	F	F	T FF	F TF	T F F

We may symbolize the motion made by Representative McClory, "to postpone for ten days further consideration of whether sufficient grounds exist for the House of Representatives to exercise constitutional powers of impeachment unless by 12 noon, eastern daylight time, on Saturday, July 27, 1974, the president fails to give his unequivocal assurance to produce forthwith all taped conversations subpoenaed by the committee. . . ." as P ∨ ~A, where P: Further consideration is postponed or ten days; A: The president gives his unequivocal assurance to produce all subpoenaed tapes.

21.

P	A	P ∨ ~A	A ⊃ P	
F	T	F F FT	T F F	Equivalent

24.

P	R	P ⊃ R	P	
F	F	F T F	F	
T	F	T F F	T	Independent

27.

P	R	P ⊃ R	R ⊃ P	
T	F	T F F	F T T	
F	T	F T T	T F F	Independent

30.

P	R	P ⊃ R	~R ⊃ ~P	
T	F	T F F	TF F FT	Equivalent

7.3 Negation, Disjunction, and Conjunction

3.

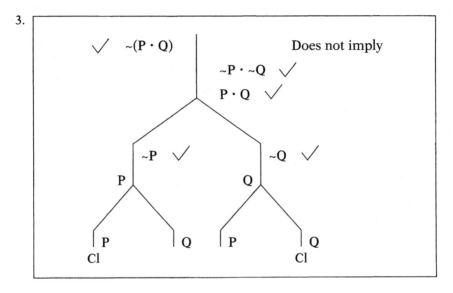

6.

9.

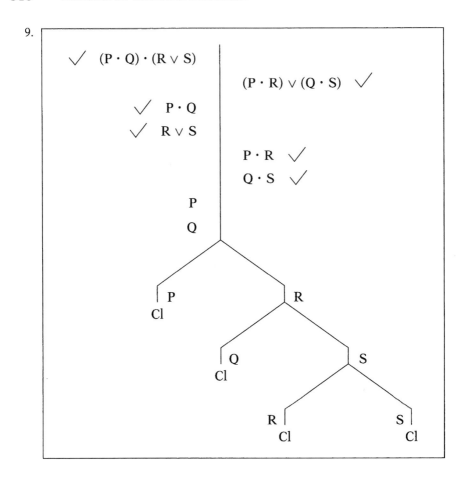

12. G: A man serves God; M: A man serves Mammon; S: A man starves; U: A man survives.

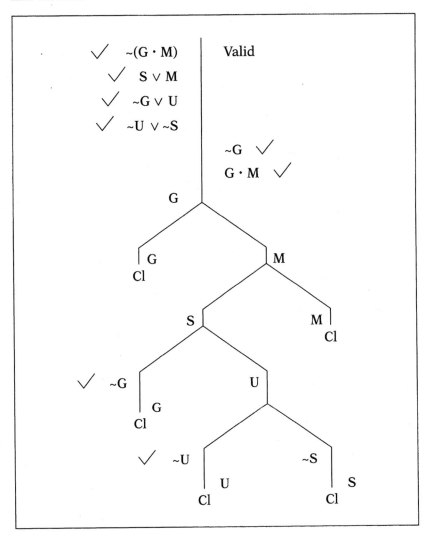

15. M: Laura is in love with Mel; N: Laura is in love with Ned; L: Ned loves Laura; O: Ned loves Olive; S: Olive is miserable; H: Laura is happy.

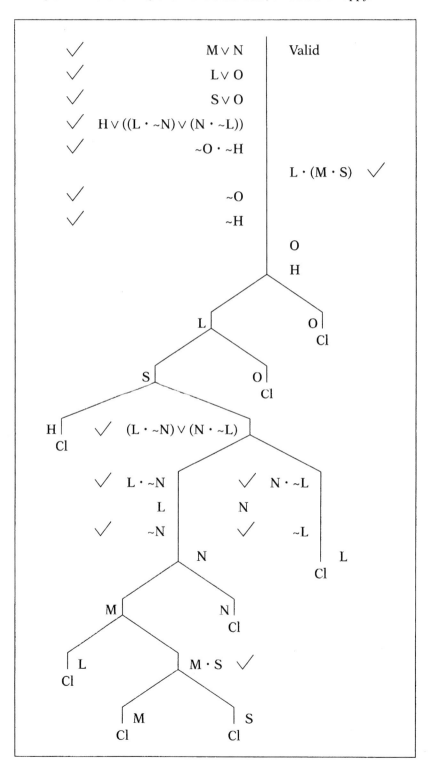

7.5   The Conditional and biconditional

3.

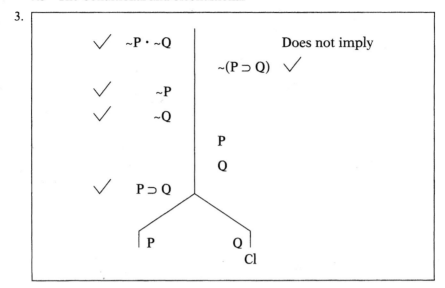

6.

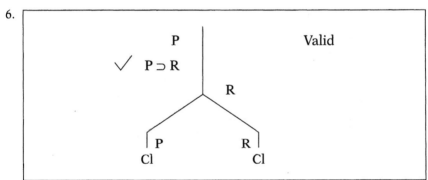

9.

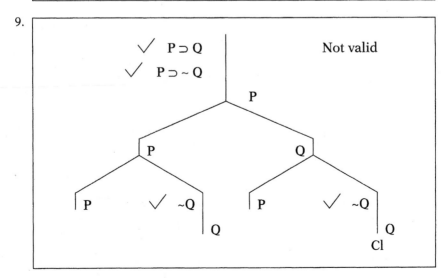

12.

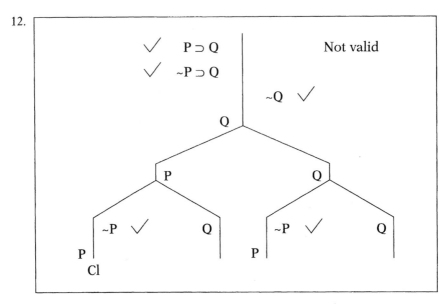

15.

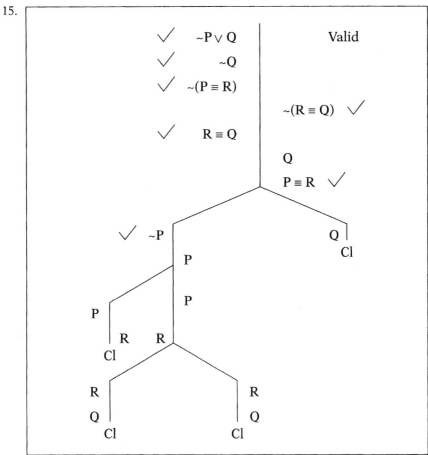

18.

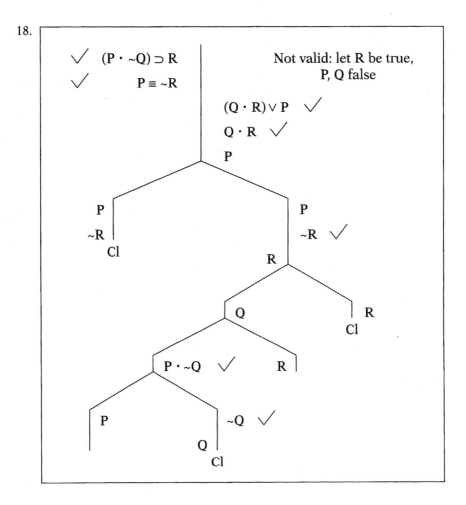

21. Not valid; let P be false, and Q and R have opposite truth values

24.

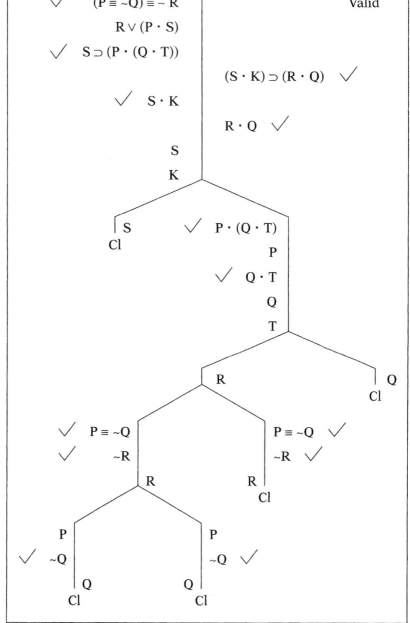

27. P: God is all–powerful; R: God can make a rock so heavy he cannot lift it. Valid.

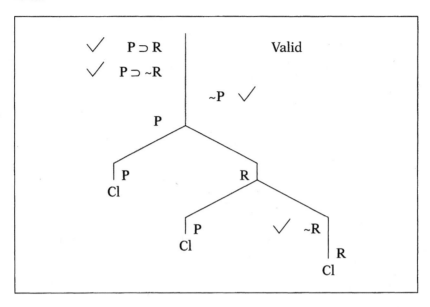

30. J: Johnson leaves the firm; N: Jones leaves the firm; R: Robinson leaves the firm. Valid.

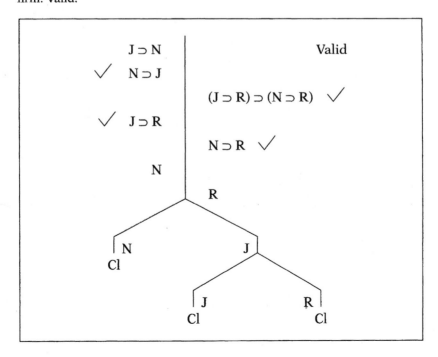

33. C: My cat will sing opera; L: The lights are out; I: I am very insistent; H: I howl at the moon. Valid.

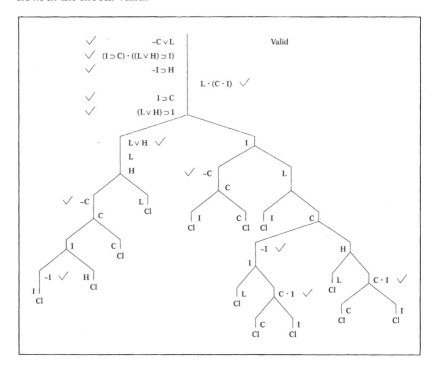

36. A: John attends the meeting; C: John will convince at least a few people to vote against the development proposal; S: Most of the neighborhood signs a petition; D: The developers change their proposal. Valid.

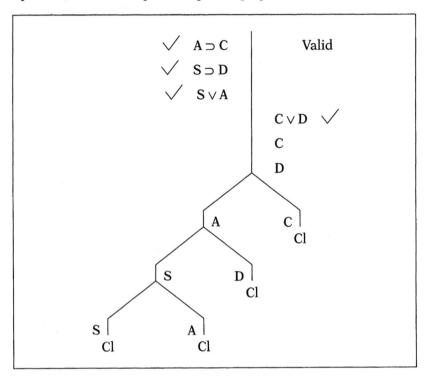

39. P: The party maintains its current economic policy; C: There is a flight of
    capital to other countries; T: The party tightens its control over the econ-
    omy; F: The nation will have to pay large amounts of foreign debt in hard
    currency. Not valid: T true, P, F false.

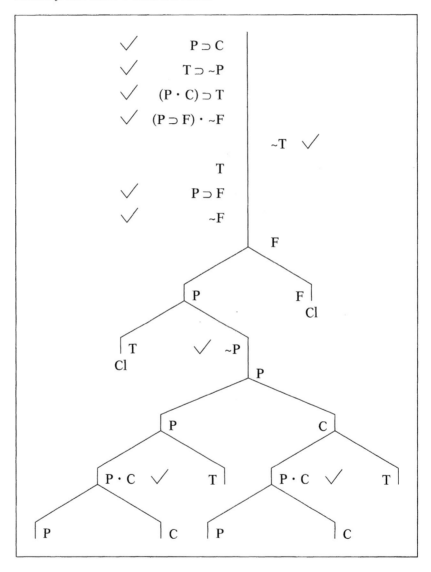

42. A: A ship is an American vessel; N: It is numbered in the U.S.; R: It is registered in the U.S.; F: It is registered in a foreign country; C: Its crew members are all U.S. citizens; E: Its crew members are all employees of corporations based in the U.S.

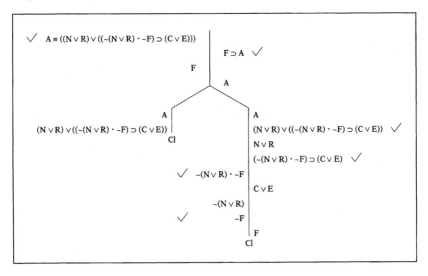

7.6 Other Applications

3.

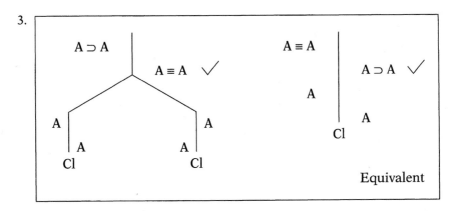

6.

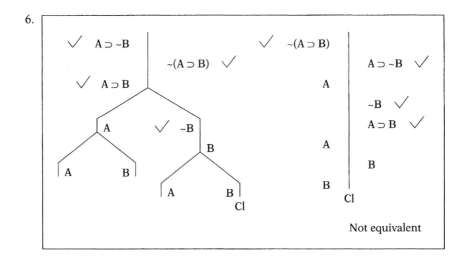

9.

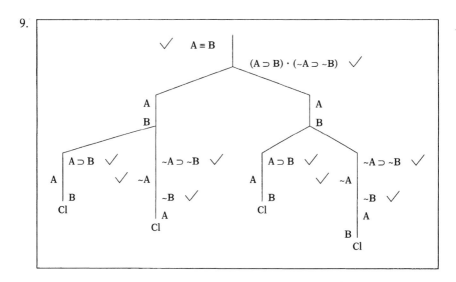

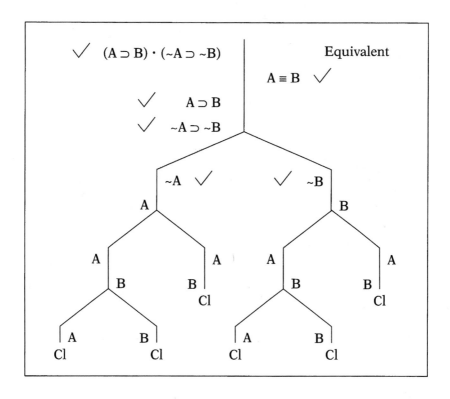

12.

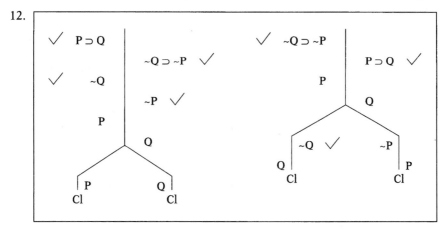

15.

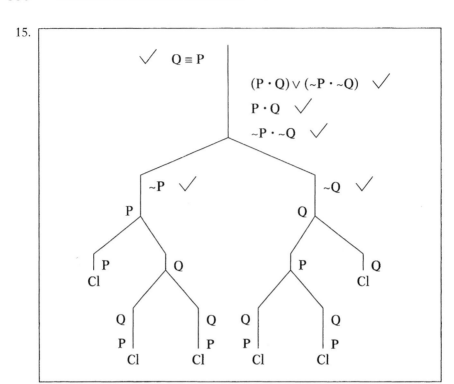

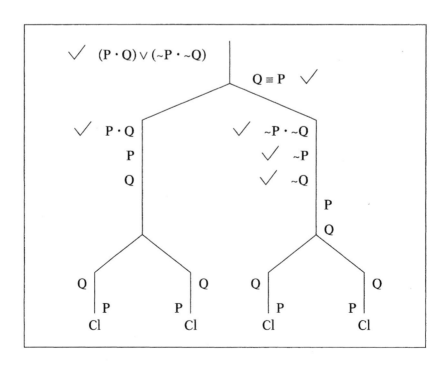

18.

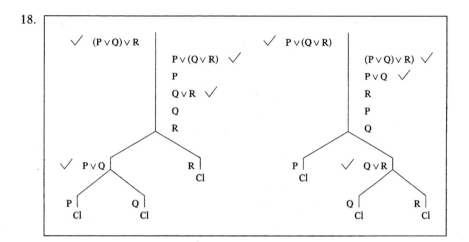

21. Tautologous:

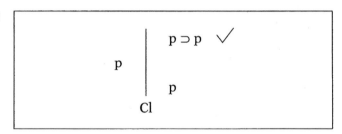

24. Contradictory:

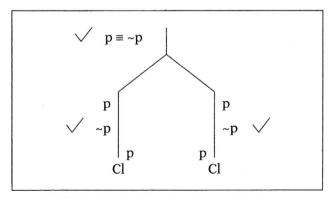

27. Contradictory:

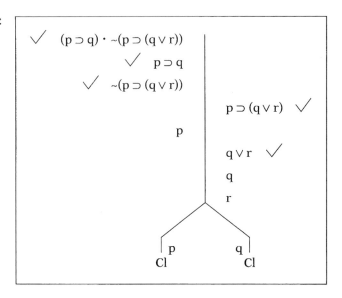

30. X: Xenia will graduate; I: She improves her GPA. Equivalent.

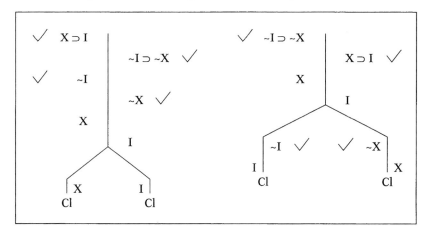

33. S: Sandra leaves by 11:00; H: Harold leaves by 11:00; I: I'll be amazed. Not equivalent; (b) implies (a).

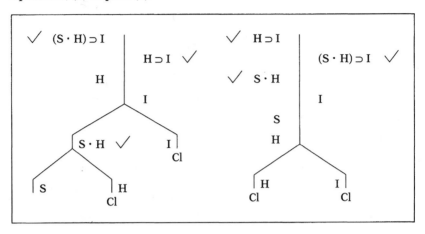

Throughout the following, P, J, C, M, and L represent that Punch, Judy, Curly, Moe, and Larry, respectively, are knights. Their negations represent that they are knaves. In a few cases, the tableaux are too long to present here; I have simply given the conclusions in those cases.

36.

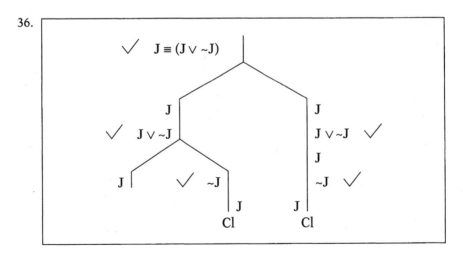

Judy is a knight.

39.

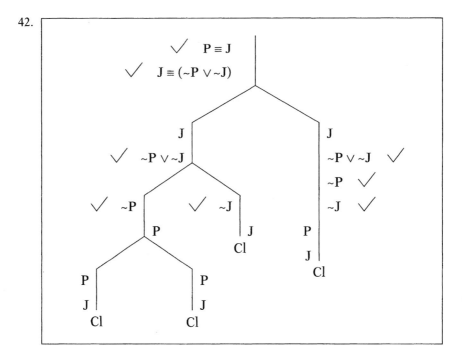

Punch and Judy are both knights.

42.

This is an impossible situation. It violates the rules of the land of knights and knaves.

45.

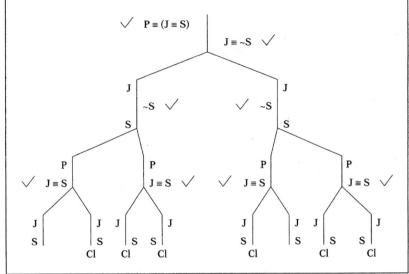

Punch is a knight. Judy and her sister are both knaves or both knights.

48. Curly, Moe, and Larry are all knights.

8. Proof

   8.2 Rules of Implication: Conjunctions and Conditionals

3. 1. ~A ⊃ B
   2. ~A · C   / ∴ C · B
   3. ~A   Simp, 2
   4. C   Simp, 2
   5. B   MP, 1, 3
   6. C · B   Conj, 4, 5

6. 1. A ⊃ (B ⊃ C)
   2. A · ~C   / ∴ ~B
   3. A   Simp, 2
   4. ~C   Simp, 2
   5. B ⊃ C   MP, 1, 3
   6. ~B   MT, 5, 4

9. 1. A ⊃ B
   2. ~A ⊃ ~C
   3. ~C ⊃ ~D
   4. ~B   / ∴ ~D
   5. ~A   MT, 1, 4
   6. ~C   MP, 2, 5
   7. ~D   MP, 3, 6

12. 1. D ⊃ G
    2. C · D   / ∴ G
    3. D   Simp, 2
    4. G   MP, 1, 3

    G: Georgia will lose the case; D: The court bases its decision on *Davis;*
    C: The composition of the court is more conservative than it was a few
    years ago.

15. 1. F ⊃ ~A
    2. F ⊃ ~P
    3. ~A ⊃ V
    4. G · F   / ∴ V · ~P
    5. F   Simp, 4
    6. F ⊃ V   HS, 1, 3
    7. ~P   MP, 2, 5
    8. V   MP, 6, 5
    9. V · ~P   Conj, 8, 7

    F: This product will fail; A: It is given adequate advertising support; The
    company will show a profit for the year; V: The vice president believes that
    another product has more potential; G: The product is great.

18. 1. (A ⊃ B) · (~A ⊃ C)
    2. ~B   / ∴ C
    3. A ⊃ B   Simp, 1
    4. ~A ⊃ C   Simp, 1
    5. ~A   MT, 3, 2
    6. C   MP, 4, 5

8.3   Rules of Implication II: Disjunction

3. 1. r ∨ p
   2. ~r · s   / ∴ p ∨ q
   3. ~r   Simp, 2
   4. p   DS, 1, 3
   5. p ∨ q   Add, 4

6. 1. ~p ∨ q
   2. ~q ∨ r
   3. ~r   / ∴ ~p
   4. ~q   DS, 2, 3
   5. ~p   DS, 1, 4

9. 1. p ⊃ r
   2. ~s
   3. p ∨ ~q
   4. ~q ⊃ s   / ∴ r
   5. ~~q   MT, 4, 2
   6. p   DS, 3, 5
   7. r   MP, 1, 6

12. 1. p ∨ q
   2. r ∨ s
   3. ~p · ~s   / ∴ (q · r) ∨ t
   4. ~p   Simp, 3
   5. ~s   Simp, 3
   6. r   DS, 2, 5
   7. q   DS, 1, 4
   8. q · r   Conj, 7, 6
   9. (q · r) ∨ t   Add, 8

15. 1. p ∨ (r ∨ q)
   2. (r ⊃ m) · (q ⊃ n)
   3. (m ∨ n) ⊃ (p ∨ q)
   4. ~p   / ∴ q
   5. r ∨ q   DS, 1, 4
   6. r ⊃ m   Simp, 2
   7. q ⊃ n   Simp, 2
   8. m ∨ n   CD, 5, 6, 7
   9. p ∨ q   MP, 3, 8
  10. q   DS, 9, 4

18. M: Management remains resolute; S: Management can stop the hostile takeover attempt; C: The company can improve its cash flow position; U: Management understands the consequences of the acquisition for its own position.
   1. M ⊃ S
   2. C ∨ ~S
   3. M ⊃ U
   4. U · ~C   / ∴ ~S
   5. ~C   Simp, 4
   6. ~S   DS, 2, 5

21. D: Peter will doubt me; C: You come along; E: You explain things to him; M: Margaret will tell him about our idea; U: Peter will understand; A: What Margaret tells Peter will just make him angry; O: You chicken out.
   1. D ∨ (C · E)
   2. ~M
   3. E ⊃ U
   4. (D ⊃ M) · A   / ∴ U ∨ O
   5. D ⊃ M   Simp, 4
   6. ~D   MT, 5, 2
   7. C · E   DS, 1, 6
   8. E   Simp, 7
   9. U   MP, 3, 8
  10. U ∨ O   Add, 9

24. M: The party maintains its current economic policy; F: There will be a flight of capital to other countries; T: The party tightens its control over the economy; C: The nation will have to pay large amounts of foreign debt in hard currency.

   1. M ⊃ F
   2. T ⊃ M
   3. (M · F) ⊃ T
   4. (M ⊃ C) · ~C   / ∴ ~T
   5. M ⊃ C   Simp, 4
   6. ~C   Simp, 4
   7. ~M   MT, 5, 6
   8. ~T   MT, 2, 7

27. 1. A ⊃ B
   2. ~C ⊃ ~D
   3. A ∨ ~C
   4. (B ∨ ~D) ⊃ ~A   / ∴ ~C
   5. B ∨ ~D   CD, 3, 1, 2
   6. ~A   MP, 4, 5
   7. ~C   DS, 3, 6

30. 1. (A ⊃ B) ⊃ (A ⊃ C)
   2. A ∨ C
   3. ~E ⊃ (A ⊃ B)
   4. ~D · ~E
   5. C ⊃ D   / ∴ C · ~C
   6. ~D   Simp, 4
   7. ~C   MT, 5, 6
   8. A   DS, 2, 7
   9. ~E   Simp, 4
 10. A ⊃ B   MP, 3, 9
 11. B   MP, 10, 8
 12. A ⊃ C   MP, 1, 10
 13. C   MP, 12, 8
 14. C · ~C   Conj, 13, 7

8.4 Rules of Replacement: Connectives

 3. 1. r ⊃ p
   2. ~r ⊃ q   / ∴ ~q ⊃ p
   3. ~r ⊃ ~~q   DN, 2
   4. ~q ⊃ r   Trans, 3
   5. ~q ⊃ p   HS, 4, 1

 6. 1. A ⊃ B
   2. ~(A · B)   / ∴ ~A
   3. A ⊃ (A · B)   Abs, 1
   4. ~A   MT, 3, 2

 9. 1. ~(A ⊃ B)   / ∴ A · ~B
   2. ~(~A ∨ B)   Impl, 1
   3. ~~A · ~B   DM, 2
   4. A · ~B   DN, 3

12. 1. ~C ∨ ~B   / ∴ ~(C · ~~B)
   2. ~(C · B)   DM, 1
   3. ~(C · ~~B)   DN, 2

15. 1. B ⊃ C
   2. ~B ⊃ A
   3. A ⊃ D   / ∴ C ∨ D
   4. ~~B ∨ A   Impl, 2
   5. B ∨ A   DN, 4
   6. C ∨ D   CD, 5, 1, 3

18. 1. C ⊃ ~A
   2. B
   3. B ⊃ (A ∨ ~D)   / ∴ D ⊃ ~C
   4. A ∨ ~D   MP, 3, 4
   5. ~~A ∨ ~D   DN, 4
   6. ~A ⊃ ~D   Impl, 5
   7. C ⊃ ~D   HS, 1, 6
   8. ~~C ⊃ ~D   DN, 7
   9. D ⊃ ~C   Trans, 8

21. 1. C ≡ A   / ∴ ~C ≡ ~A
   2. (C ⊃ A) · (A ⊃ C)   Equiv, 1
   3. C ⊃ A   Simp, 2
   4. A ⊃ C   Simp, 2
   5. ~A ⊃ ~C   Trans, 3
   6. ~C ⊃ ~A   Trans, 4
   7. (~C ⊃ ~A) · (~A ⊃ ~C)   Conj, 6, 5
   8. ~C ≡ ~A   Equiv, 7

24. 1. C ∨ (A ∨ D)
   2. E · ~G
   3. ~(~E ∨ G) ⊃ ~C
   4. (D ⊃ B) · ~B   / ∴ A
   5. D ⊃ B   Simp, 4
   6. ~B   Simp, 4
   7. ~D   MT, 5, 6
   8. ~~E · ~G   DN, 2
   9. ~(~E ∨ G)   DM, 8
  10. ~C   MP, 3, 9
  11. A ∨ D   DS, 1, 10
  12. A   DS, 11, 7

27. 1. R ⊃ F
   2. I ⊃ ~R
   3. F ⊃ R   / ∴ ~(F · I)
   4. ~~I ⊃ ~R   DN, 2
   5. R ⊃ ~I   Trans, 4
   6. F ⊃ ~I   HS, 3, 5
   7. ~F ∨ ~I   Impl, 6
   8. ~(F · I)   DM, 7
  R: I'm right; F: You're a fool; I: I'm a fool.

30. 1. ~(G · M)
   2. ~M ⊃ S
   3. S ⊃ ~G   / ∴ ~G

4. ~M ⊃ ~G   HS, 2, 3
5. G ⊃ M   Trans, 4
6. G ⊃ (G · M)   Abs, 5
7. ~G   MT, 6, 1
G: A man serves God; M: He serves Mammon; S: He starves.

33. 1. ~C
    2. (I · ~R) ⊃ C   / ∴ I ⊃ R
    3. ~(I · ~R)   MT, 2, 1
    4. ~I ∨ ~~R   DM, 3
    5. I ⊃ ~~R   Impl, 4
    6. I ⊃ R   DN, 5
    C: Something can be conceived as greater than God; I: God exists in our imaginations; R: God exists in reality.

36. 1. A ≡ (A ∨ B)
    2. ~A   / ∴ ~B
    3. (A ⊃ (A ∨ B)) · ((A ∨ B) ⊃ A)   Equiv, 1
    4. (A ∨ B) ⊃ A   Simp, 3
    5. ~(A ∨ B)   MT, 4, 2
    6. ~A · ~B   DM, 5
    7. ~B   Simp, 6

39. 1. ~C ⊃ (~A ∨ ~B)
    2. (A · D) ∨ (A · B)
    3. ~(C · D)
    4. (D ∨ E) · ~E   / ∴ D
    5. ~E   Simp, 4
    6. D ∨ E   Simp, 4
    7. D   DS, 6, 5
    8. ~C ∨ ~D   DM, 3
    9. ~~D   DN, 7
    10. ~C   DS, 8, 9
    11. ~A ∨ ~B   MP, 1, 10
    12. ~(A · B)   DM, 11
    13. A · D   DS, 2, 12
    14. D   Simp, 13

8.5   Rules of Replacement: Algebra

3. O: God is omnipotent; E: God can do everything; S: God can make a stone so heavy that He can't lift it; X: God exists.
    1. O ≡ E
    2. ~S ⊃ ~E
    3. S ⊃ ~E   / ∴ ~O ∨ ~X
    4. E ⊃ S   Trans, 2
    5. E ⊃ ~E   HS, 4, 3
    6. ~E ∨ ~E   Impl, 5
    7. ~E   Taut, 6
    8. (O ⊃ E) · (E ⊃ O)   Equiv, 1

9. O ⊃ E	Simp, 8
10. ~O	MT, 9, 7
11. ~O ∨ ~X	Add, 10

6. C: My cat sings opera; L: All the lights are out; I: I am very insistent; H: I howl at the moon.

1. ~C ∨ L	
2. I ⊃ C	
3. (L ∨ H) ⊃ I	
4. ~I ⊃ H	/ ∴ (L · I) · C
5. ~~I ∨ H	Impl, 4
6. (~~I ∨ H) ∨ L	Add, 5
7. ~~I ∨ (H ∨ L)	Assoc, 6
8. ~~I ∨ (L ∨ H)	Comm, 7
9. ~I ⊃ (L ∨ H)	Impl, 8
10. ~I ⊃ I	HS, 9, 3
11. ~~I ∨ I	Impl, 10
12. I ∨ I	DN, 11
13. I	Taut, 12
14. C	MP, 2, 13
15. C ⊃ L	Impl, 1
16. L	MP, 15, 14
17. L · I	Conj, 16, 13
18. (L · I) · C	Conj, 17, 14

9. S: There will be scorpions in the house this week; B: There's building going on in the neighborhood; D: The depression in the housing market ends; K: The cats will kill scorpions; T: The cats will bring scorpions to me as trophies.

1. B ⊃ S	
2. D ⊃ B	
3. S ⊃ (K · T)	/ ∴ D ⊃ T
4. D ⊃ S	HS, 2, 1
5. D ⊃ (K · T)	HS, 4, 3
6. ~D ∨ (K · T)	Impl, 5
7. (~D ∨ K) · (~D ∨ T)	Dist, 6
8. ~D ∨ T	Simp, 7
9. D ⊃ T	Impl, 8

12. A: Applications to universities will fall; C: Universities counteract demographic trends; F: Universities teach fewer students; L: Universities become less selective; S: Universities shrink; E: Universities become even more expensive.

1. A ∨ C	
2. A ⊃ F	
3. F ⊃ (S · E)	/ ∴ C ∨ E
4. C ∨ A	Comm, 1
5. ~~C ∨ A	DN, 4
6. ~C ⊃ A	Impl, 5

7. ~C ⊃ F	HS, 6, 2
8. ~C ⊃ (S · E)	HS, 7, 3
9. ~~C ∨ (S · E)	Impl, 8
10. C ∨ (S · E)	DN, 9
11. (C ∨ S) · (C ∨ E)	Dist, 10
12. C ∨ E	Simp, 11

15. E: The earth has been visited recently by extraterrestrial beings; G: The government has kept the information silent; A: Some reported UFO sightings have been authentic; U: Our current understanding of our place in the universe is seriously mistaken.

1. E ⊃ (G · E)	
2. A ⊃ E	
3. A ⊃ U	/ ∴ U ∨ ((E · G) ⊃ ~A)
4. ~A ∨ U	Impl, 3
5. U ∨ ~A	Comm, 4
6. (U ∨ ~A) ∨ ~(E · G)	Add, 5
7. U ∨ (~A ∨ ~(E · G))	Assoc, 6
8. U ∨ (~(E · G) ∨ ~A)	Comm, 7
9. U ∨ ((E · G) ⊃ ~A)	Impl, 8

18.
1. A ∨ (B ∨ C)	
2. ~B	/ ∴ A ∨ C
3. A ∨ (C ∨ B)	Comm, 1
4. (A ∨ C) ∨ B	Assoc, 3
5. A ∨ C	DS, 4, 2

21.
1. A ≡ B	
2. ~(H ⊃ B)	/ ∴ ~A
3. ~(~H ∨ B)	Impl, 2
4. ~~H · ~B	DM, 3
5. ~B	Simp, 4
6. (A ⊃ B) · (B ⊃ A)	Equiv, 1
7. A ⊃ B	Simp, 6
8. ~A	MT, 7, 5

24.
1. A · B	/ ∴ ~(~B · ~C) · A
2. B	Simp, 1
3. B ∨ C	Add, 2
4. ~(~B · ~C)	DM, 3
5. A	Simp, 1
6. ~(~B · ~C) · A	Conj, 4, 5

27.
1. (~A ⊃ B) ∨ C	/ ∴ ~A ⊃ (B ∨ C)
2. (~~A ∨ B) ∨ C	Impl, 1
3. ~~A ∨ (B ∨ C)	Assoc, 2
4. ~A ⊃ (B ∨ C)	Impl, 3

30.
1. A ≡ B	
2. A ⊃ C	/ ∴ A ⊃ (B · C)
3. (A ⊃ B) · (B ⊃ A)	Equiv, 1
4. A ⊃ B	Simp, 3

5. ~A ∨ B   Impl, 4
6. ~A ∨ C   Impl, 2
7. (~A ∨ B) · (~A ∨ C)   Conj, 5, 6
8. ~A ∨ (B · C)   Dist, 7
9. A ⊃ (B · C)   Impl, 8

33. 1. A ⊃ (B ⊃ ~C)
    2. ~~B ⊃ ~A
    3. C · A   / ∴ B ⊃ (~A · ~C)
    4. A   Simp, 3
    5. ~~A   DN, 4
    6. ~~~B   MT, 2, 5
    7. ~B   DN, 6
    8. ~B ∨ (~A · ~C)   Add, 7
    9. B ⊃ (~A · ~C)   Impl, 8

36. 1. A ∨ B
    2. C ≡ B   / ∴ C ∨ A
    3. (C ⊃ B) · (B ⊃ C)   Equiv, 2
    4. B ⊃ C   Simp, 3
    5. ~~A ∨ B   DN, 1
    6. ~A ⊃ B   Impl, 5
    7. ~A ⊃ C   HS, 6, 4
    8. ~~A ∨ C   Impl, 7
    9. A ∨ C   DN, 8
    10. C ∨ A   Comm, 9

39. 1. A · D
    2. B ∨ C
    3. A ≡ ~B   / ∴ C · D
    4. (A ⊃ ~B) · (~B ⊃ A)   Equiv, 3
    5. A   Simp, 1
    6. A ⊃ ~B   Simp, 4
    7. ~B   MP, 6, 5
    8. C   DS, 2, 7
    9. D   Simp, 1
    10. C · D   Conj, 8, 9

42. 1. ~(p ≡ q)   / ∴ p ≡ ~q
    2. ~((p ⊃ q) · (q ⊃ p))   Equiv, 1
    3. ~(p ⊃ q) ∨ ~(q ⊃ p)   DM, 2
    4. ~(~p ∨ q) ∨ ~(q ⊃ p)   Impl, 3
    5. ~(~p ∨ q) ∨ ~(~q ∨ p) Impl, 4
    6. (~~p · ~q) ∨ ~(~q ∨ p) DM, 5
    7. (~~p · ~q) ∨ (~~q · ~p) DM, 6
    8. (p · ~q) ∨ (~~q · ~p) DN, 7
    9. (p · ~q) ∨ (q · ~p)   DN, 8
    10. p ≡ ~q   Equiv, 9

45. 1. p ≡ q   / ∴ q ≡ p
    2. (p ⊃ q) · (q ⊃ p)   Equiv, 1

    3. $(q \supset p) \cdot (p \supset q)$ Comm, 2

    4. $q \equiv p$ Equiv, 3

48.
   1. $p \supset q$
   2. $p \supset r$
   3. $\sim q \vee \sim r$ / ∴ $\sim p$
   4. $q \supset \sim r$ Impl, 3
   5. $p \supset \sim r$ HS, 1, 4
   6. $\sim\sim p \supset \sim r$ DN, 5
   7. $r \supset \sim p$ Trans, 6
   8. $p \supset \sim p$ HS, 2, 7
   9. $\sim p \vee \sim p$ Impl, 8
   10. $\sim p$ Taut, 9

51.
   1. $\sim(\sim C \cdot A) \cdot (C \equiv \sim A)$ / ∴ $C \equiv (A \supset B)$
   2. $\sim(\sim C \cdot A)$ Simp, 1
   3. $C \equiv \sim A$ Simp, 1
   4. $(C \supset \sim A) \cdot (\sim A \supset C)$ Equiv, 3
   5. $\sim\sim C \vee \sim A$ DM, 2
   6. $C \supset \sim A$ Simp, 4
   7. $\sim C \vee \sim A$ Impl, 6
   8. $(\sim C \vee \sim A) \vee B$ Add, 7
   9. $\sim C \vee (\sim A \vee B)$ Assoc, 8
   10. $C \supset (\sim A \vee B)$ Impl, 9
   11. $C \supset (A \supset B)$ Impl, 10
   12. $\sim A \supset C$ Simp, 4
   13. $\sim C \supset \sim A$ Impl, 5
   14. $\sim C \supset C$ HS, 13, 12
   15. $\sim\sim C \vee C$ Impl, 14
   16. $C \vee C$ DN, 15
   17. C Taut, 16
   18. $(A \cdot \sim B) \vee C$ Add, 17
   19. $(\sim\sim A \cdot \sim B) \vee C$ DN, 18
   20. $\sim(\sim A \vee B) \vee C$ DM, 19
   21. $\sim(A \supset B) \vee C$ Impl, 20
   22. $(A \supset B) \supset C$ Impl, 21
   23. $(C \supset (A \supset B)) \cdot ((A \supset B) \supset C)$ Conj, 11, 22
   24. $C \equiv (A \supset B)$ Equiv, 23

54.
   1. $(A \equiv (B \cdot A)) \equiv (A \equiv C)$
   2. $A \supset B$
   3. $C \cdot D$ / ∴ $A \cdot (B \cdot (C \cdot D))$
   4. C Simp, 3
   5. $((A \equiv (B \cdot A)) \supset (A \equiv C)) \cdot ((A \equiv C) \supset (A \equiv (B \cdot A)))$ Equiv, 1
   6. $(A \equiv (B \cdot A)) \supset (A \equiv C)$ Simp, 5
   7. $A \supset (B \cdot A)$ Abs, 2
   8. $\sim A \vee (B \cdot A)$ Impl, 7
   9. $(\sim A \vee B) \cdot (\sim A \vee A)$ Dist, 8
   10. $\sim A \vee A$ Simp, 9
   11. $\sim B \vee (\sim A \vee A)$ Add, 10

12. (~B ∨ ~A) ∨ A	Assoc, 11
13. ~(B · A) ∨ A	DM, 12
14. (B · A) ⊃ A	Impl, 13
15. (A ⊃ (B · A)) · ((B · A) ⊃ A)	Conj, 7, 14
16. A ≡ (B · A)	Equiv, 15
17. A ≡ C	MP, 16, 6
18. (A ⊃ C) · (C ⊃ A)	Equiv, 17
19. C ⊃ A	Simp, 18
20. A	MP, 19, 4
21. B · A	MP, 7, 20
22. B	Simp, 21
23. B · (C · D)	Conj, 22, 3
24. A · (B · (C · D))	Conj, 20, 23

57.

1. ~(A ⊃ ~B)	
2. C ⊃ (~A ∨ ~B)	
3. (C ∨ D) ≡ E	/ ∴ E ≡ D
4. ((C ∨ D) ⊃ E) · (E ⊃ (C ∨ D))	Equiv, 3
5. (C ∨ D) ⊃ E	Simp, 4
6. ~(C ∨ D) ∨ E	Impl, 5
7. (~C · ~D) ∨ E	DM, 6
8. (~C ∨ E) · (~D ∨ E)	Dist, 7
9. ~D ∨ E	Simp, 8
10. D ⊃ E	Impl, 9
11. E ⊃ (C ∨ D)	Simp, 4
12. ~E ∨ (C ∨ D)	Impl, 11
13. ~E ∨ (D ∨ C)	Comm, 12
14. (~E ∨ D) ∨ C	Assoc, 13
15. ~(~A ∨ ~B)	Impl, 1
16. ~C	MT, 2, 15
17. ~E ∨ D	DS, 14, 16
18. E ⊃ D	Impl, 17
19. (D ⊃ E) · (E ⊃ D)	Conj, 10, 18
20. D ≡ E	Equiv, 19

8.6 Categorical Proofs

3. (p ∨ p) ⊃ p

1. p ⊃ p	SI
2. (p ∨ p) ⊃ p	Taut, 1

6. ((p ∨ q) ∨ ~r) ⊃ ~~(p ∨ (~~q ∨ ~r))

1. ((p ∨ q) ∨ ~r) ⊃ ((p ∨ q) ∨ ~r)	SI
2. ((p ∨ q) ∨ ~r) ⊃ ~~((p ∨ q) ∨ ~r)	DN, 1
3. ((p ∨ q) ∨ ~r) ⊃ ~~((p ∨ ~~q) ∨ ~r)	DN, 2
4. ((p ∨ q) ∨ ~r) ⊃ ~~(p ∨ (~~q ∨ ~r))	Assoc, 3

9. q ⊃ (p ∨ q)

1. q ⊃ q	SI
2. ~q ∨ q	Impl, 1
3. p ∨ (~q ∨ q)	Add, 2

4. $(p \lor \neg q) \lor q$	Assoc, 3
5. $(\neg q \lor p) \lor q$	Comm, 4
6. $\neg q \lor (p \lor q)$	Assoc, 5
7. $q \supset (p \lor q)$	Impl, 6

12. $p \supset (\neg p \supset q)$

1. $p \supset p$	SI
2. $\neg p \lor p$	Impl, 1
3. $\neg p \lor \neg\neg p$	DN, 2
4. $(\neg p \lor \neg\neg p) \lor q$	Add, 3
5. $\neg p \lor (\neg\neg p \lor q)$	Assoc, 4
6. $p \supset (\neg\neg p \lor q)$	Impl, 5
7. $p \supset (\neg p \supset q)$	Impl, 6

15. $(p \supset q) \supset ((p \supset r) \supset (p \supset (q \cdot r)))$

1. $((p \supset q) \cdot (p \supset r)) \supset ((p \supset q) \cdot (p \supset r))$	SI
2. $((p \supset q) \cdot (p \supset r)) \supset ((\neg p \lor q) \cdot (p \supset r))$	Impl, 1
3. $((p \supset q) \cdot (p \supset r)) \supset ((\neg p \lor q) \cdot (\neg p \lor r))$	Impl, 2
4. $((p \supset q) \cdot (p \supset r)) \supset (\neg p \lor (q \cdot r))$	Dist, 3
5. $((p \supset q) \cdot (p \supset r)) \supset (p \supset (q \cdot r))$	Impl, 4
6. $(p \supset q) \supset ((p \supset r) \supset (p \supset (q \cdot r)))$	Exp, 5

18. $p \supset (q \supset p)$

1. $p \supset p$	SI
2. $\neg p \lor p$	Impl, 1
3. $\neg q \lor (\neg p \lor p)$	Add, 2
4. $(\neg q \lor \neg p) \lor p$	Assoc, 3
5. $(\neg p \lor \neg q) \lor p$	Comm, 4
6. $\neg p \lor (\neg q \lor p)$	Assoc, 5
7. $p \supset (\neg q \lor p)$	Impl, 6
8. $p \supset (q \supset p)$	Impl, 7

21. $(((p \cdot q) \lor (p \cdot \neg q)) \lor (\neg p \cdot q)) \lor (\neg p \cdot \neg q)$

1. $p \supset p$	SI
2. $q \supset q$	SI
3. $\neg p \lor p$	Impl, 1
4. $\neg q \lor q$	Impl, 2
5. $(\neg p \lor p) \cdot (\neg q \lor q)$	Conj, 3, 4
6. $(p \lor \neg p) \cdot (\neg q \lor q)$	Comm, 5
7. $(p \lor \neg p) \cdot (q \lor \neg q)$	Comm, 6
8. $(p \cdot (q \lor \neg q)) \lor (\neg p \cdot (q \lor \neg q))$	Dist, 7
9. $((p \cdot q) \lor (p \cdot \neg q)) \lor (\neg p \cdot (q \lor \neg q))$	Dist, 8
10. $((p \cdot q) \lor (p \cdot \neg q)) \lor ((\neg p \cdot q) \lor (\neg p \cdot \neg q))$	Dist, 9
11. $(((p \cdot q) \lor (p \cdot \neg q)) \lor (\neg p \cdot q)) \lor (\neg p \cdot \neg q)$	Assoc, 10

24.

1. $A \supset B$	
2. $\neg D \supset \neg B$	
3. $\neg A \supset C$	
4. $\neg D \lor E$	$/ \therefore C \lor E$
5. $A \supset A$	SI
6. $\neg A \lor A$	Impl, 5

7. B ⊃ D	Trans, 2
8. A ⊃ D	HS, 1, 7
9. D ⊃ E	Impl, 4
10. A ⊃ E	HS, 8, 9
11. C ∨ E	CD, 6, 3, 10

8.7 Indirect Proofs

3. A: Allen is out of the shop; B: Baker is out of the shop; C: Carr is out of the shop.

1. A ⊃ B	
2. C ⊃ (A · ~B)	/ ∴ (~C · ~B) ⊃ ~A
3. ~((~C · ~B) ⊃ ~A)	AIP
4. ~(~(~C · ~B) ∨ ~A)	Impl, 3
5. ~~(~C · ~B) · ~~A	DM, 4
6. ~~(~C · ~B)	Simp, 5
7. ~C · ~B	DN, 6
8. ~B	Simp, 7
9. ~~A	Simp, 5
10. ~A	MT, 1, 8
11. ~A · ~~A	Conj, 10, 9

6.

1. A ≡ B	/ ∴ ~(G ⊃ ~B) ⊃ A
2. ~(~(G ⊃ ~B) ⊃ A)	AIP
3. ~(~~(G ⊃ ~B) ∨ A)	Impl, 2
4. ~~~(G ⊃ ~B) · ~A	DM, 3
5. ~A	Simp, 4
6. ~~~(G ⊃ ~B)	Simp, 4
7. ~~~(~G ∨ ~B)	Impl, 6
8. ~(~G ∨ ~B)	DN, 7
9. ~~G · ~~B	DM, 8
10. ~~B	Simp, 9
11. (A ⊃ B) · (B ⊃ A)	Equiv, 1
12. B ⊃ A	Simp, 11
13. ~B	MT, 12, 5
14. ~B · ~~B	Conj, 13, 10

9.

1. D ⊃ (B · C)	
2. A ⊃ (~B · ~E)	/ ∴ (A · D) ⊃ F
3. ~((A · D) ⊃ F)	AIP
4. ~(~(A · D) ∨ F)	Impl, 3
5. ~~(A · D) · ~F	DM, 4
6. ~~(A · D)	Simp, 5
7. A · D	DN, 6
8. A	Simp, 7
9. D	Simp, 7
10. B · C	MP, 1, 9
11. B	Simp, 10
12. ~B · ~E	MP, 2, 8
13. ~B	Simp, 12
14. B · ~B	Conj, 11, 13

12. 1. ~(~A · B)
    2. (A ∨ C) ⊃ ~(M · D)
    3. B ≡ D   / ∴ M ⊃ ~(B ∨ D)
    4. ~(M ⊃ ~(B ∨ D))   AIP
    5. ~(~M ∨ ~(B ∨ D))   Impl, 4
    6. ~~M · ~~(B ∨ D)   DM, 5
    7. ~~M   Simp, 6
    8. ~~(B ∨ D)   Simp, 6
    9. M   DN, 7
    10. ~(~B · ~D)   DM, 8
    11. (B · D) ∨ (~B · ~D)   Equiv, 3
    12. B · D   DS, 11, 10
    13. D   Simp, 12
    14. M · D   Conj, 9, 13
    15. ~~(M · D)   DN, 14
    16. ~(A ∨ C)   MT, 2, 15
    17. ~A · ~C   DM, 16
    18. ~A   Simp, 17
    19. B   Simp, 12
    20. ~A · B   Conj, 18, 19
    21. (~A · B) · ~(~A · B)   Conj, 20, 1

15. 1. (B ≡ ~A) ⊃ ~C
    2. (~B · D) ∨ (A · M)
    3. (D ∨ M) ⊃ C   / ∴ A ⊃ B
    4. ~(A ⊃ B)   AIP
    5. ~(~A ∨ B)   Impl, 4
    6. ~~A · ~B   DM, 5
    7. ~B · ~~A   Comm, 6
    8. (B · ~A) ∨ (~B · ~~A)   Add, 7
    9. B ≡ ~A   Equiv, 8
    10. ~C   MP, 1, 9
    11. ~(D ∨ M)   MT, 3, 10
    12. ~D · ~M   DM, 11
    13. ~D   Simp, 12
    14. ~M   Simp, 12
    15. ~A ∨ ~M   Add, 14
    16. ~(A · M)   DM, 15
    17. ~B · D   DS, 2, 16
    18. D   Simp, 17
    19. D · ~D   Conj, 18, 13

9. Predicate Logic

    9.1 Quantifiers

3. $(\exists x)Wx$
6. $\sim Hc \supset (\exists x)Px$
9. $(\exists x)Dx \supset \sim(x)Px$

    9.2 Categorical Statement Forms

3. $\sim(\exists x)(Ex \cdot Lx)$ (or $(x)(Ex \supset \sim Lx)$)
6. $\sim(\exists x)(Jx \cdot Cx)$ (or $(x)(Jx \supset \sim Cx)$)
9. $(\exists x)(Sx \cdot Ex) \cdot (\exists x)(Sx \cdot \sim Ex)$

    9.3 Symbolization

3. nonintersective: one could be a fast learner and a runner without being a fast runner.
6. intersective (Though one could argue that an interesting documentary, for example, is interesting for a documentary, but not interesting for a movie. I'm not convinced, however.)
9. nonintersective (Daffy might be an intelligent duck but not an intelligent animal.)
12. intersective
15. nonintersective
18. $(x)(Px \supset Gx)$
21. $(x)(Kx \supset Rx)$
24. $\sim(\exists x)(Rx \cdot Ix)$ (or $(x)(Ix \supset \sim Rx)$)
27. $(x)((Bx \cdot Wx) \supset Lx)$
30. $\sim(\exists x)((Mx \cdot Wx) \cdot Gx)$ (or $(x)((Mx \cdot Wx) \supset \sim Gx)$)
33. $(x)((Mx \cdot \sim Tx) \supset \sim Ax)$ (or $\sim(\exists x)((Mx \cdot \sim Tx) \cdot Ax)$)
36. $\sim(\exists x)((Gx \cdot Wx) \cdot (Ax \cdot \sim Px))$ (or $(x)(((Gx \cdot Wx) \cdot Ax) \supset \sim\sim Px)$)
39. $(x)(Mx \supset ((Ux \cdot Px) \lor Cx))$
42. $(x)((Wx \cdot Lx) \supset Px)$; $(x)((Tx \cdot \sim Cx) \supset \sim Mx)$
45. $(x)Lx \cdot (\exists x)Sx$
48. $(x)(Mx \lor Wx) \cdot (x)\sim(Mx \cdot Wx)$
51. $(x)((Ex \cdot Px) \supset Gx)$
54. $(x)(Sx \supset (Dx \cdot Ax))$ (or $(x)(Sx \supset Dx) \cdot (x)(Sx \supset Ax)$)
57. $\sim(\exists x)(Hx \cdot \sim Ix) \cdot (x)(Px \supset Ux)$
60. $(x)((Mx \cdot (Wx \cdot \sim Ax)) \supset G)$
63. $(\exists x)(Wx \cdot (Fx \cdot Ix)) \cdot (\exists x)(Wx \cdot (Fx \cdot Px))$
66. $(x)((Px \cdot Ox) \supset Fx) \supset \sim(\exists x)Ax$
69. $(x)((Gx \cdot Vx) \supset ((Sx \cdot Ax) \supset Rx))$

9.4   Quantified Tableaux

3.

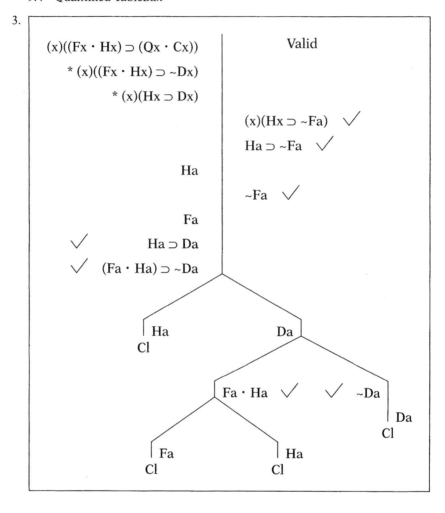

$(x)((Fx \cdot Hx) \supset (Qx \cdot Cx))$ Valid

* $(x)((Fx \cdot Hx) \supset {\sim}Dx)$

* $(x)(Hx \supset Dx)$

$(x)(Hx \supset {\sim}Fa)$ ✓

$Ha \supset {\sim}Fa$ ✓

Ha

${\sim}Fa$ ✓

Fa

✓ $Ha \supset Da$

✓ $(Fa \cdot Ha) \supset {\sim}Da$

Ha   Da
Cl

Fa · Ha ✓   ✓ ${\sim}Da$

Da
Cl

Fa   Ha
Cl   Cl

6.

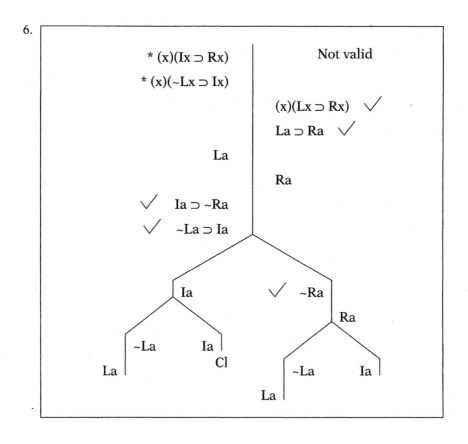

9.

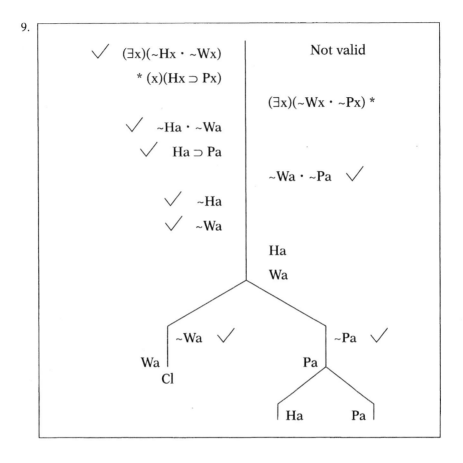

12.

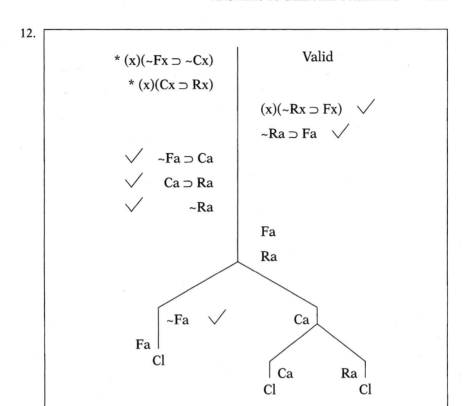

15.

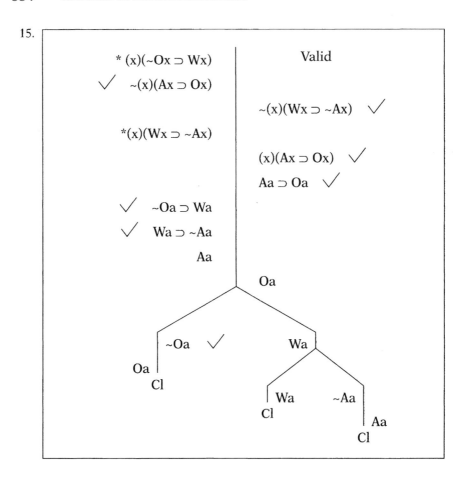

18.

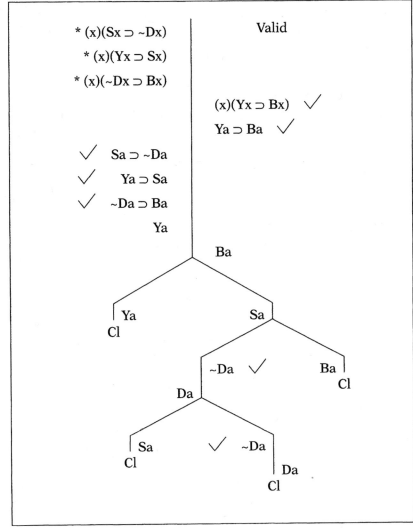

21.

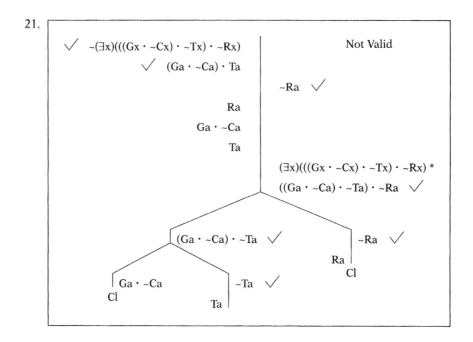

24.

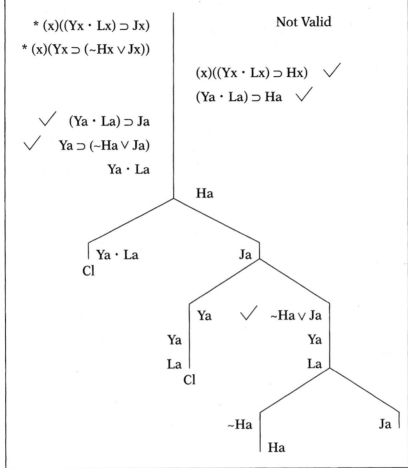

27.

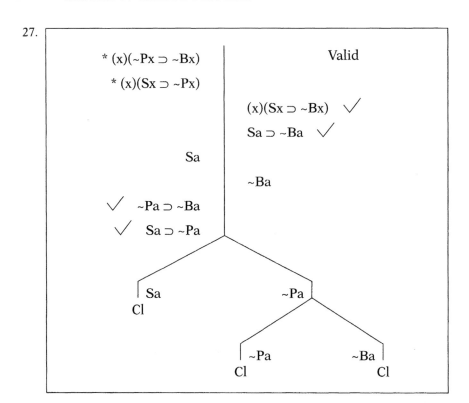

30.

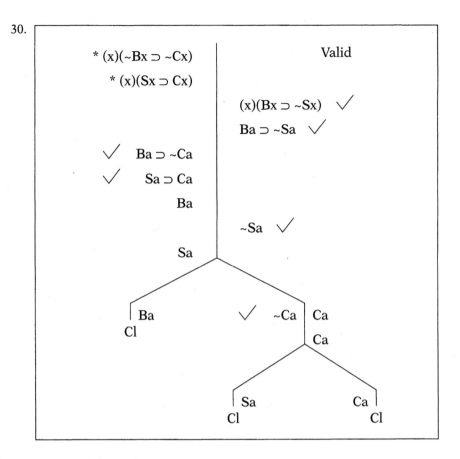

33. Contingent:

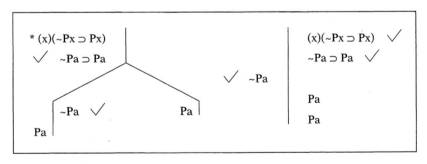

36.

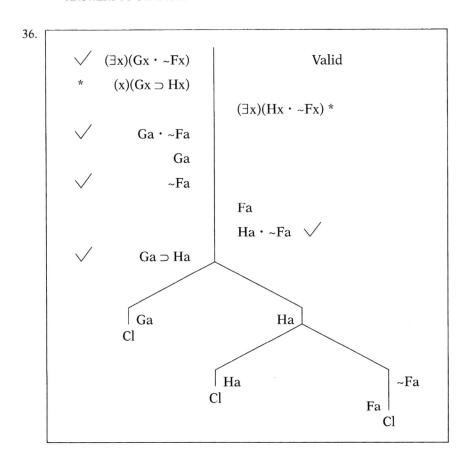

39.

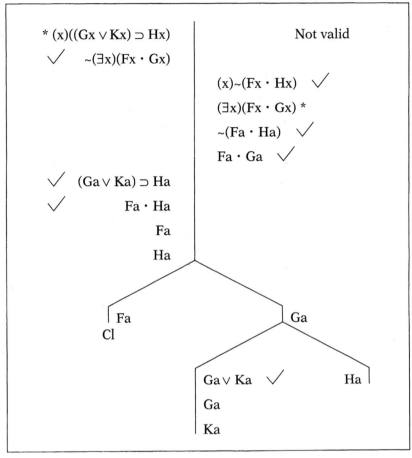

42.

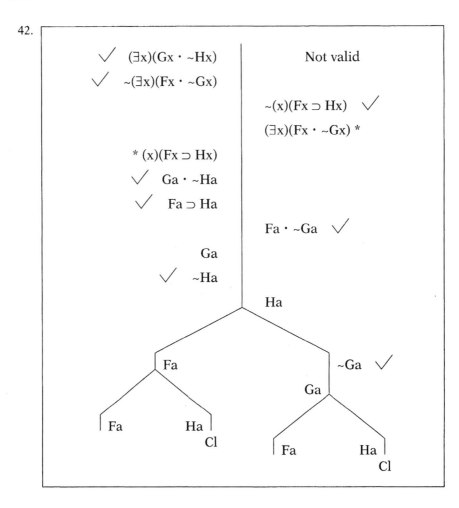

45.

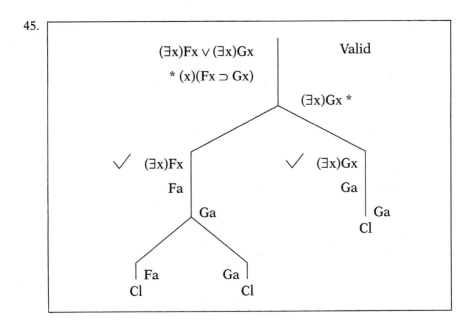

48.

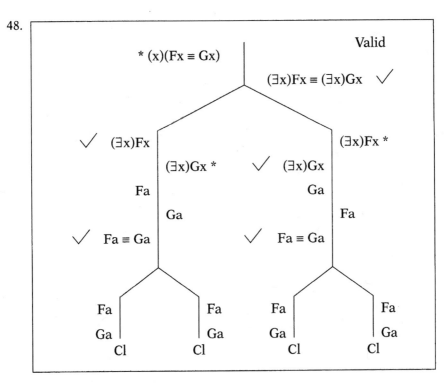

9.5   Quantified Proof

3. 1. (x)Cx
    2. (∃x)Ex   / ∴ (∃x)(Cx · Ex)
    3. Ea   EI, 2
    4. Ca   UI, 1
    5. Ca · Ea   Conj, 4, 3
    6. (∃x)(Cx · Ex)   EG, 5

6. 1. (∃x)((Nx · Hx) · (Lx · ~Wx))
    2. (x)((Lx · Hx) ⊃ Ix)   / ∴ (∃x)((Hx · Ix) · ~Wx)
    3. (Na · Ha) · (La · ~Wa)   EI, 1
    4. (La · Ha) ⊃ Ia   UI, 2
    5. Na · Ha   Simp, 3
    6. La · ~Wa   Simp, 3
    7. Ha   Simp, 5
    8. La   Simp, 6
    9. ~Wa   Simp, 6
    10. La · Ha   Conj, 8, 7
    11. Ia   MP, 4, 10
    12. Ha · Ia   Conj, 7, 11
    13. (Ha · Ia) · ~Wa   Conj, 12, 9
    14. (∃x)((Hx · Ix) · ~Wx)   EG, 13

9. 1. (x)(Cx ⊃ ~(Wx ∨ Rx))
    2. (∃x)((Dx · Cx) · Ix)   / ∴ (∃x)((Ix · Dx) · ~Rx)
    3. (Da · Ca) · Ia   EI, 2
    4. Ca ⊃ ~(Wa ∨ Ra)   UI, 1
    5. Da · Ca   Simp, 3
    6. Ia   Simp, 3
    7. Da   Simp, 5
    8. Ca   Simp, 5
    9. ~(Wa ∨ Ra)   MP, 4, 8
    10. ~Wa · ~Ra   DM, 9
    11. ~Ra   Simp, 10
    12. Ia · Da   Conj, 6, 7
    13. (Ia · Da) · ~Ra   Conj, 12, 11
    14. (∃x)((Ix · Dx) · ~Rx)   EG, 13

12. 1. (∃x)(Bx · Tx)
    2. (x)(Tx ⊃ ~Rx)   / ∴ (∃x)(Bx · ~Rx)
    3. Ba · Ta   EI, 1
    4. Ba   Simp, 3
    5. Ta   Simp, 3
    6. Ta ⊃ ~Ra   UI, 2
    7. ~Ra   MP, 6, 5
    8. Ba · ~Ra   Conj, 4, 7
    9. (∃x)(Bx · ~Rx)   EG, 8

15. 1. (x)(Mx ⊃ Lx)
    2. (∃x)(Sx · Mx)   / ∴ (∃x)(Sx · Lx)
    3. Sa · Ma   EI, 2

	4. Sa	Simp, 3
	5. Ma	Simp, 3
	6. Ma ⊃ La	UI, 1
	7. La	MP, 6, 5
	8. Sa · La	Conj, 4, 7
	9. (∃x)(Sx · Lx)	EG, 8

18. 1. (x)(Lx ⊃ Mx)
    2. (∃x)(Sx · ~Mx)   / ∴ (∃x)(Sx · ~Lx)
    3. Sa · ~Ma   EI, 2
    4. Sa   Simp, 3
    5. ~Ma   Simp, 4
    6. La ⊃ Ma   UI, 1
    7. ~La   MT, 6, 7
    8. Sa · ~La   Conj, 4, 7
    9. (∃x)(Sx · ~Lx)   EG, 8

21. 1. (∃x)(Mx · Lx)
    2. (x)(Mx ⊃ Sx)   / ∴ (∃x)(Sx · Lx)
    3. Ma · La   EI, 1
    4. Ma ⊃ Sa   UI, 2
    5. Ma   Simp, 3
    6. La   Simp, 3
    7. Sa   MP, 4, 5
    8. Sa · La   Conj, 7, 6
    9. (∃x)(Sx · Lx)   EG, 8

24. 1. (x)(Mx ⊃ ~Lx)
    2. (∃x)(Mx · Sx)   / ∴ (∃x)(Sx · ~Lx)
    3. Ma · Sa   EI, 2
    4. Ma ⊃ ~La   UI, 1
    5. Ma   Simp, 3
    6. Sa   Simp, 3
    7. ~La   MP, 4, 5
    8. Sa · ~La   Conj, 6, 7
    9. (∃x)(Sx · ~Lx)   EG, 8

27. 1. (x)(Lx ⊃ ~Mx)
    2. (x)(Sx ⊃ Mx)
    3. (∃x)Sx   / ∴ (∃x)(Sx · ~Lx)
    4. Sa   EI, 3
    5. Sa ⊃ Ma   UI, 2
    6. La ⊃ ~Ma   UI, 1
    7. Ma   MP, 5, 4
    8. ~~Ma   DN, 7
    9. ~La   MT, 6, 8
    10. Sa · ~La   Conj, 4, 9
    11. (∃x)(Sx · ~Lx)   EG, 10

30. 1. (x)(Mx ⊃ Lx)
    2. (∃x)(Sx · Mx)   / ∴ (∃x)(Lx · Sx)

	3.	Sa · Ma	EI, 2
	4.	Ma ⊃ La	UI, 1
	5.	Sa	Simp, 3
	6.	Ma	Simp, 3
	7.	La	MP, 4, 6
	8.	La · Sa	Conj, 7, 5
	9.	(∃x)(Lx · Sx)	EG, 8

33.	1.	(x)Fx	/ ∴ (∃x)Fx
	2.	Fa	UI, 1
	3.	(∃x)Fx	EG, 2

36.	1.	(x)(Gx ⊃ Hx)	
	2.	(∃x)(Fx · Gx)	/ ∴ (∃x)(Fx · Hx)
	3.	Fa · Ga	EI, 2
	4.	Ga ⊃ Ha	UI, 1
	5.	Fa	Simp, 3
	6.	Ga	Simp, 3
	7.	Ha	MP, 4, 6
	8.	Fa · Ha	Conj, 5, 7
	9.	(∃x)(Fx · Hx)	EG, 8

39.	1.	(x)(~Gx ⊃ ~Hx)	
	2.	(∃x)(Fx · ~Gx)	/ ∴ (∃x)(Fx · ~Hx)
	3.	Fa · ~Ga	EI, 2
	4.	~Ga ⊃ ~Ha	UI, 1
	5.	~Ga	Simp, 3
	6.	Fa	Simp, 3
	7.	~Ha	MP, 4, 5
	8.	Fa · ~Ha	Conj, 6, 7
	9.	(∃x)(Fx · ~Hx)	EG, 8

42.	1.	(∃x)(Gx · Hx)	
	2.	(x)(Gx ⊃ Fx)	/ ∴ (∃x)(Fx · Hx)
	3.	Ga · Ha	EI, 1
	4.	Ga	Simp, 3
	5.	Ha	Simp, 3
	6.	Ga ⊃ Fa	UI, 2
	7.	Fa	MP, 6, 4
	8.	Fa · Ha	Conj, 7, 5
	9.	(∃x)(Fx · Hx)	EG, 8

45.	1.	(∃x)(Hx · Gx)	
	2.	(x)(Gx ⊃ Fx)	/ ∴ (∃x)(Fx · Hx)
	3.	Ha · Ga	EI, 1
	4.	Ga ⊃ Fa	UI, 2
	5.	Ha	Simp, 3
	6.	Ga	Simp, 3
	7.	Fa	MP, 4, 6
	8.	Fa · Ha	Conj, 7, 5
	9.	(∃x)(Fx · Hx)	EG, 8

48.   1. (x)(Fx ∨ ~Gx)
      2. (x)(Fx ⊃ Hx)
      3. (x)(~Gx ⊃ Jx)
      4. (∃x)~Jx   / ∴ (∃x)Hx
      5. ~Ja   EI, 4
      6. Fa ∨ ~Ga   UI, 1
      7. Fa ⊃ Ha   UI, 2
      8. ~Ga ⊃ Ja   UI, 3
      9. ~~Ga   MT, 8, 5
     10. Fa   DS, 6, 9
     11. Ha   MP, 7, 10
     12. (∃x)Hx   EG, 11

   9.6   Universal Generalization

3.   1. (x)(Lx ⊃ ~Mx)
     2. (x)(Sx ⊃ Mx)   / ∴ (x)(Sx ⊃ ~Lx)
     3. Sa ⊃ Ma   UI, 2
     4. La ⊃ ~Ma   UI, 1
     5. ~~Ma ⊃ ~La   Trans, 4
     6. Ma ⊃ ~La   DN, 5
     7. Sa ⊃ ~La   HS, 3, 6
     8. (x)(Sx ⊃ ~Lx)   UG, 7

6.   1. (x)(Dx ⊃ Sx)
     2. (x)(Ix ⊃ ~Sx)   / ∴ (x)(Dx ⊃ ~Ix)
     3. Da ⊃ Sa   UI, 1
     4. Ia ⊃ ~Sa   UI, 2
     5. ~~Sa ⊃ ~Ia   Trans, 4
     6. Sa ⊃ ~Ia   DN, 5
     7. Da ⊃ ~Ia   HS, 3, 6
     8. (x)(Dx ⊃ ~Ix)   UG, 7

9.   1. (x)(Px ⊃ ~Ix)
     2. (x)(Vx ⊃ Ix)   / ∴ (x)(Px ⊃ ~Vx)
     3. Pa ⊃ ~Ia   UI, 1
     4. Va ⊃ Ia   UI, 2
     5. ~Ia ⊃ ~Va   Trans, 4
     6. Pa ⊃ ~Va   HS, 3, 5
     7. (x)(Px ⊃ ~Vx)   UG, 6

12.   1. (x)(Ex ⊃ ~~Cx)
      2. (x)(Sx ⊃ Bx)
      3. (x)(Cx ⊃ ~Bx)   / ∴ (x)(Sx ⊃ ~Ex)
      4. Ea ⊃ ~~Ca   UI, 1
      5. Sa ⊃ Ba   UI, 2
      6. Ca ⊃ ~Ba   UI, 3
      7. ~~Ba ⊃ ~Ca   Trans, 6
      8. Ba ⊃ ~Ca   DN, 7
      9. ~~~Ca ⊃ ~Ea   Trans, 4
     10. ~Ca ⊃ ~Ea   DN, 9

11. Ba ⊃ ~Ea   HS, 8, 10
12. Sa ⊃ ~Ea   HS, 5, 11
13. (x)(Sx ⊃ ~Ex)   UG, 12

15. 1. (x)(Sx ⊃ Tx)
    2. (x)(Wx ⊃ ~Bx)
    3. (x)(Tx ⊃ Bx)   / ∴ (x)(Sx ⊃ ~Wx)
    4. Sa ⊃ Ta   UI, 1
    5. Wa ⊃ ~Ba   UI, 2
    6. Ta ⊃ Ba   UI, 3
    7. Sa ⊃ Ba   HS, 4, 6
    8. ~~Ba ⊃ ~Wa   Trans, 5
    9. Ba ⊃ ~Wa   DN, 8
   10. Sa ⊃ ~Wa   HS, 7, 9
   11. (x)(Sx ⊃ ~Wx)   UG, 10

18. 1. (x)(Hx ⊃ ~Dx)
    2. Ra
    3. (x)(Px ∨ Hx)
    4. (x)(Dx ∨ Sx)
    5. (x)(Rx ⊃ ~Px)   / ∴ Sa
    6. Ra ⊃ ~Pa   UI, 5
    7. ~Pa   MP, 6, 2
    8. Pa ∨ Ha   UI, 3
    9. Ha   DS, 8, 7
   10. Ha ⊃ ~Da   UI, 1
   11. ~Da   MP, 10, 9
   12. Da ∨ Sa   UI, 4
   13. Sa   DS, 12, 11

21. 1. (x)(Fx · Gx)   / ∴ (x)Fx · (x)Gx
    2. Fa · Ga   UI, 1
    3. Fa   Simp, 2
    4. (x)Fx   UG, 3
    5. Ga   Simp, 2
    6. (x)Gx   UG, 5
    7. (x)Fx · (x)Gx   Conj, 4, 6

24. 1. (x)(Fx ≡ (Gx · Hx))   / ∴ (x)(Fx ⊃ Gx) · (x)(Fx ⊃ Hx)
    2. Fa ≡ (Ga · Ha)   UI, 1
    3. (Fa ⊃ (Ga · Ha)) · ((Ga · Ha) ⊃ Fa)   Equiv, 2
    4. Fa ⊃ (Ga · Ha)   Simp, 3
    5. ~Fa ∨ (Ga · Ha)   Impl, 4
    6. (~Fa ∨ Ga) · (~Fa ∨ Ha)   Dist, 5
    7. ~Fa ∨ Ga   Simp, 6
    8. Fa ⊃ Ga   Impl, 7
    9. (x)(Fx ⊃ Gx)   UG, 8
   10. ~Fa ∨ Ha   Simp, 6
   11. Fa ⊃ Ha   Impl, 10
   12. (x)(Fx ⊃ Hx)   UG, 11
   13. (x)(Fx ⊃ Gx) · (x)(Fx ⊃ Hx)   Conj, 12

27. 1. $(x)(Fx \equiv (\sim Gx \cdot \sim Fx))$ / ∴ $(x)Gx$
     2. $Fa \equiv (\sim Ga \cdot \sim Fa)$ UI, 1
     3. $(Fa \supset (\sim Ga \cdot \sim Fa)) \cdot ((\sim Ga \cdot \sim Fa) \supset Fa)$ Equiv, 2
     4. $Fa \supset (\sim Ga \cdot \sim Fa)$ Simp, 3
     5. $\sim Fa \vee (\sim Ga \cdot \sim Fa)$ Impl, 4
     6. $(\sim Fa \vee \sim Ga) \cdot (\sim Fa \vee \sim Fa)$ Dist, 5
     7. $\sim Fa \vee \sim Fa$ Simp, 6
     8. $\sim Fa$ Taut, 7
     9. $(\sim Ga \cdot \sim Fa) \supset Fa$ Simp, 3
    10. $\sim(\sim Ga \cdot \sim Fa)$ MT, 9, 8
    11. $\sim\sim Ga \vee \sim\sim Fa$ DM, 10
    12. $Ga \vee \sim\sim Fa$ DN, 11
    13. $Ga \vee Fa$ DN, 12
    14. Ga DS, 13, 8
    15. $(x)Gx$ UG, 14

30. 1. $(x)(Fx \supset (Gx \vee Hx))$
     2. $(x)((Jx \cdot Fx) \supset \sim Gx)$
     3. $(x)(\sim Fx \supset \sim Jx)$ / ∴ $(x)(Jx \supset Hx)$
     4. $\sim Fa \supset \sim Ja$ UI, 3
     5. $Ja \supset Fa$ Trans, 4
     6. $Fa \supset (Ga \vee Ha)$ UI, 1
     7. $(Ja \cdot Fa) \supset \sim Ga$ UI, 2
     8. $Ja \supset (Ja \cdot Fa)$ Abs, 5
     9. $Ja \supset \sim Ga$ HS, 8, 7
    10. $Ja \supset (Ja \cdot \sim Ga)$ Abs, 9
    11. $Ja \supset (Ga \vee Ha)$ HS, 5, 6
    12. $Ja \supset (\sim\sim Ga \vee Ha)$ DN, 11
    13. $Ja \supset (\sim Ga \supset Ha)$ Impl, 12
    14. $(Ja \cdot \sim Ga) \supset Ha$ Exp, 13
    15. $Ja \supset Ha$ HS, 10, 14
    16. $(x)(Jx \supset Hx)$ UG, 15

10. Generalizations and Analogies

   10.1   Inductive Strength

From Chapter 1, section 1.4 (pp. 45–47):

   3. Deductively valid.
   6. Deductively valid.
   9. Deductively valid.
  12. Not valid; inductively strong (the conclusion holds of some things aren't easy).
  15. Deductively valid.
  18. Deductively valid.
  21. Deductively valid.

   10.2   Enumeration

   3. a. (much) less—little variation
      b. less—little variation
      c. more—sample variation increased

    d. less—less variation
    e. less—less variation (the population may have changed)

10.3 Statistical Generalizations

3. a. Little variation; representatives in the student government may not be representative of the student body as a whole.
    b. Biased: not all students are equally likely to be in the Student Union at lunchtime. (On some campuses, of course, they might be, in which case this is a fairly good method.)
    c. Probably a good method, though students who are full-time are more likely to be included than those who are part-time. (Full-time students will be in more classes.) This may be appropriate for some purposes and inappropriate for others. Whether it makes a difference on the question of raising admissions standards depends in part on whether full- and part-time students are admitted under the same standards.
    d. Biased; not all students are equally likely to be in the library. The studious will be more likely to be included than the slackers, and humanities majors may be more likely to be included than science majors.
    e. Self-selected: highly biased.

6. Problems: Few men will want to answer such a personal question; many of those who do answer are likely to lie. Those asked with their wives present are even more likely to lie. Probably no method would be very reliable.

9. Problems: By definition, the homeless have no fixed address or telephones; they are unlikely to have cars, be registered to vote, etc. Shelters and other organizations aiding the homeless are possible sources of information, but many homeless people may not use shelters for food or lodging, and organizations have incentives to inflate both the scope of the problem they are addressing and the number of people they are helping.

12. Amounts of bleach may be measured by weight or volume, but that probably makes little difference here. This claim doesn't specify the other term of the comparison—"20 percent more bleach" than what?—but probably compares the new with the old version of the product. If it does, this claim doesn't indicate whether there is now 20 percent more bleach than before, per unit—per ounce, for example—or whether there is simply 20 percent more bleach in the container, which could result from having a bigger container. (Of course, if the product is simply bleach, there can be 20 percent more only by having a bigger container. Whether the claim is impressive then depends on whether the price has changed.)

15. Children in affluent schools learning 27 percent more than whom? Than children in less affluent schools? Than they were before? Also, learning is not so easily quantified: Was there a standardized test on which these children scored 27 percent higher? Covering what material?

18. The length of a person's life can be measured precisely. But what is the other term of the comparison? Unmarried men? Married women?

21. This argument depends on the claims (a) that those who actually vote are not a representative sample of the eligible electorate, and (b) that the rich, well-educated, white and conservative are over-represented, while the poor, uneducated, nonwhite, and liberal are under-represented. The term "eligible electorate" is unclear; does it mean "registered voters" or "those who could

register to vote"? If those registered are not representative of those qualified to register, the claim that actual voters do not represent the registered
voters may be false while the corresponding claim about those qualified
to register is true. Whether the actual voters were a representative sample
of the registered voters could be determined by a random sampling of
those registered. Even this presents problems, however; people often lie
to pollsters, not wanting to admit preferring controversial candidates or
propositions, and they may change their minds after the fact, wanting to
appear to have preferred the winner. Whether the actual voters were a representative sample of those who could register is much more difficult, for
the unregistered frequently do not want to be contacted; they may be alienated from the political process, lack information, or simply have moved
recently.

Evaluating (b) is also difficult. It is well-known that the old are more
likely to vote than the young; it may be possible to determine whether whites
are more likely to vote than minorities. But judging whether the rich are
more likely to vote than the poor is difficult, since it is hard to gather accurate economic information on actual or potential voters, and since "rich"
and "poor" are vaguely defined notions that may easily be distorted by being
made precise. (Politicians often call households earning over $75,000 a year
"rich"—something that surely would be news to them—while studies have
shown that surprising percentages of those labelled "poor" by the government own their own homes, have two cars, and even have swimming pools.
Part of the difficulty is that income and wealth are different notions; one
might have a high net worth but a low income (as many elderly people do)
or a high income but low net worth (as many families with children do).
Also, "rich" and "poor" imply some stability of economic circumstance over
time, but many of those showing up in a given year with a high or low income may be in such a situation only temporarily. A graduate student may
have a low income and a low net worth, but be about to graduate and be
employed at a good-paying job; a retiree may have a high income for a year
as a result of selling stock or a business.

"Liberal" and "conservative" raise even more problems, since these
terms too are hard to define precisely, and many people are liberal on some
issues and conservative on others.

10.4   Analogies

3. a. Men are compared to moths, mountains to tufts of wool. b. Men and
mountains will be scattered. c. Men and mountains appear to be fixed and
sturdy, while moths and tufts of wool don't.

6. a. The world is compared to a stage and people to players. b. People have
exits and entrances and play many parts. c. The world is larger; players can
leave the stage and then return, but death is permanent.

9. a. The evening is compared to a nun. b. Both are quiet, calm, free, filled
with adoration. c. The nun is a person; the evening is not. More importantly,
the nun is the one who adores, while the evening is the object of adoration.

12. a. The world is compared to a land of dreams and to a darkling plain.
b. The world seems various, beautiful, and new, like a land of dreams; but
it is really swept with confusion and struggle. c. The world contains novelty
and confusion, beauty and struggle.

15. a. Seeing Yankee fans is related to waking up in a Brazilian jail. b. Presumably, both are shocking and very unpleasant. c. Yankee Stadium is in the Bronx, not Brazil; baseball is played there. No one is forced to go to Yankee Staduim or remain there for any set length of time. d. No argument.

18. a. Television is compared to bumper stickers, news to philosophy. b. Television and bumper stickers convey little information, and convey it in a crude form, while news and philosophy are complex and subtle. c. Television shows series of pictures, while bumper stickers are static; news is concerned with issues of the day, while philosophy is concerned with the issues of perennial concern. d. No argument.

21. a. Judgments are related to watches. b. They differ, but each person believes his own. c. Watches, unlike judgments, operate mechanically, can be checked against a fixed standard, and give information only about time. d. No argument.

24. a. Light and other forms of radiation are compared to water ripples or waves. b. Both distribute energy from a central source. c. Radiation isn't wet. d. No argument.

27. a. A clock is related to society. b. Both are human constructions that can be reconstructed. c. There are people who understand clocks. d. No argument.

30. a. A system is compared to the tail of a lizard, the truth to a lizard. b. The truth, like a lizard, is hard to grasp; catching the tail is easy but doesn't lead to catching the lizard. c. A lizard is an animal; a system is abstract. d. No argument.

33. a. Mistletoe is related to fighting animals, and its attraction of birds to a struggle. b. Mistletoe and fighting animals are engaged in competition for survival. c. Mistletoe competes with other species; individual animals fight other individual animals. d. No argument.

36. a. The Federal Reserve is compared to a doctor, the economy to a patient. b. The Federal Reserve is in charge of the health of the economy the way a doctor is in charge of the health of a patient. c. The patient is an individual, while the economy is the product of the activities of many individuals. (Some economists believe that growth that is too rapid creates inflation; presumably there are no problems caused by patients that recover too quickly.) d. Conclusion: The Federal Reserve should not clamp down too tightly on interest rates.

39. a. Contracting muscles around the eyes while crying is related to animals drawing back their ears to fight. b. In both cases, something began as a voluntary movement has become involuntary. c. Animals draw back their ears when they feel threatened; people contract their eye muscles when they cry.

42. a. weaker: additional difference.
    b. weaker: additional difference.
    c. weaker: additional difference.
    d. weaker: additional (highly relevant) difference.
    e. stronger: additional similarity.

11. Causes

     11.1 Kinds of Causes

3. a. No—there could have been many other sources of delay. b. Probably—it depends on the forms in which the information was stored. c. Everything

else was set up properly to transmit the information, but it was available only on the computer that crashed. d. Remote.

6. a. No—there are other ways of melting ice. b. No—salt won't melt ice if the temperature is low enough. c. Nothing else was done to melt the ice, and the temperature was in the right range. d. Proximate. (Though one could count the dissolving of the salt as a separate event, in which case the cause is remote.)

9. a. Depends on whether the bread contained any other rising agents (such as eggs or baking powder). b. No. c. We must assume that the right ingredients were in the dough and that the dough rose, and the bread baked, at appropriate temperatures, with appropriate humidity, pressure, and so on. d. Proximate.

12. a. No. b. No. c. We must assume that objects capable of being destroyed were in the flood plain and couldn't be moved in time. d. Proximate.

15. a. No. (Hank could have been injured.) b. No. (Hank might have been replaced by somebody even better.) c. Hank was valuable and was not replaced with an equally valuable player. d. Remote.

18. a, b, c, d.   Neither necessary nor sufficient.
In appropriate contexts, the scandal, the electoral defeat, and the change in leadership could each be said to have caused widespread turmoil.

11.2   Agreement and Difference

3. Difference: East and West Germany differed only in form of political and economic organization. This then must be the cause of their different economic performance.

6. Difference: Lulu survived because she got less heroin, and less toxin, than the others.

9. Two uses of agreement: All areas with malaria are infested with a particular kind of mosquito, and all areas infested with that kind of mosquito have malaria. The two methods are much more powerful than one, for they show that the mosquitoes are necessary and sufficient for malaria.

12. Joint method of agreement and difference: 250 liters were pasteurized, 250 were not. The heated bottles agreed in being fine 10 months later; the 250 unheated agreed in being sour. This is strong grounds for inferring that the heating process caused the wine to resist degeneration.

15. (a) By the joint method of agreement and difference, we can attribute the difference to wearing a tie—assuming that there are no other relevant differences. (b) Probably, the results would be similar, or even more striking, for a jacket, shirt, shoes, pants, or grooming. This would suggest that none is itself the cause of being treated with trust and respect, but that each is a part of the cause, and that failing to wear such an item, or failing to groom one's hair, and so on, causes one to be treated with a lack of trust and respect.

11.3   Residues and Concomitant Variation

3. A humorous allusion to something like the method of residues. Given that the effect of a Senate is the same in both cases, what additional effects will the House have?

6. This seems to rely on the method of residues. Children are often sick; a single difference, adding garlic, led to a record of uninterrupted health. So,

the method indicates, the garlic must have been responsible. This argument is weak, since there are bound to be many other genetic and environmental differences between Mr. Buscaglia and other people. (Note: because this passage discusses only a single case, no other method applies.)

9. This is a denial of an inference someone might be tempted to make based, presumably, on the method of residues (since we have only a single case).

12. MBAs and PhDs are advanced-degree professionals and have much in common. Yet MBAs are more satisfied sexually. The method of difference implies that we should examine differences between MBAs and PhDs to find the cause, or a substantial part of the cause, of MBAs' greater satisfaction. The passage points to one such difference: MBAs have money, while PhDs have learning. Of course, there are probably many other relevant differences.

15. This relies on the methods of agreement and concomitant variation. Note, however, that nothing in the argument supports the conclusion that both alphabets descended from the Phoenician; at most, we can conclude that they have some causal connection.

18. The method of residues suggests that, once all other factors have been accounted for, radiation must be the explanation of the missing matter.

21. Agreement: Compare the U.S. to other great nations. Difference: Contrast the U.S. with other wealthy nations.

24. Agreement: Compare nonquantitative questions to see whether they agree in being important. Difference: Contrast important with unimportant questions, quantitative with nonquantitative questions. Concomitant variation: See whether degree of importance correlates negatively with degree of susceptibility to quantitative solution.

27. Agreement: Consider cases in which people have been shielded from the (harmful) effects of their actions and see whether they agree in being more likely to elicit those actions. Difference: Contrast cases in which people have been shielded with those in which they have not.

12. Explanations

12.1 Generalizations and Laws

3. Generalization: Anybody who wants to be the best at anything in the world must spend 80 percent of their waking time on it. And you lose the pursuit of other things.
Conclusion: I never want to be the best at anything.

6. Generalization: People in capitalist countries do not earn enough money to buy products. Soviet people can buy what they desire.
Conclusion: Soviet shelves are empty, while shelves in capitalist countries are full.
Covering law (implicit): If people can buy what they want, nothing will remain on the shelves.

9. Generalization: In eclipses the dividing line is always rounded. The eclipse is due to the interposition of the earth.
Conclusion: The earth is spherical.
Covering law: If the eclipse is due to the interposition of the earth, the rounded line results from its spherical shape.

12. Conclusion: Rays of light are very small bodies emitted from shining substances.
    Covering law: Such bodies will pass through uniform mediums in right lines without bending into the shadow. They will also be capable of several properties and be able to conserve their properties unchanged in passing through several mediums.
15. Generalization: The density of wood is lower than that of water.
    First Conclusion: Any body floats if its weight is less than the weight of an equal volume of water.
    Second Conclusion: Wood floats on water.
    Covering law: A liquid supports a body immersed in it with a force equal to the weight of the liquid the body displaces.
18. Generalization: Inflation has changed prices by a factor of 10; it moved the entire workforce into higher tax brackets, which reduced after-tax purchasing power.
    Conclusion: The good economy went.
21. Generalization: Somehow people have gotten the idea that they won't mind being old so much if they can turn on the television and see what they were like when they were young. Not true. The best memories are ones that have been allowed to evolve unhindered by documentary proof.
    Conclusion: Video cameras are popular.
    Covering law: Memory is better than a video camera, because, in addition to being free, it doesn't work very well.

12.2   The Hypothetico-Deductive Method

3. Hypothesis: Anybody who wants to be the best at anything in the world must spend 80 percent of their waking time on it. No explicit hypothetical reasoning.
6. Hypotheses: People in capitalist countries do not earn enough money to buy products. Soviet people can buy what they desire. If people can buy what they want, nothing will remain on the shelves. No explicit hypothetical reasoning.
9. Confirming evidence: In eclipses the dividing line is always rounded. The eclipse is due to the interposition of the earth.
    Hypothesis: The earth is spherical.
12. Hypothesis: Rays of light are very small bodies emitted from shining substances.
    Confirming evidence: Such bodies will pass through uniform mediums in right lines without bending into the shadow. They will also be capable of several properties and be able to conserve their properties unchanged in passing through several mediums.
15. No real hypothesis; all generalizations assumed known. Possible hypothesis: The density of wood is lower than that of water.
    Confirming Evidence: Wood floats on water.
18. Hypothesis: Inflation reduced after-tax purchasing power.
    Confirming evidence: Contrast between 1955 or 1963 and today.
21. Hypothesis: People want video cameras to make them less worried about aging. No explicit hypothetical reasoning.

12.3   Confirmation and Auxiliary Assumptions

3. Electricity can exert force (by creating an electric or magnetic field) and perform work (by driving a motor, for example); it can be converted to other forms of energy (to light by a lamp; to heat by a stove or furnace; to sound by a radio; etc.).

6. Many people eat tomatoes in many forms, raw and cooked, and suffer no apparent ill effects.

9. People who eat high-fat diets are more likely to gain weight and suffer from weight-related illnesses such as diabetes, high blood pressure, and heart disease.

12. Hypothesis: everything in the heavens revolved around the earth. Disconfirming evidence: The moons of Jupiter revolve around Jupiter, not the earth.

15. Hypothesis: The general theory of relativity. Prediction 1: Light should bend, or be deflected, in the vicinity of massive objects like the sun, to a greater extent than Newtonian physics allows. Confirming evidence: A British expedition in Principe confirmed this experimentally by photographing a solar eclipse. Prediction 2: Mercury's orbit should turn, its perihelion advancing very slightly with each trip around the sun. Confirming evidence: Observations have indicated that this prediction is correct. Prediction 3: A reddening of light from strong gravitational fields. Confirming evidence: Observations of "white dwarfs," small, very dense stars, have borne out the theory's prediction.

18. Imperial Hypothesis: An eclipse would occur on the morning of June 21, 1629 at 10:30 and would last for two hours. Jesuit Hypothesis: The eclipse would not come until 11:30 and would last for only two minutes. Evidence (confirming for the Jesiuts, disconfirming for the Imperial astronomers): At 11:30, the eclipse began and lasted for a brief two minutes

12.4   Evaluating Explanations

It is difficult to evaluate all the explanations by all the criteria. Only the most important for that particular explanation are mentioned.

3. Reliable, relevant, accurate, consistent, well-confirmed, powerful, simple.

6. Relevant, powerful, simple. Not, however, well-confirmed, reliable, or even consistent. (If the store shelves are empty, how can you buy everything you desire?)

9. Reliable, relevant, accurate, consistent, well-confirmed, powerful, simple.

12. Reliable, relevant, accurate, consistent, powerful, simple. Light does exhibit many properties of particles, but also many properties of waves; confirmation here is thus mixed.

15. Reliable, relevant, accurate, consistent, well-confirmed, powerful, simple.

18. Reliable, relevant, consistent, well-confirmed, powerful, simple.

21. Relevant, consistent, powerful, simple. But points about the connections between memory, accuracy, and happiness aren't very well confirmed.

GLOSSARY

Abusive ad hominem argument
Argument purporting to discredit a
position by insulting those who hold
it. Page 93

Accent Fallacy of trying to justify a
conclusion by relying on presuppositions arising from a change in
stress in a premise. Page 133

Accident Fallacy of trying to justify
a conclusion by treating an accidental feature of something as essential. Page 117

Accuracy If, on the basis of a sample of Fs n percent of which are observed to be G, it is argued that
between $n - e$ percent and $n + e$
percent of the total population of
Fs are G, the generalization has an
accuracy of $1 - e$ percent. Page 398

**Ad hominem (or ad personam) ar-
gument** An argument "to the man
(or person)," attempting to refute
positions by attacking those who
hold or argue for them. Page 93

Ambiguity Possession of more than
one meaning. Page 70

Amphiboly Fallacy of trying to justify a conclusion by relying on an
ambiguity in sentence structure.
Page 130

Analytical definition Definition that
tries to explain the meaning of a
term by indicating what the things
to which it applies have in common. Page 74

Appeal to authority Argument that
tries to justify its conclusion by
citing the opinions of authorities.
Page 116

Appeal to common practice Argument that tries to justify a kind of
action by appealing to the common
practice of the community. Page 107

Appeal to force Argument that tries
to justify a kind of action by threatening the audience. Page 108

Appeal to ignorance Argument that
tries to justify its conclusion by appealing to what is not known. Page
114

Appeal to pity Argument that tries
to justify a kind of action by arousing sympathy or pity in the audience over the consequences of the
action. Page 109

Appeal to the people Argument
that tries to justify its conclusion
by appealing to the audience's emotions. Page 105

Argument A finite string of statements, called premises, together
with another statement, the conclusion, which the premises are taken
to support. Page 2

Argument form Argument pattern
in the metalanguage, such as $p \lor q$;
$\sim p$; $\therefore q$. Page 237

Aristotle's assumption The hypothesis that no terms have empty extensions. Page 174

Atomic formula Formula that contains no connectives. Page 238

Atomic statement Statement that
has no other statements as components. Page 219

Auxiliary assumption An item of
knowledge, hypothesis, principle,
or fact assumed to be true for the

purpose of testing a hypothesis. Page 458

Begging the question A fallacy in which the premises include or presuppose the conclusion. Page 88

Biased sample Unrepresentative sample. Page 400

Categorical proof Proof using no assumptions. Page 298

Categorical syllogism A syllogism made up of categorical propositions and containing three terms, each appearing in two propositions. Page 181

Cause An event A causes an event B in a context C if and only if A, together with C, is a necessary and sufficient condition for B. Page 14

Circumstantial ad hominem argument Argument purporting to discredit a position by appealing to the circumstances or characteristics of those who hold it. Page 93

Closed tableau branch Branch on which the same formula appears live on both sides. Page 267

Closed tableau Tableau every branch of which is closed. Page 267

Cogent argument Inductively strong argument with true premises. Page 44

Complement Term that applies to exactly those things (in the universe of discourse) to which a given term does not apply. Page 166

Complex argument Argument that contains other arguments. Page 24

Complex question A fallacy in which the premises include a question that presupposes the conclusion. Page 89

Component Proper part of a statement; replacing it with another statement yields something meaningful. Page 218

Composition Fallacy of attributing something to a whole or group be-

cause it can be attributed to the parts or members. Page 133

Compound Statement that has components. Page 219

Conclusion The thesis an argument tries to justify. Page 3

Confidence interval If, on the basis of a sample of Fs n percent of which are observed to be G, it is argued that between n − e percent and n + e percent of the total population of Fs are G, n − e percent to n + e percent is called the generalization's confidence interval. Page 398

Confidence level In a statistical generalization, the likelihood of the actual percentage of the population having the characteristic under study falling within the confidence interval. Page 399

Connective Word or phrase that forms a single, compound statement from one or more component statements. Page 219

Contingent statement Statement that could be true or false. Page 240

Contradictory statement Statement that cannot be true. Page 240

Contradictory Statement that always disagrees in truth value with a given statement. Page 163

Contrapositive Proposition that results from another categorical proposition by (1) switching the order of the terms and (2) replacing each term with its complement. Page 169

Converse A proposition that results from switching the predicates of a categorical proposition. Page 166

Counterexample An instance contradicting a general statement; especially, an argument with true premises and a false conclusion that shows an argument form to be invalid. Page 54

Deductively invalid argument Argument the premises of which could all be true in a circumstance in which its conclusion is false. Page 42

Deductively valid argument Argument whose premises cannot all be true while its conclusion is false. Page 42

Definiendum The term being defined. Page 79

Definiens The defining expression. Page 80

Definition per genus et differentiae Definition "by genus and difference," that tries to specify a species or kind of thing by indicating (a) what genus, or more general classification, it falls under, and (b) what distinguishes the species from others of the same genus. Page 74

Degree of confidence In a statistical generalization, the likelihood of the actual percentage of the population having the characteristic under study falling within the confidence interval. Page 399

Demonstrative argument Argument whose premises are believed to be true. Page 45

Description A group of statements about a situation that relays information about some circumstance. Page 11

Descriptive definition Definition that tries to describe the meaning of a term as it is commonly used. Page 69

Dialectical argument Argument whose premises are not believed to be true; that tries to show that a certain statement is false by using it as a premise to reach an outrageous or absurd conclusion, or that adopts a premise purely as a hypothesis, to see what would follow if it were true. Page 45

Distribution A proposition distributes a term occurring in it if and only if its truth depends on the en-

tire extension of the term. In general, a categorical proposition distributes a term if it asserts of every member of the term's extension that it is included in or excluded from the other term's extension. Page 198

Division Fallacy of attributing something to members or parts because it can be attributed to the whole. Page 134

Enthymeme Argument that relies on unstated premises. Page 28

Equivalence One statement is equivalent to another if and only if it is impossible for them to disagree in truth value. Page 48

Equivocation Fallacy of trying to justify a conclusion by relying on an ambiguity in a word or phrase. Page 128

Extended argument Argument that contains other arguments. Page 24

Extension The set of objects of which a term is true. Page 149

Fallacy A kind of bad argument. Page 87

Figure Property of a syllogism that depends on the placement of its middle term. This chart summarizes the figures according to the position of the middle term, M, in standard form (where P is the major, and S the minor, term): Page 183

	FIRST FIGURE		SECOND FIGURE	
Major premise	M	P	P	M
Minor premise	S	M	S	M
Conclusion	S	P	S	P

	THIRD FIGURE		FOURTH FIGURE	
Major premise	M	P	P	M
Minor premise	M	S	M	S
Conclusion	S	P	S	P

Finished tableau Open tableau on which only atomic formulas are live. Page 268

Formal fallacy A bad (invalid) argument form. Page 87

Generalization $(v)\mathfrak{J}(v)$ and $(\exists v)\mathfrak{J}(v)$ are generalizations of $\mathfrak{J}(c)$. Page 449

Hypothesis Statement not yet believed but to be adopted tentatively for purposes of testing. Page 457

Hypothetical proof Proof that begins with premises (also called assumptions, or hypotheses), on which the conclusion depends. Page 298

Hypothetical reasoning The process of proposing hypotheses, testing them, and accepting, rejecting, or modifying them in light of evidence. Page 457

Ignoratio elenchi An irrelevant argument ignorant of its own goal or purpose. Page 92

Implication A set of statements implies a statement if and only if, whenever every statement in the set is true, that statement must also be true. One statement implies another if and only if, whenever the first is true, the second must be true as well. Page 47

Incomplete enumeration A fallacy presupposing a disjunction that does not include all available possibilities. Page 90

Inductive definition Definition with three parts: the basis of the definition, stating that the term being defined applies to certain objects or categories of objects; inductive clauses, having a conditional form, specifying that if certain objects satisfy the term being defined, then so do certain others; and a closure condition, saying that the term being defined applies to nothing else: the term applies only to the objects to which the basis and inductive clauses of the definition force it to apply. Page 81

Inductively strong argument Argument in which the truth of the premises does not guarantee the truth of the conclusion, but does make the truth of the conclusion probable. Page 43

Inferential relation A relation of intended support between premises and conclusion. When indicators are used to mark a premise or conclusion, the inferential claim is explicit. Many arguments, however, lack indicators. In such cases, the inferential claim is implicit. Page 10

Informal fallacy A kind of argument typically violating considerations of evidence, relevance, or clarity in particular contexts. Page 87

Instance Say that $\mathfrak{J}(c)$ is the result of substituting constant c for every occurrence of variable v throughout the statement function $\mathfrak{J}(v)$. If $(v)\mathfrak{J}(v)$ and $(\exists v)\mathfrak{J}(v)$ are statements, then $\mathfrak{J}(c)$ is an instance of them. Page 55

Invertible rule A rule of equivalence; a rule of inference working in both directions, expressing an equivalence. Page 300

Listing definition Definition that tries to specify the meaning of a word by listing things or kinds of things to which it applies. Page 74

Main connective The connective occurrence in a given formula with the largest scope. Page 238

Major premise Premise of a syllogism that contains the major term. Page 182

Major term The conclusion's predicate. Page 182

Margin of error If, on the basis of a sample of Fs n percent of which are observed to be G, it is argued that between n − e percent and n + e percent of the total population of Fs are G, the percentage e is the margin of error of the generalization. Page 398

Metalanguage Language in which another language is discussed. Page 237

Middle term Term that appears in both premises. Page 182

Minor premise Premise of a syllogism that contains the minor term. Page 182

Minor term The conclusion's subject. Page 182

Misapplication Fallacy of trying to justify a conclusion about a particular case by appealing to a rule that is generally sound but inapplicable or outweighed by other considerations in that case. Page 118

Mood List of three letters signifying the form of the major premise, minor premise, and conclusion of a syllogism. Page 183

Narrative A group of statements that tells a story and relays information about a course of events. Page 11

Natural deduction system A set of rules of inference. Page 207

Necessary condition Condition such that a given event cannot occur when it does not hold. Page 422

Object language Artificial, symbolic logical language. Page 237

Obverse Proposition that results from another categorical proposition by (1) changing its quality (from affirmative to negative, or vice versa) and (2) replacing its predicate term with its complement. Page 169

Open tableau Tableau with at least one open branch. Page 267

Open tableau branch Branch on which the same formula does not appear live on both sides. Page 267

Ostensive definition Definition that tries to clarify the meaning of a term by pointing to examples of things to which it applies. Page 73

Persuasive definition Definition that tries to explain the meaning of a term, but contentiously, expressing not only meaning but attitude as well. Page 70

Post hoc ergo propter hoc Fallacy ("after this, therefore because of this") of drawing a causal conclusion simply from the temporal ordering of events. Page 119

Precising definition Definition that tries to remain close to the meaning of a term in natural language, but also to make that meaning more precise, or more useful for a particular purpose. Page 70

Premises The initial assertions of an argument. Page 3

Proof An extended argument, each statement of which is an assumption or follows from previous statements by a rule of inference. Page 298

Proposition Expressed by statements with the same meaning in the same context; the basic things that are true or false; and the things people believe, know, think, doubt, etc. Page 4

Propositional logic Logic that takes statements or propositions as basic units and examines relationships between them that pertain to reasoning. Page 148

Random sample Sample such that every member of the population has an equal chance of being included in it. Page 403

Recursive definition Definition with three parts: the basis of the definition, stating that the term being defined applies to certain objects or categories of objects; inductive clauses, having a conditional form, specifying that if certain objects satisfy the term being defined, then so do certain others; and a closure condition, saying that the term being defined applies to nothing else: the term applies only to the objects to which the basis and inductive clauses of the definition force it to apply. Page 81

Red herring An irrelevant point introduced into an argument or debate; an argument that tries to undermine an opponent's argument by introducing a point irrelevant to the issue at hand. Page 100

Reliability In a statistical generalization, the likelihood of the actual percentage of the population having the characteristic under study falling within the confidence interval. Page 399

Reliable argument Argument in which the truth of the premises does not guarantee the truth of the conclusion, but does make the truth of the conclusion probable. Page 43

Replacement The principle that substituting for any subformula in a formula a logically equivalent subformula results in a formula equivalent to the original. Page 300

Representative sample Sample that mirrors the total population in its relevant characteristics. Page 395

Response rate The proportion of people who respond to a survey as a fraction of all of those asked to respond. Page 404

Rule of equivalence Rule of inference working in both directions, expressing an equivalence. Page 300

Rule of implication Rule of inference working in one direction only. Page 300

Rules of inference Rules that deduce propositions from other propositions. Page 298

Satisfiable statement Statement that could be true. Page 240

Schema Expression in the metalanguage ranging over object language statements having a certain form. Page 237

Schematic letter Variable in the metalanguage ranging over object language statements. Page 237

Scope A given connective together with the subformulas (and any parentheses) it links. Page 238

Simple argument Argument that does not contain other arguments. Page 24

Simple statement Statement that has no other statements or components. Page 219

Sound argument Valid argument with true premises. Page 44

Standard form Syllogism stated in this form: Page 4

> Major premise
> Minor premise
> $\downarrow$
> $\therefore$ Conclusion

Statement Sentence that can be true or false. Page 3

Statement form Expression in the metalanguage ranging over object language statements having a certain form. Page 237

Statement variable Variable in the metalanguage ranging over object language statements. Page 237

Stipulative definition Definition that assigns a meaning to a term without regard to its ordinary use. Page 69

Straw man The fallacy of trying to justify rejecting a position by attacking a different, and usually weaker, position; also, that weaker position. Page 101

Subformula Any formula constructed in building a given formula. Page 238

Sufficient condition Condition such that a given event must occur when it holds. Page 422

Syllogism An argument containing three statements: two premises and one conclusion. Page 181

Symbolic argument   Argument symbolized in the object language. Page 250

Synonymous definition   Definition giving a synonym for the word being defined. Page 74

Tautology   Statement that cannot be false. Page 240

Term   An expression that applies to objects taken individually. Page 149

Truth function   Function from truth values into truth values. Page 222

Truth table   A computation of the truth value of a formula or set of formulas under each combination of truth values for its statement letters. Page 244

Truth–functional connective   Connective such that the truth values of the component statements the connective joins always completely determine the truth value of the compound statement formed by the connective; two compounds formed from it have the same truth value whenever their corresponding components match in truth value. Page 218

Tu quoque   Argument purporting to discredit a position by charging those who hold it with inconsistency or hypocrisy. Page 93

Vagueness   Imprecision in meaning. Page 70

INDEX

RULES OF INFERENCE

Conjunction

Simplification (Simp)

| n. | p · q | |
| n + m. | p | (or q) | Simp, n |

Conjunction (Conj)

n.	p	
m.	q	
p.	p · q	Conj, n, m

The Conditional

Modus Ponens (MP)

n.	p ⊃ q	
m.	p	
p.	q	MP, n, m

Modus Tollens (MT)

n.	p ⊃ q	
m.	~q	
p.	~p	MT, n, m

The Conditional (*continued*)

Hypothetical Syllogism (HS)

n.	p ⊃ q	
m.	q ⊃ r	
p.	p ⊃ r	HS, n, m

Disjunction

Addition (Add)

| n. | p | (or q) |
| n + p. | p ∨ q | Add, n |

Disjunctive Syllogism (DS)

n.	p ∨ q	
m.	~p	(or ~q)
p.	q	(or p)	DS, n, m

Constructive Dilemma (CD)

n.	p ∨ r	
m.	p ⊃ q	
p.	r ⊃ s	
q.	q ∨ s	CD, n, m, p

RULES OF REPLACEMENT

Double Negation (DN)

p ⇔ ~~p   DN

DeMorgan's Laws (DM)

~(p · q) ⇔ (~p ∨ ~q)   DM
~(p ∨ q) ⇔ (~p · ~q)   DM

Material Implication (Impl)

p ⊃ q ⇔ ~p ∨ q   Impl

Transposition (Trans)

p ⊃ q ⇔ ~q ⊃ ~p   Trans

Absorption (Abs)

p ⊃ q ⇔ p ⊃ (p · q)   Abs

Exportation

(p · q) ⊃ r ⇔ p ⊃ (q ⊃ r)   Exp

Material Equivalence (Equiv)

p ≡ q ⇔ (p ⊃ q) · (q ⊃ p)   Equiv
p ≡ q ⇔ (p · q) ∨ (~p · ~q)   Equiv

Commutativity (Com)

p ∨ q ⇔ q ∨ p   Com
p · q ⇔ q · p   Com

Associativity (Assoc)

(p ∨ q) ∨ r ⇔ p ∨ (q ∨ r)   Assoc
p · (q · r) ⇔ p · (q · r)   Assoc

Tautology (Taut)

p ⇔ p ∨ p   Taut
p ⇔ p · p   Taut

Distribution (Dist)

p · (q ∨ r) ⇔ (p · q) ∨ (p · r)   Dist
p ∨ (q · r) ⇔ (p ∨ q) · (p ∨ r)   Dist

For Categorical Proofs

Self Implication (SI)

n.   p ⊃ p   SI

Indirect Proofs

Premises

Negation of Conclusion   AIP
⋮
Contradiction